HISTORY OF BIOLOGY

Third Edition

HISTORY

OF

BIOLOGY

ELDON J. GARDNER

Utah State University
Logan, Utah

BURGESS PUBLISHING COMPANY ● MINNEAPOLIS, MINNESOTA

Copyright©1972, 1965, 1960 by Burgess Publishing Company
Printed in the United States of America
Library of Congress Catalog Card Number 75-188575
Standard Book Number 8087-0702-7

4   5   6   7   8   9   0

PREFACE

The purpose of this book is to present the main landmarks and themes in the history of biology. Although many contributors are characteristically involved in establishing the major principles and concepts of any science, time tends to erode credit from all but those who arrived at broad, enduring generalizations. This book necessarily adheres to such a process of oversimplification. Space limitations and the use for which the book is intended seem to justify the plan that has been followed. The book is based upon experience in the classroom and represents an attempt to answer questions that students ask about the history of biology. It is designed for use as a text for a one quarter or one semester junior-senior college course that enrolls science majors, premedical and predental students and those interested in broad cultural aspects of biological science. The materials provided will constitute only pillars and superstructure for those who want an understanding of the detailed history of biology. Even the casual student, however, should supplement the book by library reading which may be drawn from the references and readings given at the ends of chapters and in general literature sections. These lists include primary sources, some of which are not written in English and are not readily available, along with a number of readings that are available in most college and university libraries.

The new edition includes a short introductory chapter designed to show what science is, how it has developed, where biology fits into the general pattern among the sciences, and underlying factors that led to the development of basic and applied biology. From this beginning, succeeding chapters present information about methods employed, main accomplishments, and people responsible for major contributions. Chronological order is followed in Chapters 2-6. Special tools and methods are highlighted in Chapters 7-12, while the major themes and concepts of eighteenth, nineteenth, and twentieth century biology are developed in Chapters 13-17. Chapter 18 is a brief summary statement on the past and present of biology. A summary or chronology has been included with Chapters 2-17. This new edition benefits from additional illustrations in several chapters and the inclusion of tables to summarize and supplement correlated information that is discussed.

Many people, including my colleagues, students, and those who have used earlier editions of the book in other institutions, have made suggestions for the improvement of this manuscript. My thanks to all who have assisted in any way. Portraits of biologists presented throughout the book were drawn by Professor Everett Thorpe. Most of these portraits were prepared originally for this book. Those that appeared previously in other works are reproduced here with the permission of John Wiley and Sons, Inc., New York, and Limusa-Wiley, S. A., Mexico City. Credits for illustrations are given in the legends of individual items according to the wishes of the owners. Miss Lois Cox edited the manuscript and made valuable suggestions. Mrs. Barbara Thomas typed the manuscript with great care and skill. Mr. Stanley Irvine proofread the completed manuscript. Along with my thanks to those who have assisted with the manuscript, I express a sincere hope that those who use the book will continue to suggest improvements.

Eldon J. Gardner

Logan, Utah
May, 1972

CONTENTS

CHAPTER I

INTRODUCTION
THE SCIENCE OF BIOLOGY

BIOLOGY is the science that deals with living things. As a science, it is
based on objective observations of natural phenomena and the results
of controlled experiments. Interest in biological subjects, however, is
much older than experimental science.

Two profound and universal human characteristics lie behind the
origin and growth of science: the urge to control the workings of
nature to promote man's perceived welfare, and the simple irrepres-
sible desire to understand both the world about us, and ourselves.
We, today, are sometimes inclined to draw too sharp a distinction
between these two attitudes toward the world, the one practical and
utilitarian, the other theoretical and contemplative, and to talk of pure
research as something quite distinct from applied science. Yet both
motives have been (and are) operative, sometimes apart and some-

1

times together. The great progress of astronomy during the seventeenth century was prompted at least to some degree by the needs of the art of navigation. Much biological knowledge of a fundamental kind originated in man's search for ways to conquer illness. The great contributions that Pasteur made to science came about as a result of his interest in mundane problems such as what made wine turn sour or what could be done about a plague that was destroying the silkworms of France. On the other hand, many investigators had (and have) little concern with the possible utility of their work. Fortunately, a good deal of scientific research which seemed to have no possible practical value at first, proved later to be extremely useful.

Man has followed various routes in trying to understand his world. Certainly religion, art, literature, history, and philosophy represent attempts to account for ourselves and our environment, to specify the unknown, and to diminish both the mystery and the burden of existence. Magic and the superstitious invoking of religion, art, and rituals have also been tried as ways to control nature. Often enough, useful knowledge and magic have been mixed together. Some of the Indians of North America would accompany the planting of corn with elaborate prayers and rituals, but they made sure they planted a dead fish along with the corn. Indians of South America treated malaria with a tea brewed from the bark of the cinchona tree, which does in fact contain an effective agent, quinine, until recent times the only treatment for this disease. But they knew no more about the causes of the disease than their North American relatives knew about the nitrogen cycle.

Nevertheless, primitive peoples made substantial progress in learning enough about nature to help themselves manage it. Although their understanding of what they were doing was based largely on trial and error and transmitted as myth, it is chastening to us to realize that fire, the wheel, the domestication of animals, agriculture, and other discoveries or inventions of first importance were made by prehistoric man. Were these people scientists? If not, how did they differ from today's scientists?

Science, as currently defined, is made up of bodies of knowledge about the natural world (chemistry, physics, geology) and about ourselves as a part of nature (anthropology, economics, physiology). Each of these bodies of knowledge, moreover, is systematically organized and derived entirely from observation and experimentation. Scientific knowledge, in principle and often in fact, allows us to predict specific events and control them. It is, finally, open to verification by others through repetition of the original work. A philosophically sophisticated reader might take exception to this loose definition, but it serves to specify some of the most distinguishing features of scientific knowledge.

The term biology was first introduced at the beginning of the nineteenth century to refer to the science concerned with living things. We now know that living things may take us into geology, chemistry, or physics, and, in the case of man, into history, anthropology and psychology. Whether an investigation is biological often depends on motivation. For example, a geologist usually studies fossils in order to find oil while a biologist will study them to understand more about biological evolution.

What, from the standpoint of modern biology, are the most important working principles of biological inquiry? Perhaps the most important is the conviction that living organisms obey the laws of physics and chemistry. This principle was by no means universally accepted in the past, and any history of biology recounts innumerable conflicts between mechanists who thought that living matter was governed by the same laws that govern inorganic matter, and vitalists of one kind or another who argued that living things were endowed with special features that could not be explained by the laws of nature. Certainly the most productive approach to biological subjects has utilized the scientific method and the mechanistic premise that natural laws are consistent and that natural phenomena can be discovered by objective methods.

CHAPTER 2

BEGINNINGS IN BIOLOGY

IT SEEMS safe to assume that at some remote time early man became consciously aware of the differences between living and nonliving objects. He then may have begun to think about special features of living and recently living organisms. In preparing animals for food or for religious ceremonies, he became acquainted with their external and internal characteristics, but apparently made no attempt to learn about particular animals or plants or to compare their anatomies. Sometimes this haphazardly acquired information formed the basis for superstitious beliefs. The fate of a king or a nation, for example, was on occasion forecast by the shape of a single ram's liver. Progress in accumulating biological knowledge remained extremely slow as long as capricious demons were believed to control all events. And if any of our earliest ancestors had time for study and reflection, they were apparently inclined to devote themselves not to visible objects in the natural world but rather to invisible and presumed controlling forces beyond. But even so, some observations were made before the written record, and these were perpetuated by signs and symbols as part of the traditions of primitive tribes of people.

Most of what we claim to know about early man has been derived by inference from mummies, archaeological "digs," and present-day primitive people who are assumed to have much in common with ancient primitives. Thus, evaluating the biological understandings and concepts of early man is extremely difficult because our substantiated information about prehistoric periods ranges from meager to non-existent. At best this unwritten record is "spotty."

It seems reasonable to assume, though, that early man necessarily recognized, appreciated, and respected the laws of nature. Through trial and error, and inventive management, he adapted some of these laws of his physical and biological worlds to a personal advantage. He learned to use fire, to make tools from stone and later from bronze, to improvise eating and cooking utensils, and to create pottery. Of particular interest to the biologist, he evidently also domesticated virtually all amenable animals and plants that were accessible to him.

Communication

There can be no effective "science" until there is effective, lasting communication of information. This truism holds for biology as for all science.

Man first learned to communicate his thoughts by hand signs and by inventing meaningful symbols. Eventually he formulated vocal symbols; that is, he learned to speak. Even so, man still could communicate only as far as his voice would carry. He next devised sketches and drawings that could be set on stone or other physical material. This greatly increased his facility for exchanging and preserving observations and ideas.

Once man began to write, his own imaginative skill and the material he had to write upon were the limiting factors. Early Chinese wrote on strips of bamboo. This material was fragile and deteriorated unless stored in a very special way. Probably only a small proportion of the treatises written on such material was preserved long enough to be useful to later generations. The very few that are available at the present time have provided virtually all the information we have concerning the early Chinese civilization.

Babylonians and other Mesopotamians used cumbersome, inflexible blocks of clay and stones as writing material. The usefulness of these materials was limited, and only under special conditions were the records preserved for the enlightenment of succeeding generations. The Babylonian civilization, thus, was limited in its potential by a complex system of writing and an awkward material.

The Egyptians also used complex symbols, but they developed a much improved writing material, papyrus. This crude paperlike material was made from the pitch of a tall sedge, *Cyperus paprus*, native to the Nile regions. Papyrus was flexible and could be rolled on a scroll for filing and safekeeping. It can be assumed that at least one reason why the Egyptians recorded their observations more extensively than the Mesopotamians was that their writing medium was more practical. Many early Egyptian manuscripts have been discovered and translated and have added to our understanding of how biology developed.

Egyptian papyrus was the best writing medium in existence until the Chinese invented paper (ca. A.D. 105). This was one of China's greatest contributions to world culture and civilization. Ts'ai Lun, a Chinese court attendant, is credited as being the pioneer papermaker. Legend has it that he learned the principle from the first papermakers, wasps, that used this material for their nests. Later, papermaking became standardized in China. The inner bark of the mulberry tree was macerated in liquid to a fibrous slop, dehydrated to a thin, tough film, and cut into uniform sheets. Not only was this a biological derivative, but it proved to be the light, convenient medium on which biology's own story continued to be recorded.

Medicine

Medicine, founded on empirical knowledge and experience, seems to have formed the central core of most early biological developments. Because the specific causes of life and death were such enigmas to early man, his medical practices were often quite complex, mystical, and superstition ridden. Nevertheless, a medicine man practicing his art in those days was probably relatively successful despite severely limited resources and a lack of objective data. Some of his few effective drugs are standbys of modern medicine. Physical therapy, in the forms of massage, heat applications, and bathing had a place even in prehistoric times. Psychological factors such as confession and suggestion were basic components of the healing arts then, as they are now.

The specific ailments with which the earliest medicine men were most concerned, however, differed from those that often confront a modern practitioner. Epidemic diseases, such as smallpox, typhoid, cholera, measles, syphilis, and diphtheria did not become common until mankind developed a complex community life. Although contagious diseases were known many years before the time of Christ, their spread was sporadic and localized. Degenerative diseases such as cancer and heart disease were rare because few people lived to old age. Most people who reached maturity died from causes such as accidents and warfare before they reached the "heart disease age" or the "cancer age." So far as we can deduce, early medicine men were mainly concerned with digestive disturbances, respiratory infections, rheumatic fever, skin diseases, irregularities at childbirth, and diseases characteristic of early infancy.

Reproductive processes are known to have stimulated interest from prehistoric to modern times. The oldest *recorded* idea of procreation was spontaneous generation; that is, plants and animals were believed to arise from mud in the bottom of a pool or from various other materials, rather than through reproductive processes of parent organisms. Demonstrations could easily be made to show that worms and insects became prevalent at certain times of the year for no known reason except that conditions were propitious, and that mice accumulated in places where dirty rags and food materials were available. The apparently obvious explanation was spontaneous generation.

It is remarkable, therefore, that early man so effectively applied cause and effect reasoning to human birth. We have reason to believe that among primitive peoples sex was understood and usually was surrounded with beauty. The menstrual cycle was likened to the cycle of the moon, and rituals at puberty showed logic and honesty of thinking. Festivals with some biological significance were perpetuated to celebrate important events such as the rebirth of spring.

Agriculture

Agriculture, one of the "applied" forms of biology, developed independently in several widely separated regions of the world where the need existed and where environmental conditions were favorable. The need came as people accumulated in cities. Food-gathering or hunting economies could not sustain concentrated populations living in cities.

A proper combination of climate, soil, edible plants and domesticated animals equates with agriculture. The first civilizations character-istically developed in the great valleys of the major river systems of the world. Cultural stability was enhanced in relatively dry regions of the world, particularly the valley of the Indus in India, the Nile in Egypt, and Mesopotamia (between the Tigris and Euphrates Rivers) in Asia. Fertility is soon leached from the soil in humid regions but may be retained and replaced even by crude methods in more arid regions so long as adequate moisture is available from rivers. Nearly all animals and plants that have proved capable of domestication were controlled at an early period, long before the beginning of the written record. Trial and error, coupled with good husbandry, apparently made this accomplishment possible.

Because of communication and transportation problems, the source areas for potentially domesticated plants and animals and the locations where they were cultivated must have been in some proximity with each other. For example, rice, citrus, beans, some varieties of yams, and sugar cane originated in the northern Malay Peninsula and were not raised in other parts of the world until they were transported by people to other locations.

From Persia and the surrounding area came many fruits such as apples, pears, quinces, plums, cherries, almonds; also the cucurbits including gourds, cucumbers, small grains such as wheat, barley, oats, and millet. The cabbage family, including cabbage, cauliflower, and sprouts was also cultivated in the area of Mesopotamia. Among the domesticated animals of the Old World were the Asiatic elephant, sheep, goats, swine, the ass, the horse, cattle, buffalo, reindeer, camels, pigeons, chickens, ducks, geese, and swans.

In the New World domestication of native plants and animals took place primarily within a rather narrow area of South and Central America. In Peru the white potato, maize, beans, squash, gourds, sweet potato, the yam, red pepper, tobacco, pumpkins, the avo-cado, chocolate, vanilla, tomato, and pineapple had their origin. The New World produced the turkey, the llama, and the alpaca in northern South America and Central America.

Biology in Some Early Civilizations

In Eastern Asia, independent advances in biological understanding occurred among the Chinese and Hindus. The developments in Western Asia, however, provided the foundations for the biology we know today.

In China, intellectual achievements that had biological implications were sporadic and utilitarian, and for the most part had little in common with modern biology. Some Chinese applications of biological principles, however, are known to date from the distant past and are worthy of note. Their medical literature in particular began to accumulate early and became extensive.

Medicine in Ancient China

As early as 3000 B.C. the Chinese ascribed medicinal values to several different plants. These plants became the central objects of the plant lore as well as the medicine that developed early in human history. Beginnings in Chinese medicine associated with plant lore were attributed to three legendary emperors. The first was Fu Hsi (ca. 2900 B.C.) who talked of medicine in terms of the Taoist legend in which the god Pan Ku formed the universe by dividing the chaos into two vital principles of Yang and Yin. This sense of dualism exists in early Chinese philosophy, folklore, and religion, as well as in medicine. In Chinese lore, Yang and Yin were the two vital forces of man and represented opposing qualities. Yang, the masculine force, represented the left side and all positive qualities (good, heaven, day, tall, hot, acid). Yin, the feminine force, represented the right side and all the passive qualities (bad, earth, night, short, cold, base). The Chinese believed that when all factors were in perfect equilibrium, the body would be healthy and tranquil. Disease was attributed to disharmony in balance of the two vital forces.

Five basic elements, fire, earth, water, wood, and wind, were interwoven into Chinese lore. Five sensations of taste were related to the elements. Behind the Chinese dish, chop suey, is Chinese philosophy. Its mixture of elements and corresponding tastes were believed to prolong life because it embodied an appropriate balance of Yang and Yin.

Shen Nung, the second legendary emperor, lived sometime between 2838 and 2700 B.C. He is credited with the authorship of a 20-chapter medical herbal that included descriptions of the important food item, the soybean, and of numerous plants having medicinal value such as rhubarb, ginseng, opium, and pomegranate root, which were part of the Chinese pharmacopedia. The medical herbal also included descriptions for cures that the author developed by experimenting on himself.

The third legendary emperor of ancient China was Huang-Ti (ca. 2600 B.C.), who is said by many to be the author of *Neiching*, a classic dealing with internal diseases. This treatise, written some 4,500 years ago on strips of bamboo, represents the foundation of Chinese medical literature. It contains a passage suggesting that the early Chinese may have discovered the continuous circulation of blood many centuries before the time of William Harvey: "All the blood in the body is under the control of the heart. . . . The blood current flows continuously in a circle and never stops." If the statement concerning continuous circulation of the blood has been properly interpreted from the original, the accomplishment was particularly remarkable inasmuch as the Chinese avoided animal dissection whenever possible because of their horror of death and dead things.

The Yang and Yin dualism dealt specifically with five basic organs of the body: lungs, liver, kidneys, heart and spleen. In the descriptive part of *Neiching*, the source of thought was located in the spleen, and the center for will power was placed in the kidneys. According to this ancient Chinese lore, the lungs represented the seat of sorrow. They produced skin and hair, formed the kidneys, and controlled the heart. Blood, the seat of anger containing the soul, was stored in the liver. The liver was the controlling organ or general of the body. It was connected with the eyes and became involved in the production of tears. The kidneys controlled fright and provided willpower. They also produced bone marrow, were connected with the ears as well as the urogenital system, and they produced saliva. It was noted that the heart tasted bitter, smelled like toast, formed the spleen, and was the source of sweat. The heart acted as king of the body and was described as having three chambers. As the source of thought, the spleen was associated with earth, controlled reproduction, and created the five taste sensations.

Although many authors have specifically associated smallpox vaccination with early Chinese, the date of its invention is not known. Methods were crude and there was no underlying understanding of the pathogen or of the degrees of its virulence, but the philosophy of "better to introduce it than to wait for it to strike" was accepted, and vaccination evidently was practiced with some success. Various methods of introducing the virus into the body were employed. An old crusty scab could be introduced into the wound of a healthy person, or powdered material from scales could be inhaled. Less desirable practices were those of inserting a pus-soaked rag into the nose of the person being vaccinated or his voluntary wearing of the patient's contaminated clothing.

Chinese Contributions

Among the lasting contributions made by the early Chinese in the fields of medicine and agriculture were: ashes of sponges used to relieve goiter, opium from poppy juice used to relieve pain, ephedrine employed for treating asthma (a fact recorded in the medical herbal presumably written by Shen Nung), and the ephedrine plant (ma huang, *Ephedra vulgaris*) also recognized for its properties as a diaphoretic and circulatory stimulant. Agar-agar, the vegetable solidifying matter derived from marine algae, was used by the Chinese for food and medicine, and later became a laboratory substance of considerable importance in microbiology for solidifying the nutrient medium used in plate cultures. The narcotic, marijuana, from the hemp plant *Cannabis indica*, was used by the Chinese as an anesthetic.

The biologic control concept was employed on a macroscopic level by the early Chinese. Farmers learned to pit one organism against another as a control measure. Entomophagous ants, for example, were "attached to trees" to destroy a "boring insect." Also during an early period in the history of ancient China, fowl eggs were artificially incubated. Chinese ingenuity found a way to check the temperature at which eggs were being incubated. Man's eyelids were employed as sensitive thermometers to check the temperature during incubation.

A biology-related development of great economic significance in China was the silk industry. Because the caterpillar of *Bombyx mori* can produce silk, it is one of the most important beneficial insects in history. When man first used silk is unknown, but the Chinese Empress, See Ling, had it in 2697 B.C. Paintings, woodwork, and embroidery dating back to early periods of Chinese history show a great appreciation for insects and particularly the silkworm. This worm cannot live without man's help at the present time and must have been nurtured purposely by the early Chinese for the value of its product (silk).

Other biological applications in ancient China were associated with food getting. Wild plants and animals native to China were used as food, and some were domesticated and made to serve man on a more controlled basis. The saliva of birds was considered a delicacy. Nests of the Asian swift which were composed of mud cemented with bird saliva were boiled and made into a choice dish, bird's nest soup. Sea foods such as fish, snails, crabs, cuttlefish, and sea cucumbers were obtained from China's long seashore and used extensively. The Chinese made good use of the soil and began early their intensive cultivation of tea, oranges, soybeans, rice, barley, and bamboo for practical purposes.

Ornamental gardens where the morning glory, chrysanthemum, and rose were grown and improved became closely associated with botanical gardens in which plants with medicinal values were raised. Chinese plant lore was never widely dispersed, but was developed to a large extent as a racial culture. Some botanical historians have indicated that the early Chinese were as well acquainted with the flora of their region in eastern Asia as was Dioscorides at a later date (A.D. 60) with the flora of the Mediterranean region (Chapter 4).

Early Chinese medicine attained a high level of achievement, as compared with other contemporary peoples, but it declined after the time of Christ. Several possible reasons for the decline have been suggested by students of Chinese culture:

1. Their religious beliefs discouraged anatomical studies.

2. A complacent attitude prevailed. People asked, "Why go beyond this point?" and said, "We know enough now."

3. Their system was not elastic enough to accommodate new circumstances and ideas. They believed that what had become established must continue and man must adapt himself to nature. They tried to make new developments conform to old patterns. New developments that could not be made to fit were discarded.

The Japanese, who influenced Chinese medicine in the ninth century A.D., did not show the same conservatism. During the sixteenth century A.D., the Japanese developed a direct clinical approach which reflected more reliance on the healing forces of nature. Japanese surgeons made progress under Western influence during the seventeenth century, and in the eighteenth century they made great strides especially in obstetrics and anatomy. In the latter half of the nineteenth century they assimilated Western medicine as a whole with ease and skill.

Contributions in India

Early Hindu medicine is divided into two periods. The first, continuing from the earliest beginnings until about 800 B.C., is called the Vedic period because the information we have is derived mainly from the Vedas, a collection of four Sanskrit books. The second or Brahmanic period (roughly 800 B.C. to A.D. 1000), was based on a culture ideologically dominated by the Brahmanic caste of Hindu priests. After A.D. 1000, large parts of India came under Islamic rule and Arab doctors took over medical practice in the country.

Vedic medicine was primitive and archaic. Sin was viewed as the cause of disease and confession as the healing rite. Unsanitary conditions provided a breeding ground for plague and cholera, and malaria was also prevalent. Vedic books reflect the existence of many diseases and show a marked predilection for purifying treatments with water.

During the early Brahmanic period doctors belonged to a special caste. They were not educated at priest schools as were contemporary Greeks, but were trained through apprenticeships. At this time Hindu medical education was of a high caliber, with a wholesome balance between theory and practice and a wholesome relation between medical and surgical knowledge. A solemn oath much like the Hippocratic oath (Chapter 3) was taken as an entrance ceremony into the profession.

In the sixth century B.C. two great universities existed in India; Kasi in the East and Takasila in the West. Atreya, the physician, taught in the latter university, and his younger contemporary, Susruta, the surgeon, taught in the former. Susruta, a true scientist who combined direct observation with sound theory, described several operations relating to cataract, hernia, the caesarean section, and plastic surgery. He also gave an account of the surgical instruments used.

The greatest overall accomplishment of the Hindus was in the field of plastic surgery. Because mutilation was current judiciary procedure, there was ample opportunity to patch ears and noses. The technique of hypnosis was applied frequently in India to facilitate painless operations.

The three great classics of Brahmanic medicine are the books of (1) Charaka (written at the beginning of the Christian Era), (2) Susruta (about A.D. 500), and (3) Vagbhata (about A.D. 600). These books were based on much older Vedic material. Hindu writings state that the human body has several skins and list some 300 bones, 107 joints, 900 tendons, 700 blood vessels, and 500 nerves. Although the numerical lists of structures are impressive, the Hindus had only primitive ideas concerning the functions of these structures. Body parts, various fluids, and different kinds of air which provided for the body's renewal were of more interest to the Hindus from the numerical than the functional point of view. Hindu medicine, like Chinese medicine, froze into dogmatism before the time of Christ and continued in a static form through subsequent centuries. Developments in agriculture were parallel and, to some extent, interwoven with those in medicine. Plant studies in India, as in other parts of the world, had utilitarian purposes. Plants were a source of food and were classified as wholesome or injurious. In a medical sense, some plants were shown to alleviate pain, and/or heal wounds and sores, and others were found to give pleasurable excitement or have a narcotic effect. Plants were studied in relation to medicine, agriculture, and horticulture, and early treatises on botany such as "Ayurveda," "Charaka-samhita," and "Susruta-samhita" were therefore primarily practical in orientation.

Evidence from early Vedic literature (2500 B.C.?), Ramyana (1900 B.C.?), and Mahabharata (1400 B.C.?) indicate that the science of agriculture was well developed in those periods. The beginnings of Indian agriculture can be traced back to the Mohenjo-daro period (3250-2700 B.C.); it reached maturity in the Vedic period. Specimens of wheat and barley unearthed in the ruins of Mohenjo-daro indicate that these grains were cultivated at this time. *Triticum compactum* and *Hordeum vulgare* have been identified. Millets, dates, melons, and other fruits and vegetables and cotton were all known to the ancient Hindus.

Plants were propagated by seed, cuttings, grafting, and budding during the Vedic period. The ancient Hindus were aware of sexuality in plants; for example, male and female types of screw pine *Pandanus odoratissimus* were recognized by different names. Plants were classified on the basis of external morphology, medicinal properties, and environmental associations.

Mesopotamian Culture

Two cultures that had developed in Africa and western Asia during an early period of history were flourishing around 3000 B.C. One was located in the valley of the Nile River and gave rise to the Egyptian civilization; the other was between the Tigris and Euphrates river systems (Fig. 2.1) and became known as the Mesopotamian (between two rivers) culture. The Egyptian and Mesopotamian civilizations paralleled each other in many ways, but the Mesopotamian culture developed to a high level at an earlier period than the Egyptian civilization. Both left some evidence of their biological knowledge and interest in biological subjects.

In Mesopotamia an extensive knowledge of zoology was developed. This was especially true of one highly civilized group, the Babylonians, who occupied the Euphrates valley. They gained information about anatomy by preparing sacrifices to the Babylonian gods and from their associated habit of seeking omens. Omens were sought through various methods, including examination of the livers of sacrificed animals. Such omens might indicate whether a man would live or die, whether there would be rain or drought, or whether a battle would be won or lost. Practically anything found to be "out of the ordinary" was considered an omen of impending occurrences. Omens were used to determine the prognosis and diagnosis of an illness and were also considered to be important preventative measures. While the liver was being carefully examined for omens, substantial knowledge concerning other aspects of internal anatomy was also obtained. Some

Figure 2.1
Location of Mesopotamia,
between the Euphrates
and Tigris river systems.

of this information carried over to medicine and was recorded in Babylonian medical texts that are now available for examination.

Our most pertinent information about early Babylonian medicine comes from the Emperor Hammurabi (1955-1913 B.C.), a near contemporary of the Biblical patriarch, Abraham. The *Code of Hammurabi*, the oldest surviving code for physicians, includes laws specifying the maximum fees for medical practice and punishments in case of malpractice. The principle "an eye for an eye" was the basis for the code. If a physician injured a patient in the course of his treatment, the physician was punished by having the same injury inflicted upon himself. According to an excerpt from the code:

> If the doctor shall treat a gentleman and shall open an abscess with the knife and shall preserve the eye of the patient, he shall receive ten shekels of silver. If the doctor shall open an abscess with a blunt knife and shall kill the patient or shall destroy the sight of the eye, his hands shall be cut off or his eye shall be put out.

Modern laws and precedents concerning the legal aspects of malpractice may be traced to the Babylonians, but they have usually been tempered over the years by a better understanding of the physician's problems.

Much of the early Mesopotamian literature was preserved in the enormous library of Ashurbanipal (668-626 B.C.), a great king and scholar. Approximately 25,000 clay tablets collected in Nineveh represent one of the best sources of information concerning the Mesopotamian medicine during this period. Some of these tablets are in the British Museum. They illustrate symptoms of diseases and indicate pharmacological treatment. Others describe demons, their ways, and their characteristics. Prayers and incantations also are included on the tablets since religion was closely related with medical practice at this time.

Although Herodotus (484-425 B.C.), a Greek historian, stated that there were no physicians in Babylon, other evidence indicates that a Babylonian school, which is now believed to have been a school of medicine, was in existence during the seventh century B.C. In addition, letters have been found that were presumably written by a Babylonian physician and addressed to an Assyrian king in the seventh century B.C. These letters, intended for the king's chief physician, gave directions for treatment of a nosebleed from which someone in the kingdom was apparently suffering and also commented on the probable recovery of a citizen whose eyes were diseased. Other letters written in the same general period indicate that physicians were present, but they evidently were restricted to the various courts of rulers.

Early Mesopotamians used some effective drugs in their medical practice, including the fruit, leaf, flower, bark, and root parts from various plants such as the lotus, olive, myrtle, and garlic. Organs of animals were used as remedies, and minerals occurred frequently in recipes for the cure of eye and skin diseases. Various sulphur compounds, arsenic, copper, mercury, and sour milk were prescribed for diseases of the eyes. Salves and ointments were frequently swallowed through a tube which was also used to blow medicaments into the nose or ear. Tablets on which medical information was written were divided into three columns. The first listed the drugs, the second the diseases for which each drug was to be used, and the third the form in which the drug must be applied.

In observing the course of disease, the Babylonians noted that the general sequence from illness to health in an individual often was followed by the occurrence of the same illness in someone else who had been associated with the ill person. They decided that the demon of the disease must be passed from one person to another and that if they gave the demon a medium in which to pass, it would leave the household in peace. A lamb was placed next to the patient to make the transfer of the disease effective. If the lamb had a diseased liver or lung when it eventually was killed and opened up, the priests would infer that the transfer had been successful. This had a tremendous psychological effect on the patient because it meant that he would recover, and in this sense the ancient "magical" formula for cure was comparable with the psychology used by the modern physician.

Cleanliness was practiced widely in ancient Mesopotamia. Effective sewage systems were developed in this civilization some four thousand years ago. Notions of contagious diseases, isolation of lepers, and regular rest days are also indicated in Babylonian medical literature. Even though these practices have come into modern culture from the ancient Hebrews, they seem to have originated in Mesopotamia.

Egyptian Accomplishments

The early Egyptian civilization (Fig. 2.2) accomplished many things in biology, particularly in the "applied" field of medicine, but detailed evidence is fragmentary. Many questions have been left unanswered and open to speculation by members of later generations. Besides creating a writing system and learning to make bricks and fashion pottery, the Egyptians devised ways to make bread, brew beer, and prepare leather. In connection with extensive domestication of plants and animals, the very early dwellers in the Nile Valley developed a high level of agriculture. Records of these accomplishments, however, have been difficult to piece together. When Herodotus explored the Nile River in the fifth century B.C., most of the evidence of the earlier civilization had already been destroyed or carried to the sea. Even the documents that were cast in bronze, and thus preserved, do not tell much of the story, but they do indicate an important development in the refining and the use of metal. The use of bronze was a major step forward by the Egyptians because much can be done with bronze that is impossible with flint or limestone. Egypt's important contributions to biology may be classified in three categories: (1) medical and surgical practice, (2) embalming, and (3) personal hygiene.

By the time the Egyptians began to write about their medical practices, the art was so old that the writers did not attempt to describe the beginnings but attributed its origin to the gods. Medicine and supernatural healing could not be separated because the physicians were priests, and the gods who controlled various aspects of medicine were ever present. The known history of early Egyptian medicine centers primarily around one personage, Imhotep. Presumably Imhotep lived about 3000 B.C. and was a physician to one of the pharaohs. Although we have much information about him, it is difficult to be

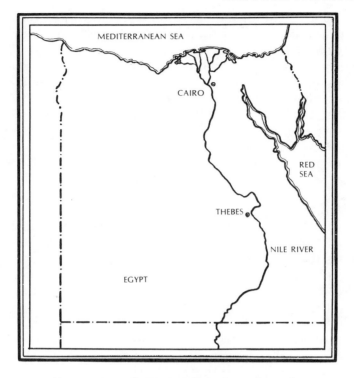

Figure 2.2.   Nile River Valley where the early
Egyptian civilization developed.

sure whether he was an actual man or a myth. Whatever the truth
may be, he eventually was glorified and finally deified. Legends
about Imhotep have persisted throughout the medical history of
Egypt.

From 3000 to 1000 B.C. Imhotep was attributed with far-reaching
powers and was invoked in prayers for the sick. By 1000 B.C. he was
considered a god; temples were erected to him and people came from
great distances to be cured from their illnesses and have their fears
alleviated. Imhotep and his emissaries were supposed to appear to
people in temples to give cures for illness. Ill people were said to have
gone to sleep in a temple and awakened well and free from their
symptoms. The influence of Imhotep persisted for many centuries.
During a period of widespread disease in Rome as late as 273 B.C.,
a special emissary was sent to the temple of Imhotep to invoke assistance for the stricken Romans.

An impressively complex Egyptian civilization persisted from about

3000 to 1000 B.C. One of the great steps that was taken during that period involved the invention of papyrus, a paperlike writing material of great flexibility and durability. This material was not only of value to the Egyptians for recording their current knowledge, but it preserved the records for present-day historians. Most of the history of the ancient Egyptians has been obtained from a few preserved papyrus records that have been found. The oldest known papyrus is the Kahun papyrus (2000 B.C.), which is concerned with gynecology and veterinary medicine. Biologically speaking, however, the two most important papyruses are the Edwin Smith papyrus (1600 B.C.), which deals with surgery, and the Ebers papyrus, entitled the *Therapeutic Papyrus of Thebes* (1550 B.C.).

The Edwin Smith papyrus, or *Surgical Papyrus*, enumerates many examples of illness and treatment, mostly surgical. Some are given in great detail, and these occasionally may include a workable checklist of questions such as: "Can he lift his arm?" "Does he shiver?" "Does he bleed from the nose?" "Can he open his mouth?" "How strong is his pulse?" Injuries to the different parts of the body were discussed separately. The eyebrow, nose, cheek, ear, jaw, temple, and forehead were included under the general headings of head and skull. Directions for bandaging wounds, restoring fractures, and burning out tumors were carefully enumerated. The papyrus obviously was meant to be a complete book on surgery and external medicine, beginning with injuries to the head and concluding with the feet. It ends abruptly, however, at the bottom of the 17th column in the middle of a line.

Each case discussion begins with the name of the ailment, followed by the symptoms, diagnosis, prognosis, and treatment. In knife-cuts, sword-slashes, and battle-axe blows of the skull, the surgeon is instructed to probe the wound. The prognosis is favorable in contusion, doubtful in a penetrating gash, and unfavorable in fracture. Feeble pulse and fever are noted among the symptoms of severe cranial injuries. In fracture of the skull under the skin, the practitioner must elevate the depression outward. A cut in the forehead is treated by a linen bandage prepared by the embalmer and known as the "physicians' skin." Nasal secretions are to be cleared away by swabs consisting of two rolls of linen dipped in ointment and inserted in the nostrils. Fracture of the temporal bone is noted as producing deafness. In dislocations of the jaw the attendant is instructed to force the bones into place. According to the text, temperature develops from a knife wound to the gullet. If the patient with such a wound drinks water, it turns aside, issuing from the wound. Loss of control of arms, legs, and excretory organs is associated with dislocation of the vertebrae and the neck, and the physician can do nothing.

The Ebers papyrus is a medical text which contains most of the biological and medical information concerning the ancient Egyptians that is now available and is the most important of the ancient papyruses. It was written in hieroglyphics on a strip of papyrus about a foot wide and 65 feet long and totaled 2,289 lines. An Arab was said to have found it in a tomb in the Nile River Valley some time before A.D. 1873. George Ebers, an Egyptologist, gained possession of the document in 1873 and it is preserved in a library at Leipzig, Germany. Ebers himself translated most of the document, some of the more obscure symbols were later deciphered by other scholars, and parts still have not been satisfactorily translated. The document was written about 1550 B.C. and was apparently a copy of a manuscript prepared many years earlier.

The information the Ebers papyrus emphasizes is the supernatural aspects of medicine, but it includes descriptions of diseases and treatments that reflect critical observation. Diseases of the viscera and organs of special sense were given particular attention. Sections on tumors and obstructions, and those on diseases of the ear and eye, from which the Egyptians suffered extensively, were given considerable space. Many ailments, from pains in the head to sore toes, were discussed; one chapter was devoted to diseases of children, another to diseases of women. Chapter 20 contained "remedies for burns and suppurative sores, gangrene, and wallops from flogging." More than 70 cosmetic prescriptions were included for sunburn, freckles, wrinkles, and other facial blemishes. Perfumes for women were recommended so their clothing and breath would be pleasant. For men, there were remedies for the care and preservation of hair. Methods for overcoming baldness and for preventing hair from turning grey were included along with directions for inducing hair to grow on scars on the scalp.

Some 700 remedies were listed in the Therapeutic Papyrus and included: opium, caster oil, copper salts, squill, acacia, calamus, saffron, gentian, pomegranate, and olive oil. Although the Egyptians used these drugs effectively, they did not know their specific values. Other agents such as fly-specks scraped from a wall and moisture from a pig's ear were sometimes recommended as highly as those that have gained a place in modern medicine. Parasites that are common today evidently were prevalent in ancient Egypt. The large intestinal worm,

Ascaris, received detailed consideration. The Egyptians knew that beetles developed from eggs, blowflys from maggots, and frogs from tadpoles.

Egypt became the medical center for the ancient world. The Biblical statement, "And Moses was learned in all the wisdom of the Egyptians," indicated that the Egyptians influenced the origin of the Mosaic laws of health. The comment of Jeremiah, "Go up into Gilead and take balm, O virgin daughter of Egypt; in vain doest thou use many medicines; there is no healing for thee," shows that Egypt was noted for numerous remedies. Homer spoke of "Egypt teeming with drugs, the land where each is a physician skillful beyond all men." Herodotus described Egypt as the home of specialists: "Medicine is practiced among them on a plan of separation; each physician treats a single disorder, and no more; thus the country swarms with medical practitioners, some undertaking to cure diseases of the eye, others of the head, others again of the teeth, others of the intestines, and some those which are not local." Diodorus Siculus explained: "The whole manner of life in Egypt was so evenly ordered that it would appear as though it had been arranged according to the rules of health by a learned physician, rather than a lawgiver." Cyrus of Persia sent for an Egyptian oculist to take care of his sick mother, and the body physician of Darius came from the Nile region.

The importance to the early Egyptians of the weather and the floods of the Nile River influenced their attitudes toward science and life in general. Supernaturalism was associated with this dependence upon nature's whims. There was little or no speculation such as that found among the Greeks in a later period.

The Egyptians gained some working understanding of anatomy from the practice of embalming the dead and forming mummies, a lost art in more recent time. Knowledge of anatomy was required for removal of the viscera and other internal organs, and particularly in removing the brain. Embalmers relied on the inherent properties of common salt, wine, aromatics, myrrh, cassia, and other preservatives available to them. The wrapping linen was smeared with gum to exclude all putrefactive agencies. Egyptians also understood the value of extreme dryness in the exercise of their antiseptic art. In spite of their cutting, probing, and sewing, however, no lasting anatomical knowledge evolved. No charts, lists, or any other notations about scientific anatomical information have been found that explain how the art of embalming attained such heights of perfection in the early Egyptian period.

Personal cleanliness, especially of the priests, was given extraordinarily detailed consideration; in fact, it was regulated by law. Priests changed regularly into freshly-laundered clothing and washed themselves twice a day and also at night. Like the Babylonians, the Egyptians evidently were conscious of the importance of cleanliness in maintaining health.

Meat of domestic animals was inspected before the populace could accept it for food. Each animal was examined for external irregularities, and the mouth and tongue were checked. Instead of using a blue stamp of acceptance Egyptian inspectors branded or notched the horn to show that the meat had passed inspection.

Public Health in the Old Testament

In spite of the absence of references to medicine and surgery in the Bible, it is a source of information on personal and social hygiene, and might even be regarded as the first textbook on public health. Leviticus, Chapters 13-15, which make up the Pentateuchal Code, give explicit instructions for the handling of lepers. They are to be isolated, their clothing must be burned, and their houses disinfected or completely destroyed. Another example of the explicit sanitation that was practiced is the use of salt to sterilize a contaminated well (Kings 2:19-22). Though the Hebrews seem to have thought disease was caused by demons, they also believed that pestilences were sent by Jehovah as punishments for wickedness. Jehovah could send disease, and he also could prevent and cure disease. The ancient Hebrews seemingly had no great physician, but they contributed much to the progress of medical science by fostering a social consciousness among the people, by instituting valuable measures for the prevention of epidemic disease, and by promoting health in the community.

In ancient Palestine, the animals were classified in two groupings, the unclean and clean. Swine were unclean animals and this condemnation seems to have had some basis in fact because, as modern parasitologists have found, swine contain a multiplicity of parasites. Some animals were avoided because of deep-seated unreasoning fear; snakes and mice fell into this category. Turtle doves, on the other hand, were the only birds permitted at sacrifices, and they were an emblem of peace, hope, and tranquility. Genesis 1:20-25 includes a crude classification of animals: creeping creatures, fowl, whales, cattle, beasts, and man.

"Like begets like" was the premise on which marriages of close relatives was made standard procedure, even expected and demanded. The same principle was applied to livestock insofar as the poorest animals were eliminated from the herds. Jacob apparently did some selective breeding in his flocks. Phenotypic selection is not necessarily the most effective, but it is still the most common improvement method practiced by animal breeders throughout the world today.

Plant classification in the Old Testament was limited to three categories: grasses, herbs with seeds, and fruit trees. The Old Testament contains 23 names for stem and branch plants. Linen, the badge of excellence in Biblical days, was made from locally grown flax. Even though the Bible contains many references to unleavened bread, the Jews used leavened bread as a staple food. They did not understand the nature and application of yeast, but kept pieces of sour dough from one baking to the next to perpetuate the gas-forming organisms.

Summary

Early man, before the time of the written record, learned much about plants and animals in his search for food and for the satisfaction of other requirements. He had already domesticated many plants and animals for his use. His desire for health and a long life led him to develop considerable empirical knowledge about himself and also led him into magic and superstition. He eventually learned to communicate vocally and with written symbols and developed a system of agriculture that provided a stable food supply. These accomplishments became major factors in the development of civilizations. Two civilizations in the East (Chinese and Hindu) and three in the West (Babylonian, Egyptian, and Hebrew) are known to have made achievements in biology during the period of about 3000 to 500 B.C.

Location	Main accomplishments in biology
China	Plant lore, herbs for medicine, plants and animals for food.
India	Medicine, surgery.
Mesopotamia	Anatomy, remedies for ailments, agricultural practices.
Egypt	Medicine, surgery, embalming, agriculture, papyrus for writing.

References and Readings

Ackerknecht, E. H. 1955. *A Short History of Medicine*, (Chapters 3 and 4). New York: Ronald Press.

Breasted, J. H. 1909. *A History of Egypt*. New York: Charles Scribner's Sons.

_____. 1930. *Translation of the Edwin Smith Surgical Papyrus*. Chicago: University of Chicago Press.

_____. 1933. *The Dawn of Conscience*. New York: Charles Scribner's Sons.

_____. 1935. *Ancient Times*, 2nd ed. Boston: Ginn and Co.

Castiglioni, A. 1947. *A History of Medicine*, 2nd ed. (Chapters 1-7). New York: Alfred A. Knopf.

Chiera, E. 1938. *They Wrote on Clay*. Chicago: University of Chicago Press.

Coonen, L. P. 1951. "The prehistoric roots of biology." *Sci. Monthly* 73: 154-165.

_____. "Biology in Ancient Egypt." *Biologist* 34:79-85.

_____. 1953. "Herodotus on biology." *Sci. Monthly* 76:63-70.

_____. 1954. "Biology in Old China." *Biologist* 36:3-12.

_____. 1955. "Biology in Ancient Palestine." *Biologist* 38:3-12.

Dawson, W. R. 1964. *The Beginnings, Egypt and Assyria*. New York: Hafner Publishing Co.

Ebbell, B. 1937. *Translation of the Papyrus Ebers*. Copenhagen: Leven and Munksgaard.

Frankfort, H. 1954. *The Birth of Civilization in the Near East*. Bloomington: Indiana University Press.

Frazer, J. G. 1935. *The Golden Bough*, vols. 1 and 2. New York: Macmillan Co.

Glanville, S. R. K., ed. 1942. *The Legacy of Egypt*. Oxford: Clarendon Press.

Gordon, C. H. 1962. *Before the Bible*. New York: Harper and Row.

Isaac, E. 1962. "On the domestication of cattle." *Science* 137:195-204.

Jastrow, M., Jr. 1915. *The Civilization of Babylonia and Assyria*. Philadelphia: J. B. Lippincott Co.

Morse, W. R. 1934. *Chinese Medicine*. New York: Paul B. Hoeber.

Needham, J. 1954. *Science and Civilization in China*, vol. 1-111. Cambridge: The University Press.

Rawlinson, G., trans. 1939. *The History of Herodotus*. New York: Tudor Publishing Co.

Reed, C. A. 1959. "Animal domestication in the prehistoric Near East." *Science* 130:1629-1639.

Reed, H. S. 1942. *A Short History of the Plant Sciences*. Waltham, Mass.: Chronica Botanica.

CHAPTER 3

THINKERS IN ANCIENT GREECE

Early Philosopher-Scientists

LITTLE is known of the specific ideas of the earliest Greek philosopher-scientists. The surviving fragments of their writings indicate that their objective was to find one basic element that would explain the functioning of the physical world. Thales was one of these early philosopher-scientists.

Thales

Thales (639-544 B.C.) of Miletus lived in the Ionian colonies on the coast of Asia Minor. He left no writings and may not have been able to write; however, he was a profound thinker who educated himself by traveling and studying in Egypt. He was a merchant, statesman, engineer, mathematician, and astronomer as well as a philosopher and original thinker. According to legend he was once asked if he was a wise man, and he modestly replied that he could not call himself wise; he was merely a lover of wisdom. Because of his wealth and high social standing, he was able to collect around him a large number of disciples. The word "philosopher" was coined by him, and he is credited with several specific observations in nature. He is said to have observed an eclipse of the sun that occurred on May 28, 585 B.C. after predicting it with the help of Babylonian astronomical tables. Thales was one of the first philosophers to promote the new European school of thought, which presumed that the whole universe was run by natural law. This school emphasized rational inquiry into nature and represented a definite break from the mythological tradition of supernatural forces governing the universe. This was the foundation for the development of natural science.

Thales observed that water was the most abundant material on the earth and that both plants and animals require moisture. Life, he thought, originated directly from the water. He believed that water produced air when expanded by evaporation and earth when congealed and contracted. Alluvial deposits at the mouth of rivers seemed to confirm the idea that water could change into earth. In his view, the earth was a solid disc floating on a sea of water.

The idea of one primary element fostered philosophical skepticism, however, because it implied that such solid materials as wood and iron were essentially the same as water. Either more elements capable of interaction must be involved, some philosophers reasoned, or the senses are untrustworthy for interpreting the physical nature of the world and its living inhabitants.

Anaximander

Anaximander (611-547 B.C.), a pupil of Thales, was also interested in the physical phenomena of the universe. He was a philosopher, astronomer, and geographer but his reputation is due mainly to his broad views of origins in nature. Thales' philosophy that water alone was involved in the formation of the earth did not satisfy Anaximander. He believed that air and earth were important and that living things also incorporated an ethereal substance, *apeiron*, that was endless, unlimited, and subject to neither old age nor decay. It was perpetually yielding fresh substance for the beings that issued from it. Apeiron was supposed to resolve itself into two different principles— hot and cold. Various intermixtures of these gave rise to the four familiar elements of which Anaximander believed the earth was made: earth, water, air, and fire. Apeiron embraced everything and directed the movements of things from which there grew up a host of shapes and differences. Out of the vague and limitless body there sprang a central mass that became the earth and all associated with it. The element earth was in the center and made up the hard core of the world. Water covered the earth, mist formed above the water, and fire potentially embraced everything. According to Anaximander's theory, fire, which heated the water and caused it to evaporate, consequently made dry land appear, but this process also increased the volume of the mist. As the pressure intensified, the fiery integument of the universe burst and took the form of wheels of fire enclosed in tubes of mist circling round the earth and sea. This was Anaximander's working model of the universe. The stars were explained as holes in the tubes through which the fire flowed. A total eclipse was explained as a complete closing of a hole, and a partial eclipse was a partial closing.

Anaximander was one of the first to describe the world as a sphere. Previously it had been visualized as a flat floor with a solid base of limitless depth. The earth was conceived by Anaximander and his Greek followers to be the stationary center of the universe. They believed the sun passed underground from one side to the other at night, instead of around the rim of the world as described in some older systems. Anaximander is said to have been the first Greek to design a map of the known world. He also recognized that the visible portion of the sky was half of a complete sphere. The results of his scientific researches were presented in a poem, *On Nature*, which was read by Aristotle and several later philosophers, but unfortunately has been lost to modern times.

Spontaneous generation of life, as visualized by Anaximander, took place in the residue of mud and mist on the earth while the water was being evaporated by the sun. The first animals were fish, produced in moisture and covered with a spiny skin. In the course of time their descendants left the water and reached dry land. When the world's integument burst and conditions changed, these descendants modified their mode of living and became adapted to the new environmental situation. According to Anaximander, different kinds of living things came into being by transmutation. Man was supposed to have come from lower species of animals, probably aquatic. A crude explanation for the sequence of evolution was thus developed by observation and reasoning.

Xenophanes

Evolutionary theory became somewhat more objective when Xenophanes (576-490 B.C.), a pupil of the mathematician, Pythagoras, identified fossils of water animals on dry land high up in the mountains. These, he declared, represented proof that the mountains were at one time under the water. He correctly interpreted fossils as forms or replicas of organisms that had lived in earlier periods of the earth's history. Aristotle (384-322 B.C.), and his contemporaries, did not believe in fossilization, and the study of fossils was neglected by later Greeks. Galen (A.D. 130-200) revived the interest in fossils and again identified them as remains of animals previously living on the earth. The speculations of Xenophanes were again viewed with favor.

Anaximenes

Anaximenes (570-500 B.C.), another philosopher from Miletus, considered air, with its variety of contents, to be the primary matter from which the world and living things were made. Everything is air, he said, in different degrees of density. He believed air had the inherent capacity to condense into water and earth. According to his philosophy, different degrees of condensation resulted in different materials with varying properties. As air became more packed into a given space, he said, it became heavier and more firm. The idea was probably suggested to him by his observations of the industrial process of felting woven materials by pressure. His condensation theory of nature seemed to be confirmed by his observation of the processes of evaporation and condensation of liquids. A diffuse gas-like substance (e.g., water vapor) could become a liquid (e.g., water), with more body and apparently different properties than the gas; or a solid (e.g., ice), in response to different temperatures. Anaximenes considered that air not only enveloped the world, but also penetrated living things and represented the life-giving and life-preserving principle.

Heraclitus

Heraclitus (556-460 B.C.), the great teacher of Ephesus, chose fire as the fundamental element. He believed the world to be governed by law, but not static law. The world was dynamic and constantly in a state of flux. Fire caused change and made worlds rise and perish. Water was antagonistic to fire. Fire was considered reconcilable with man's soul, whereas water caused disease. Intemperance (in drinking) was thought to cloud the soul, because wine made the soul humid. Heraclitus therefore concluded that "the driest soul is the wisest."

His ideas exerted great influence on the natural philosophies of Plato and Aristotle, who lived in succeeding years.

Empedocles and the Theory of Four Humors

Suggestions made by earlier philosophers that a balance among elements was associated with good health were brought together by Empedocles (504-433 B.C.), of Agrigentum. The basic elements of fire, water, earth, and air were thought to originate from combinations of four fundamental qualities: hot, dry, cold, and wet (Fig. 3.1). All changes in the world were associated with different mixtures of the four elements. Plants and animals arose through specific combinations of the elements; and mankind, according to Empedocles, originated when fire cast up out of the interior of the earth shapeless lumps that formed themselves into limbs and eventually changed into man. Growth in childhood was due to increased warmth of the body, while diminution of warmth produced the infirmities of age. Man contained all four elements, but they were represented in different degrees of refinement in different people. The more finely and evenly the elements were mixed in a man, the better he could think. If one particular part of the body contained a more perfect mixture of the elements than the other parts, that one would be highly developed. Empedocles associated the four elements with four constituent humors of the body: blood, black bile, phlegm, and yellow bile. These were supposed to originate from the heart, spleen, brain, and liver, respectively. This formed the basis for the humoral theory

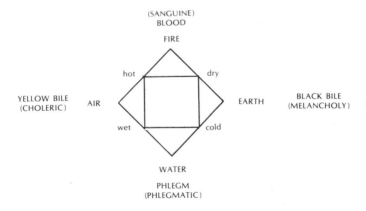

Figure 3.1.   Diagram illustrating the theory of the four humors.

of disease, later incorporated into the writings of Hippocrates and perpetuated in part until the nineteenth century A.D. An imbalance of the humors was thought to cause an individual to be sanguine, melancholy, phlegmatic, or choleric.

Fragments of the writings of Empedocles have been preserved and are available for perusal by current scholars. They represent a curious mixture of rational speculation and belief in supernatural intervention. In some places Empedocles boasts about his supernatural gifts and claims power to heal the sick, cure the infirmity of old age, raise the dead, change the direction of the wind, and bring rain or sunshine to the earth. History identifies him as a popular man of affairs and a great benefactor of mankind. He is known to have rid a town of malaria by arranging for the draining of the swampy district. Like Pythagoras, Empedocles believed in the migration of the soul and therefore forbade his disciples to kill and eat animals. Certain plants were also held to be sacred and were not used as food for this reason.

Like Anaximander, Empedocles anticipated Darwin with his speculations on natural origins of living things. He suggested in broad terms the doctrine of the survival of the fittest. He believed that when the earth was younger it produced a much greater variety of living things than existed during his time, "... many races of living things must have been unable to beget and continue their breed. For in the case of every species that now exists, either craft or courage or speed has from the beginning of its existence protected and preserved it." These words hold a clear hint of the doctrine of the survival of the fittest. The suggestion that the earth once had greater creative power than existed in his era may also be read into the statement.

Democritus and the Atomic Theory

Democritus (470-380 B.C.) of Abdura in Thrace was deeply interested in traveling and gaining knowledge. When his father died and left him a large inheritance, he traveled to Egypt and studied mathematics and physical systems. He returned to his home when the money was spent and persuaded his brother to support him in further travel. Finally, when he became established as a great scholar, he was pensioned by the province of Thrace to devote his full time and energy to travel and creative work. In his 90-some years, Democritus compiled about 70 written works. His greatest contribution was his atomic theory. According to this theory, the universe is made of atoms which move in space, and all physical changes are due to the union and separation of atoms.

The biological implications of his theory were associated with epidemics in that these were attributed to particles which originated in celestial bodies and then fell upon the earth and its people. Foreign atoms from other planets were thus invoked to explain the spread of disease. Sleep and death were also associated with atoms; sleep supposedly equated with the loss of a comparatively few atoms, whereas death presumably resulted from a great loss of atoms. Life itself was thus dependent upon the presence of atoms, and death came when the life-giving atoms departed.

Democritus studied and dissected many animals including human beings, and he found that lower animals as well as higher ones had "perfected" organs and were complex. His predecessors had considered increased complexity to be the only distinguishing feature between lower and higher animal forms. Democritus suggested that all organs present in higher animals were present in lower ones. Those which could not be seen were considered to be transparent or too small for observation with the naked eye. (To some extent this has been borne out by more recent studies. Modern tools and experimental methods have proved essential to detailed observations of minute organs.) Democritus considered the brain to be the organ of thought. In this deduction he was more accurate than Aristotle, who spoke of the heart as the center of vital activity and considered the brain to serve only in cooling off the blood. Sensations, Democritus believed, were due to the atoms that emanated from the objects with which people came in contact. Democritus considered the heart to be the organ of courage and the liver to be the organ of sensuality.

Democritus was one of the first of the Greek philosopher-scientists to attempt a classification of animals. His primary differentiation was based on the presence or absence of red blood. Aristotle's classification, developed a century later, followed the same pattern and distinguished the sanguiferous (vertebrates) from the bloodless (invertebrate) animals.

Aristotle's writings include arguments against some of the conclusions of Democritus. Indeed, it is through Aristotle's critical comments that many of Democritus' ideas have been perpetuated. These two great philosopher-scientists had much in common. In several instances modern science has proved Democritus to have been more accurate than Aristotle. In others, Aristotle's observations and interpretations are nearer presently accepted ideas. Democritus claimed, for example, that the spider's web was produced inside the spider's body, but Aristotle maintained that it was cast-off skin. In this case

Democritus was right. On the other hand, Democritus explained the sterility of the mule on the basis of a presumed contraction of the uterus. It is to his credit that he observed that mules were sterile and that, with the meager background available to him, he even attempted an explanation. A more valid answer to the mule problem had to wait until microscopes were made and chromosomes were followed in the meiotic sequence.

Medical Practice Among the Greeks

Greek medicine originated in a religious atmosphere. The Greeks had many gods, several of whom were active participants in the cause or cure of disease. Early Greeks regarded Apollo as the god of disease and healing. In the fifth century B.C. he was largely replaced by Asclepius, who, according to Greek mythology, was the son of Apollo and Cornis and the father of Hygeia. The staff and holy snake of Asclepius are still the symbols of the medical profession. Greek legend informs us that Asclepius became so proficient in the healing art that he was able to raise the dead. This brought him into conflict with other gods whose domains thus were being invaded. Greek mythology indicated that Asclepius lost his life when Pluto complained to Zeus that the prolonged life-span on earth was cutting down the population in Hades. Asclepius was slain with a "thunderbolt."

Asclepius may have been a real physician whose history followed the same pattern among the Greeks as had Imhotep's among the Egyptians. He was glorified as time went on, eventually was deified, and temples were erected to his honor. His name thus became woven into Greek mythology. Many practitioners of the healing arts, particularly those who worked with the poor and incurables, were priests of Asclepius. Healing was accomplished in the temples located in the communities and along the countryside. One treatment depended on the presumed healing powers of certain nonpoisonous serpents that were trained to lick the wounds of patients who visited the temples. In connection with the temples, sanatoria called Asclepia were established and became the first institutions provided for the care of the sick. Early methods for treating disease, even in Greece, were more supernatural than natural. It was reported that patients went to sleep in a temple and while they slept either a cure would be effected or information required for treatment would be given. Gradually the temples and Asclepia were replaced by hospitals and more modern sanatoria where more realistic treatments were developed.

Among the Asclepians, secret knowledge useful in treating patients was transmitted from father to son free of charge. Outsiders were required to pay large sums of money for the same training, if, indeed, they were allowed to participate at all. Each member of the guild took an oath in which he promised to share his knowledge freely with his colleagues but to withhold it from all others. Two impressive, but perhaps questionable "facts" emerge from the case histories of temple treatment. (1) The patients always were cured, and (2) the cures were mostly miraculous because many or all patients had been regarded as incurable. If there were failures, they were not recorded, and deaths were never mentioned.

Rational School of Medicine

Alcmaeon of Croton, who lived about 500 B.C., was one of the first Greek philosophers to write specifically on anatomy and medicine. His interest in anatomy and physiology and in their relation to medical practice prepared the way for the rational school of medicine. Among his contributions are a description of the optic nerves, a distinction between arteries and veins, and the identification of the trachea. He also identified the tongue as an organ of taste and described its function as follows: "It is with the tongue that we discern tastes. For this being warm and soft dissolves the sapid particles by its heat, while by the porousness and delicacy of its structure it admits them into its substance and transmits them to the sensorium." According to tradition, Alcmaeon was the first to "dare to undertake the excision of an eye." He learned much from this operation and from dissections of other organs, particularly the brain. He was far ahead of his time (before the time of Democritus) in designating the brain as the center of higher activities in man. His contemporary anatomists considered the brain to be merely a glandular organ for secreting phlegm, but Alcmaeon discovered that the brain had no connection with the nasal passages.

Modern medicine has taken much from the rational school of Greek medicine, including a substantial part of its medical terminology. The close relation between Greek and modern medicine in both theory and practice makes the Greek period of some 2,400 years ago seem much closer to us than more recent historical periods. Greek medicine, at its best, differed from that of all earlier periods and most of the more recent ones. It treated disease as a natural rather than a supernatural phenomenon. The Greeks assimilated the best thought and practice from previous cultures, particularly the Babylonian and Egyptian, and propelled medicine beyond its beginnings.

Figure 3.2
Hippocrates, Greek
Father of Medicine.

Hippocrates

The great physician, Hippocrates (460-377 B.C.; Fig. 3.2), who became the "father of medicine" was born and raised on the island of Cos (Fig. 3.3), in the Aegean Sea, a short distance off the coast of Asia Minor. This was a most favorable environment at that time for a prospective physician. Little is known about the early life of Hippocrates except that his father belonged to the guild of physicians. While a young man, Hippocrates traveled widely and became familiar with the cultural accomplishments of contemporary Greeks. He learned, practiced, and taught medicine on his home island and became the foremost member of the school that later became known as the "School of Hippocrates."

Hippocrates was a member of the newly-emerging rational school of Greek medicine. This school was composed of physicians, who, while not necessarily irreligious, were naturalists who separated their practice from their personal religious beliefs. A group of physicians of this school of thought established themselves on the island of Cos during the sixth century B.C. Tradition was built up among them for objectivity and critical evaluations of medical procedures. Other similar groups were located in the Greek colonies at Cnidus, Croton, Rhodes, and Cyrene. The first medical communities or schools were thus established. They were generators of medical thought and tradition rather than strictly teaching institutions in the modern sense. Teaching was done by individual masters and the students were apprentices. Schools were thus composed of mature medical men who came together for mutual assistance and improvement as well as to instruct students.

Hippocrates considered his profession to be an art—the most challenging of all the arts. He believed that even though the practice of medicine depends to a large extent on science, physicians must not lose sight of its artistic aspects. Good health, he thought, depends on emotional and social relations as well as on physical well-being. "For where there is the love of man," Hippocrates wrote in his precepts, "there is also love of the art."

In practice, the Hippocratic physicians treated the individual patient rather than the disease. The whole body was considered, not merely the part that seemed to show symptoms. These physicians knew little

Figure 3.3.   Greek lands and islands
in the Aegean Sea.

anatomy or physiology and possessed neither clinical thermometers nor stethoscopes. They knew a great deal about mankind in illness and in health, however, and used the scientific method effectively. The chief agents employed in the rational school of medical practice were sunlight, fresh air, pure water, exercise, and a proper diet.

The writings of members of the Hippocratic school are full of sound observation and logical reasoning. Methods, in general, were conservative and the healing powers of nature were properly taken into account. The physician took the attitude that he was merely helping nature provide a cure by prescribing improved diet or other environmental modifications. Only when these methods were inadequate did he resort to more violent treatments such as bloodletting to reestablish a favorable balance among the humors.

The Hippocratic theory of the four humors was basically similar to that developed by Empedocles but incorporated some modifications. Hippocrates accepted the theory of the four elements and the four qualities described by Empedocles as a basic explanation for structural and functional relations in the physical world. The theory of the four humors brought the living world into the same scheme. A logical extension was the balance theory relating the humors to conditions of illness and health. Physicians who based their medical philosophy and practice on this foundation were using the best naturalistic approach to human physiology that was available to them.

More than fifty books attributed to Hippocrates were collected at Alexandria in the third century B.C., and became the *Corpus Hippocraticum*. It is not known which, if any, of these books were actually written by Hippocrates. They represent the thinking of the school at Cos, but probably were not written by a single author. Several of the books deal with diet, one describes epidemic diseases, and another lists symptoms by which common diseases may be recognized. Several of the Hippocratic books deal with fractures, dislocations, ulcers, and wounds, and all these include excellent descriptions and conservative treatments.

Hippocrates' concern with the overall health problems afflicting his society was reflected in his book entitled *Aphorisms*, in which he made wise and pointed comments. Some examples of his aphorisms are as follows: (1) "Growing bodies have the most innate heat; they therefore require the most food, for otherwise their bodies are wasted. In old persons the heat is feeble, and therefore they require little fuel, as it were, to the flame, for it would be extinguished by much. On this account, also, fevers in old persons are not equally acute, because their bodies are cold." (2) "Old persons endure fasting more easily, next, adults; young persons not nearly so well; and most especially infants, and of them such as are of a particularly lively spirit." (3) "Such persons as are seized with tetanus die within four days, or if they pass these they recover." (4) "All diseases occur at all seasons of the year, but certain of them are more apt to occur and be exacerbated at certain seasons." (5) "In the different ages the following complaints occur: to little and newborn children, aphthae, vomiting, coughs, sleeplessness, frights, inflammation of the navel, water discharges from the ears."

One of the best known of the Hippocratic books is entitled *Airs, Waters and Places*. This treatise related disease to environmental conditions in different parts of the known world. Physicians about to set up practice in an unfamiliar town were advised to observe prevailing winds, water supply, nature of the soil, and the habits of the people. They would then, according to the book, be able to determine what diseases were the most prevalent and would thus "achieve the greatest triumphs in the practice of the art." Consideration must be given to the influence of climatic conditions upon the minds as well as the bodies of men. When the land was oppressed by winter's storms and burned by the summer sun, the author said, men became hard, stubborn, and independent, and developed more than average intelligence in the arts. In times of war, rigorous environmental conditions gave them more than average courage. Like all other Hippocratic books, this volume stressed the naturalistic approach to prevention and cure of disease. Great emphasis was placed on observing the symptoms of the patient. Disease processes in individual people were given much greater importance than broad generalizations that were made without critical observations. This book has been hailed as the first classic in medicine that emphasized medical geography or anthropology.

The famous book, *On the Sacred Disease*, a discussion of epilepsy, makes a strong case for the natural in contrast to the supernatural explanation of disease. The author begins with the statement that this dreaded disease is no more sacred than other diseases. It must have a natural cause like other diseases but the cause is not understood; therefore, it has been considered supernatural or sacred. When ignorance is dispelled and the disease is eventually understood, there will be nothing sacred about it.

Three of the Hippocratic books, *The Law, The Physician*, and *Oath*, deal with professional attributes and ethics of the physician. These writings reflect the unprotected social position of the medical man. He had to rely on himself for the establishment of a good reputation among his patients, and he could not afford failure. It was customary, therefore, for physicians to reject cases that were considered incurable in order to protect themselves in every way possible from failure or suspicion of malpractice. The following oath, taken by Greek physicians and now administered to graduating medical students, suggests the care that was taken to keep the conduct of the doctor on a high ethical plane and above reproach.

> I swear by Apollo Physician, by Asclepius, by Health, by Panacea and by all the gods and goddesses, making them my witnesses, that I will carry out, according to my ability and judgment, this oath and this indenture. To hold my teacher in this art equal to my own parents; to make him partner in my livelihood; when he is in need of money to share mine with him; to consider his family as my own brothers, and to teach them this art, if they want to learn it, without fee or indenture; to impart precept, oral instruction, and all other instruction to my own sons, the sons of my teacher, and to indentured pupils who have taken the physician's oath, but to nobody else. I will use treatment to help the sick according to my ability and judgment, but never with a view to injury and wrongdoing. Neither will I administer a poison to anybody when asked to do so, nor will I suggest such a course. Similarly, I will not give a woman a pessary to cause abortion. But I will keep pure and holy both my life and my art. I will not use the knife, not even, verily, on sufferers from stone, but I will give to such as are craftsmen therein. Into whatsoever house I enter, I will enter to help the sick, and I will abstain from all intentional wrongdoing and harm, especially from abusing the bodies of man or woman, bond or free. And whatever I shall see or hear in the course of my profession in the intercourse with men, if it be what should not be published abroad, I will never divulge, holding such things to be holy secrets. Now if I carry out this oath, and break it not, may I gain forever reputation among all men for my life and for my art; but if I transgress it and foreswear myself, may the opposite befall me.
>
> ——*Hippocratic School*

Basic Biology Among the Greeks

Aristotle

Aristotle (384-322 B.C.) was born at Stagira, on the Strymonic Gulf. He learned medicine from his father, who was physician to Amyntas II, the ruling Macedonian king. Amyntas, and his son Philip, who succeeded him, were cultured men who did much to attract scholars to their court. In this environment Aristotle had the best opportunities for education that were available in his time.

At the age of 17, Aristotle went to Plato's school, the Academy at Athens. Aristotle was associated with Plato's school for some 20 years, although he did other things during part of this period. When Plato died, Aristotle, then 37 years of age, expected to become master of the Academy. This did not materialize, and he subsequently went to the island of Lesbos (Fig. 3.3) to study marine biology.

Aristotle was one of the greatest biologists who ever lived. He was more than a biologist, however; he was interested in *all* knowledge available in his day. With his associates he undertook the tremendous task of writing down in a series of books everything known at that time. Greek catalogues list 146 books attributed to Aristotle; however, many manuscripts have since been lost. Some were copied by copyists who were not accurate, and thus the manuscripts were confounded.

Four important works on biology dealing mostly with animals and containing Aristotle's methods, observations, deductions, and theories have been preserved: *Natural History of Animals, On the Parts of Animals, On the Generation of Animals,* and *On the Psyche.* Aristotle is known to have written on botany and other biological subjects, but these books have been lost.

Perhaps Aristotle's greatest contribution to modern science was his demonstration of the scientific method. Much of his knowledge was based on his own observations and deductions. A vast amount of information was collected, classified, recorded, and evaluated. Aristotle was a keen observer and interpreted most of his observations accurately, but he was not an experimenter. Because chemistry and physics had not been developed, an accurate study of physiology was impossible. There was no substantial body of facts in any field and no consistent scientific nomenclature. Under such circumstances some inaccuracies of observations and reasoning were bound to occur.

The centuries preceding Aristotle had bequeathed little but vague speculation. In his own words, translated and quoted by Romanes (1891), we get a picture of the problems he faced: "I found no basis prepared; no models to copy. Mine is the first step and therefore a small one, though worked out with much thought and hard labor. It must be looked at as a first step and judged with indulgence. You, my readers, or hearers of my lectures, if you think I have done as much as can fairly be required for an initiatory start, as compared with more advanced departments of theory, will acknowledge what I have achieved and pardon what I have left to others to accomplish."

It is the glory of Aristotle that both his observations and reasoning on biological subjects can stand present-day comparisons as well as they do. Considering the handicaps he had to face and the enormous range of his work in biology alone, it is surprising that he accomplished as much as he did. His were original investigations, and he was, indeed, a great pioneer in biology as well as other areas of learning.

Marine fauna seemed to interest Aristotle more than land animals. His observations and interpretations of the structure and habits of sea animals were remarkably accurate. Aristotle's writings on marine biology refer frequently and with particular intimacy to places on or near the island of Lesbos. Here he spent at least two years and it is probable that most, if not all, of his observations on marine life were made during this period. No doubt much of his philosophy for which he became known also developed during his stay at the seashore at Lesbos.

Aristotle made many keen observations at Lesbos. Starfish were so abundant as to be a pest to the fishermen. Scallops had been exterminated by a period of drought and the continual working of the fishermen's dredges. There were no spiny fishes, sea crawfish, spotted or spiny dogfish. In this region all the fish, save only a little gudgeon, migrated seaward to breed. Big purple murex shells were observed at Cape Lectum, and different kinds of sponges were found on the landward and seaward side of Cape Malia.

Aristotle noted the abundance and variety of marine life and became interested in the adaptations and apparent intelligence of animals. In speaking of the sea, Aristotle said, "Now the sea is not only water but much more material than fresh water, and hot in its nature, it has a share in all living things which are produced in connection with each of these elements."

One might picture the great observer in a boat on a clear day, when the shallow water was calm, watching sea animals in their native haunts. He had no instruments and had to make all his observations with the naked eye. It is remarkable indeed that he was able to describe so accurately such small objects as eggs and developing embryos of molluscs and fish. Motivated by an intense curiosity, Aristotle watched the sea animals methodically day after day, recording what he saw. No doubt he talked frequently with the local fishermen and received enthusiastically any bits of information they were able to give.

An example of his thoroughness can be seen in his description of a certain mollusc called the "cuttlefish" or sepia. The cuttlefish was a common article of food in the Mediterranean countries and many kinds were known to fishermen. Aristotle knew most of the common kinds and described their morphology, habits, and development with far more accuracy than if he had simply gleaned a mass of fragmentary information from fishermen. He began with an outline of the general form, described the body and fins, and the eight arms with their rows of suckers, and mentioned the abnormal position of the head. He described the two great teeth that formed a beak with which the animals could cling to rocks and sway back and forth like ships at anchor. Aristotle observed the two long arms in the sepia and noted their absence in the octopus. After carefully dissecting several species and noting differences, he described the eggs and the embryos with the yolk sac attached to the head. He noted that in certain males, observed in the breeding season, one of the arms was modified into a long coiled whiplash, used in copulation.

Aristotle was interested in problems of reproduction and described in minute detail the breeding habits of certain marine organisms. He traced the development of at least two molluscs (octopus and cuttlefish or sepia) from egg to adult, and noted that the egg of each is undifferentiated. In describing the development of the octopus, he said that breeding occurs in the spring with the female laying numerous eggs and brooding over them. After about fifty days, the eggs burst and creatures like little spiders creep out in great numbers. The characteristic form of their limbs is not yet visible in detail, but their general outline is clear. They are small and helpless and many perish. The eggs of the cuttlefish, he said, look like bunches of grapes stuck together by a moist, sticky substance extruded by the female, and are not easily separated from one another. At first the eggs are white but they become larger and black after the sticky material has been extruded. A young cuttlefish is distinctly formed inside the egg, and comes out when the egg bursts.

Regarding the breeding habits of the crustacea, Aristotle said that crustacea copulated by fitting their tails to one another. Males have fine spermatic ducts and females a membraneous uterus alongside of the intestine. Aristotle made detailed comparisons of different kinds of eggs. In birds' eggs the white portion and the yellow yolk were separated, but fish eggs were of one color and the corresponding materials were completely mixed. Fish had one umbilicus like birds, connecting the embryo with the yolk. In cartilaginous fish the embryo was connected with the whole egg. The embryo, he said, gets nourishment from the yolk, and as it is being consumed the flesh encroaches upon it and grows around it.

Most fish he found to be externally oviparous, laying perfect eggs. The "frogfish" was similar to cartilaginous fish, except that the exterior of its eggs was harder and their growth was more rapid. Eggs of the frogfish, Aristotle observed, grow because they have superfluous yeasty matter in them. The eggs cannot complete their whole growth in the uterus because each frogfish produces so many. They are, therefore, small when set free and although they grow quickly, most are destroyed before hatching. Hence frogfish are prolific, for nature makes up for destruction by large numbers. Some female frogfish, Aristotle observed, actually burst as their eggs increase in size.

Speaking of fertilization of frogfish, Aristotle says, "The eggs do not arrive at completion unless the male sprinkles his milt upon them." Further, he says that the males have their milt and the females their eggs at about the same time of the year, and the nearer the female is to laying, the more abundant and the more liquid is the milt formed in the male. The union of the fishes is brief, and during Aristotle's time it had not been observed by many fishermen. Aristotle erroneously supposed that copulation generally took place in common bony fish (teleosts) before the eggs were laid, even though he knew that the eggs must be influenced by the male element after they were deposited in the water.

Aristotle was interested in the relations between an animal and its environment. He studied adaptations of sea animals, and took particular interest in the migrations of fish. Following is a summary of his account of the habits of a certain catfish, *Parasilurus aristotelis*.

> The eggs are deposited in shallow water, generally close to roots or close to reeds. They are sticky and adhere to the roots. The female, having laid her eggs, goes away, but the male stays and watches over the eggs, keeping off all other fish that might steal the eggs or fry. He thus continues for forty or fifty days, until the young are sufficiently grown to escape from the other fishes. Fishermen can tell where he is on guard, for, in warding off predators, he sometimes makes a rush in the water and gives utterance to a kind of muttering noise. Knowing his earnestness in parental duty, the fishermen catch him by the hook while he is on guard. Even if he perceives the hook, he will still stay near his charge, and will even bite the hook to pieces with his teeth.

For many centuries the only notice taken of Aristotle's account was to laugh at it. Catfish in Europe do not care for their young in this fashion, though some can make a noise with their gill covers. About the middle of the nineteenth century, Louis Agassiz studied the habits of the American catfish and observed that the male looks after its young, just as described by Aristotle. Agassiz had catfish sent to him from the Achelous River and found them to follow Aristotle's description exactly.

Aristotle also observed and described the torpedo and angler fish, both of which are aided in obtaining food by an electric discharge. Following is Aristotle's description of the habits of these two fish: "In both of these fish the breadth of the anterior part of the body is much increased. In the torpedo, the two lower fins are placed in the tail, and the broad expanse of its body serves as a fin. In the angler, the upper fins are placed behind the under fins which are close to the head. The torpedo stuns the creatures that it wants to catch, over-powering them by the force of shock in its body and then feeding upon them. It hides in the sand and mud and catches all creatures that swim within reach of its stunning power. The angler stirs up a place where there is plenty of mud and sand and hides himself there. He has a filament projecting in the front of his eyes that is long, hairlike, and rounded at the tip. It is used as bait. Little fish on which the angler feeds swim up to the filament, taking it for a bit of seaweed that they eat. Then the angler raises the filament and when the little fish strike against it, he sucks them down into his mouth. It is evident that these creatures get their living thus since, even though sluggish, they are often found with mullets in their stomachs and mullets are swift fish. Moreover, the angler is usually thin when taken after having lost the tips of its filaments."

Aristotle had no formal classification system, but he conceived the evolutionary concept and distinguished animals according to function. A classification scheme, based mainly on reproduction, can be drawn up from his definitions. His first great division separated red-blooded animals (vertebrates) from those without red blood (invertebrates). He also distinguished between land and sea animals. He noted that certain sea animals such as fish live entirely in the water, whereas others such as the otter, beaver, and crocodile spend most of their time in water but breathe air and reproduce on land. Some sea animals swim, others creep, and some are adherent. Animals can easily be divided into large divisions such as birds, fish, and whales. Aristotle observed that the large categories can be further divided into such groups as inkfish, shellfish, and crayfish. His careful discrimination in categorizing attests to his keen observation of relations. Since mammals have lungs, breathe air, are warm-blooded, and give birth to living young, he classified dolphins and whales as mammals rather than true fish.

Aristotle recognized four classes of animals without red blood. Three of these, Malakia, Malacostraca, and Ostracoderma, were made up largely of marine animals. His Malakia would be placed in today's class Cephalopoda, and his Malacostraca with the Crustacea. His Ostracoderma contained what we list as certain miscellaneous molluscs and scattered representatives of lower phyla.

At the age of 42, Aristotle was recalled to the court by King Philip to tutor the king's son and royal heir, Alexander. Aristotle influenced Alexander greatly although, as Alexander assumed additional responsibility in government, he became more interested in military pursuits than in academic accomplishments. When Philip was assassinated, young Alexander took the throne and entered upon a career of military conquest. Aristotle returned to Athens and established his school, the "Lyceum" (Gr. luminous, light bearing), in a garden. Here he walked and talked with students.

When Alexander died in 323 B.C., there was much unrest and civil strife in Greece. Aristotle had been closely associated with Alexander and the fallen government. He was under suspicion when the revolutionary forces came to power and fled from Athens to save his life, spending his last year in exile. He died in 322 B.C. at the age of 62.

Theophrastus, the Botanist

Aristotle's work on botany was lost, but fortunately the work of Theophrastus (380-287 B.C.) on plants was preserved to represent plant science of the Greek period. Theophrastus was born on the island of Lesbos, where he may have met Aristotle. He went to Athens, attended Plato's Academy, and later transferred to Aristotle's Lyceum. Theophrastus was devoted to Aristotle and succeeded him in the Lyceum, where he continued Aristotle's policies and remained for 30 years after Aristotle died.

Writings of Theophrastus on the history of plants were prepared as scrapbooks with stories of plants and folklore freely intermingled. Most biographers believe that Theophrastus traveled very little, but he somehow obtained exotic botanical specimens such as pepper, cinnamon, banyan, and frankincense which were unknown to the fields and gardens of Greece. Perhaps these came from his students, from travelers who were trained as collectors, or from military people including Aristotle's former student Alexander the Great, who traveled widely. Evidence indicates, however, that Theophrastus leaned primarily on his own observation of living plants in accumulating material for his writings. From these observations, he prepared descriptions of plants and plant parts that were quite accurate. He did not consider only the practical aspects of his studies of plants, as did most of his contemporaries and many of his followers. Instead, he displayed a predilection for basic science and went much further than just identifying plants that had medicinal values. He discussed, besides sources of plant juices and techniques for their extraction, such properties as viscosity, gumminess, odor, and color. Thus his work became not only a book for tradesmen and medical practitioners, but also a valuable document for contemporary and later botanical scholars.

His great work, *Historia Plantarum (History of Plants,* or perhaps better translated as *Enquiry into Plants),* formed the basis for the botanical section of Pliny's *Natural History* and indirectly laid the foundation for modern scientific botany. *Enquiry into Plants* was the standard botanical textbook for many centuries. It describes not only the morphological and natural history of plants, as indicated by the title, but also the use of plants in therapeutics. It differentiates among plants, catalogues their parts, and comments on their distribution, cultivation, and economic uses. Subdivisions include: (a) morphology and taxonomy; (b) propagation, especially for trees; (c) ecology and geographic botany; (d) economic botany (especially timber uses); (e) shrubs; (f) potherbs and wild herbs; (g) wild trees; (h) cereals and summer crops; and (i) juices and pharmaceuticals.

Another work of Theophrastus, *The Causes of Plants,* is more philosophical. It deals with growth, life, and reproduction. But like *Enquiry,* it contains data on weather, soil, cultivation, temperature, plant diseases, and periodicity. A section on botanical flavors and colors is included in *Causes.* Theophrastus described to some extent approximately 500 species and variations, mostly of cultivated plants.

Theophrastus explained how frankincense and myrrh were collected and how incisions could be made in the stems of plants to extract the gum. He described, largely from his own observations, the germination of seeds and included good descriptions of the sequence of events. This was the best such account available before the seventeenth century. Theophrastus was the first to differentiate between monocotyledons and dicotyledons. He listed distinguishing characteristics of trees, shrubs, and herbs, and classified plants as annuals, biennials, and perennials. Even though his descriptions of plants are now considered poor, his terminology was used by Linnaeus some 20 centuries later and has been carried over into modern botany. A number of his terms such as "carpos" for fruit, "pericarpion" or "pericarp" for seed vessel, and "metra" for core of the stem have been retained for descriptive purposes in modern botanical literature.

Theophrastus was interested in more basic aspects of plant reproduction as well as in seed germination and development, and he recognized several methods of reproduction. He believed that lower plants originated spontaneously, but that higher plants developed from seeds. He also recognized vegetative propagation. This, he said, could occur from parts of roots, pieces removed from branches, or sections from the trunk. He described cross-fertilization in one plant, the date palm. Environmental factors such as temperature extremes can influence plant development and even today may be confused with genetic variation. Theophrastus believed that the environment controlled influential variables, perhaps the ones that dictated permanent change in offspring. The direct influences of the environment thus were used to explain inherited changes. This belief was held by many observers in the centuries that followed, and it was stated by Lamarck in the eighteenth century as the theory of inheritance of acquired characteristics. A quotation from Theophrastus on the subject is as follows: "The soil seems to produce plants which resemble their parent; on the other hand a few kinds in some few places seem to undergo a change, so that wild seed gives a cultivated form, or a poor form one actually better."

Theophrastus discussed at length variations in plant habitats. On the subject of tree ecology he was particularly discerning. To Theophrastus a tree was the perfect and most important kind of plant. As Aristotle regarded the physical man as the hierarchial pinnacle of the animal kingdom, Theophrastus considered the tree to be the top of the plant world. His detailed and accurate observations of different parts of a dozen kinds of trees were catalogued and compared. In his ecological studies of trees he wrote that some trees grow only in protected areas whereas others thrive in wind-swept regions. Some trees, such as the alder and the willow, he said, grow well in either moist or dry places. He divided mountain trees on the basis of habitat into three categories: (1) those that thrive on the exposed sunward side, (2) those living on cold north declivities, and (3) those on frigid summits.

Summary

Perhaps the obvious should be emphasized at this point; it is an oversimplification to associate important developments with single individuals and particular periods of time. Most contributions to knowledge made during the Greek period, or at any time, are not made solely by the persons who may be credited. Many people, sometimes over long intervals of time, participate in most of the significant developments in biology as well as in other fields. Details tend to be sifted out by time, and landmarks are raised even higher, thus becoming increasingly accentuated for the convenience of historians and the ease of students. After many centuries, only the skeleton of history remains, stripped of much of the supporting detail that undoubtedly was an important part of the original structure.

Main contributions of Greek philosopher-scientists from about 600 to 300 B.C.

Contributor	Dates B.C.	Main contribution
Thales	636-546	Life originated from water.
Anaximander	611-547	Air, earth and *apeiron* necessary for life.
Xenophanes	576-490	Fossils, evolutionary theory.
Anaximenes	570-500	Air, primary factor in world and living things.
Heraclitus	556-460	Fire is fundamental element.
Empedocles	504-433	Theory of four humors.
Democritus	470-380	Atomic theory.
Hippocrates	460-377	Rational medicine.
Aristotle	384-322	Objective observations and theory of basic biology.
Theophrastus	380-287	Investigations on plants.

References and Readings

Aristotle. *Historia Animalium.* Translated by T. Taylor, 1809. London: Robert Wilks Co.
_____. *De Partibus Animalium.* Translated by A. L. Peck, 1937. London: W. Heinemann.
_____. *De Generatione Animalium.* Translated by A. L. Peck, 1943. London: W. Heinemann.
(Other translations of Aristotle's works and a number of books and articles about Aristotle are not listed here. These may be located through the library catalogue.)
Breasted, J. H. 1935. *Ancient Times.* 2nd ed. Boston: Ginn and Co.
Brock, A. J. 1929. *Greek Medicine.* New York: E. P. Dutton and Co.
Cohen, M. R. and I. E. Drabkin. 1948. *A Source Book in Greek Science.* New York: McGraw-Hill Book Co.
Coonen, L. P. 1957. "Theophrastus revisited." *Centennial Review* 1:404-418.

Edelstein, E. J. 1945. *Asclepius.* Baltimore: Johns Hopkins Press.
Ellegard, A. 1958. *Darwin and the General Reader.* Gôteborg: Gôteborg Universitits Årsshrift.
Farrington, B. 1947. *Greek Science.* London: Oxford University Press. (Also Penguin Books, 1953.)
Harvey-Gibson, R. J. 1919. *Outlines of the History of Botany.* London: A. and C. Black.
Hawkes, E. 1928. *The Pioneers of Plant Study.* London: Sheldon Press.
Hippocrates, trans. F. Adams. 1939. *The Genuine Works of Hippocrates.* Baltimore: Williams and Wilkins Co.
Jaeger, W. 1948. *Aristotle.* Oxford: Clarendon Press.
Lewes, G. H. 1864. *Aristotle.* London: Smith, Elder and Co.
McKeon, R., ed. 1941. *The Basic Works of Aristotle.* 4th ed. New York: Random House.
Needham, J. 1934. *A History of Embryology.* Cambridge: At the University Press.
Osborn, H. F. 1894. *From Greeks to Darwin.* New York: Macmillan Co.
Reymond, A. 1927. *History of Sciences in Greco-Roman Antiquity.* New York: E. P. Dutton and Co.
Romanes, G. J. 1891. "Aristotle as a naturalist." *Contemporary Review* 59:275-289.
Ross, W. D., ed. 1952. *Aristotle Works.* 12 vol. Oxford: Clarendon Press.
Singer, C. J. 1957. *A Short History of Anatomy from the Greeks to Harvey.* New York: Dover Publications.
Taylor, A. E. 1955. *Aristotle,* rev. ed. New York: Dover Publications.
Taylor, H. O. 1963. *Greek Biology and Medicine.* New York: Cooper Square Publishers.
Théodorides, J. 1965. *Histoire de la Biologie.* Paris: Presses Universitaires de France.
Theophrastus. *De Causis Plantarum* (On Plants). Translated by R. E. Dengkr, 1927. Philadelphia: University of Pennsylvania.
White, P. R. 1963. "Two children of Lesbos." *Phi Kappa Phi Journal* 43(4): 5-13.
Wilson, G. 1929. *Great Men of Science.* Garden City, N. Y.: Garden City Publishing Co.
Wimmer, Fridericus, and Theophrasti. 1866. *Opera Omnia.* Paris: Didot. (Reprinted: Minerva Gmbh, Frankfurt am Main, 1964.)

CHAPTER 4

PRACTITIONERS IN THE HELLENISTIC AND ARABIC CULTURES

ALEXANDER THE GREAT (356-323 B.C.) of Macedonia was an ambitious king with a strong desire for power and conquest. In his military campaign he led armies into Asia Minor, down the coast through Syria and into Egypt. He then turned eastward and marched through Mesopotamia into Persia, and as far east as India. As new areas were conquered, mostly in the Persian Empire, Alexander made certain governmental changes to keep the conquered peoples under his control. In each new governmental unit, some Persians and some Greeks were appointed to serve together as administrators under Alexander's close supervision. He hoped that the Greek and Persian peoples would become united and form one great empire.

Soon after Alexander died, his vast empire disintegrated. His generals took control and eventually divided the army and the con-

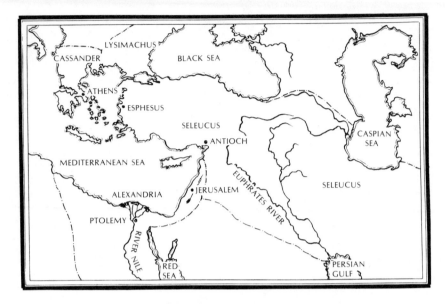

Figure 4.1.   Division of conquered territory by Alexander's
generals following the death of Alexander in 323 B.C.

quered territory among themselves (Fig. 4.1). The unity of the great empire was disrupted, but the Alexandrian conquests had laid the foundation for a new and widespread Hellenistic culture. Features of the classical Greek culture and of oriental cultures taken over by or exerting influences on the Greeks had been combined. In Greek history the Hellenistic Age is usually confined to the period from 323 B.C. to 146 B.C. (when Rome conquered Greece), but in reality Hellenistic culture was the dominant culture of the Mediterranean World from about 323 B.C. to about A.D. 500. The longer period will be considered in this discussion.

One of the most salient features of the Hellenistic culture was the continuation of scientific observation by at least a few individuals in each successive generation. Many Greek scholars moved to other areas where conditions were more conducive to their scholarly work. Some carried with them manuscripts and other tangible products of the Greek culture as well as an attitude and aptitude favorable to learning. It must be remembered that in ancient times there was no public education. Information was communicated in the forum or in the streets, and relatively few people were literate. Several cities and provinces in the Mediterranean region benefited from the dispersal of Greek scholars. Two great centers of culture developed: one at Alexandria, Egypt, and (much later) one at Rome, Italy. Alexandria in the East reached a high level in general culture and made remarkable progress in some limited areas of biology shortly after the fall of Alexander in the third century B.C. In the second century B.C., other Greek cities such as Pergamum, Rhodes, and Antioch shared with Alexandria in cultural leadership. Rome, to the west of Greece, developed much more slowly and reached its highest point around the time of Christ.

Ptolemy I (Soter, 367-285 B.C.) who became King of Egypt, was the wisest and probably the most successful of Alexander's generals. He established his Greek capital of Egypt at Alexandria and founded the Ptolemic dynasty that lasted for three centuries (323-30 B.C.). Ptolemy Soter was not only a general but a cultured man and an author as well. When he took over Egypt he decided that the world must look upon Alexandria not only as a seat of power and social order, a mart of trade and commerce, and the possessor of great wealth, but also as a treasury of learning and science where poets, philosophers, and scientists of the world could have a sanctuary.

Alexandrian Museum and Library System

Fired by enthusiasm for scholarly activity, Ptolemy I established the "Alexandrian Museum" and attracted Greek scholars to Alexandria with scholarships or grants for research and with attractive working conditions. The word museum was derived from the "Temples of the Muses," the religious dwellings named in Greek mythology after the patron goddesses of the humanities. As the museum system grew, botanical and zoological gardens as well as study halls were established. The main Alexandrian Museum was not merely a show place for specimens or a storage area for manuscripts, although it did have ten great halls filled with manuscripts, with each hall representing a different division of Hellenist knowledge. More importantly, it sheltered distinguished scholars whose work imparted lasting significance to the museum and greatly influenced the progress and direction of ancient science. Some scholars were engaged in translation, others in criticism of earlier and contemporary works, and others were writing new manuscripts. The closest current parallel to the Alexandrian Museum would probably be a government research laboratory complex or a progressive university.

The museum area included a public walk, study halls, library halls where scrolls were stored, and an enclosed meeting place. Living quarters were provided for perhaps one hundred or more scholars, with a common dining room where they were served their meals. The museum offered the scholars of that age the incomparable resources of a library which included the greatest collection of intellectual materials ever assembled in ancient times, and in their investigations. Many men of the highest rank, such as Euclid, a Greek geometrician, and Herophilus, a Greek anatomist and physician, accepted Ptolemy's invitation and came to Alexandria.

An Athenian orator, Demitrius Phalereus (Demitrius of Phaleron, 345-283 B.C.), a follower of Aristotle and a friend of Theophrastus, was appointed first chief librarian. He had already proven himself as a scholar and administrator and he gathered the first manuscripts for the library. The library benefited from Demitrius' access to many of Aristotle's writings and other Greek works, and from his vision of its potentials, which Ptolemy was wise enough to implement. Demitrius is said to have collected some 200,000 manuscript rolls. During his tenure as chief librarian, he informed Ptolemy that no record of the Jewish law was included in the library. King Ptolemy became interested in these records and asked the High Priest at Jerusalem to send translators with their laws to Alexandria. Some 70 translators were provided and each was given a cell in which to work at the museum. They translated not only laws, but also other Jewish manuscripts which had already been collected at Alexandria.

The "mother" library or Alexandriana became established in connection with the museum in the Greek section of Alexandria. In keeping with Alexander's hope of blending Greek and Egyptian cultures, a "daughter" library was established in the area surrounding the Temple of Serapsis, a place of worship in the Egyptian part of Alexandria. Some 42,000 rolls, which represented perhaps 100,000 separate works, were included in this library (the Serapeiana).

Ptolemy II (Philadelphus, 309-247 B.C.) was even more enthusiastic about scholarship and culture than his father, Ptolemy I. He attracted other distinguished librarians as well as scholars to Alexandria. One of the most noted chief librarians during the long history of the library was Erastosthenes, who served from 230 to 195 B.C. He was one of the most famous of the mathematicians and geographers of antiquity.

The wealth and power of the Ptolemies, backed by the mighty name of the deified Alexander, gave book dealers a special incentive as they ransacked the Hellenic Mediterranean area, and Asian cities, towns, and countrysides for literary manuscripts and records of every kind. Manuscripts were acquired in every way, honestly and otherwise, by private purchase or unscrupulous force. Greek and Latin writers give conflicting estimates as to the number of books (rolls and scrolls) in the libraries of Alexandria, but the conflicts may be due to different but unspecified periods of time. It should also be noted that a papyrus roll rarely represented a complete or single work in prose or verse.

A famous Latin scholium indicates that a century before the time of Christ the inner or main library contained about 400,000 unsorted rolls and 90,000 rolls arranged in order. The "daughter" library contained some 42,800 classified rolls. For nearly two and one-half centuries (about 295 to 48 B.C.) the Alexandrian Library was recognized as the greatest store of learning in the world. Under good or bad rulers of this Hellenist city of Egypt, the library maintained its importance and continuously increased its stores of manuscripts. It not only held the greatest collection of manuscripts in all antiquity, but was probably the most important library throughout history before the invention of printing.

In 48 B.C. the museum contained some 700,000 manuscripts. It had miraculously escaped the elemental hazards of nature and the vicissitudes of war. It was in its highest glory when Julius Caesar came to Alexandria. Caesar was a scholar of the first rank as well as a distinguished statesman and for a long time had wanted to visit the world-famous museum. He had several reasons for making the trip— political, diplomatic, and personal. Accompanying him were some 3,200 soldiers.

On Caesar's arrival at Alexandria in 47 B.C. his hostess, Queen Cleopatra, in a high spirit of generosity, offered Caesar many of the valuable manuscripts in the museum. Caesar, after some persuasion, gladly accepted. The scrolls were carefully packed and sent to the docks to be loaded on Caesar's ships to be sent to Rome. The people of Alexandria, however, were not as generous as their queen and they rose in arms against the removal of the manuscripts that had been laboriously accumulated over the centuries. A conflict between the Roman soldiers and the Egyptians in the streets of Alexandria forced Caesar and his men to the docks. In the meantime the Alexandrian fleet had surrounded the city. Caesar thus had to use force in order to escape. His soldiers set fire to the Roman ships to prevent them from falling into the hands of the Egyptians. The fire spread from the ships to the docks and many of the books waiting to be carried to Rome were burned. Ancient and modern historians have estimated the number of burned volumes at about 40,000. Thus began the decline of the Alexandrian museum, a great tragedy of history.

Accumulated written materials promote stability in human culture, permitting man to profit by the knowledge and accomplishments of past generations. Without written records, man might never have developed the material facilities or the cultural background necessary for modern civilization. The value of accumulating knowledge can be realized by succeeding generations only if each generation can build upon what its predecessors learned. It is true that information could be passed on orally from generation to generation, but oral transmission is generally inaccurate and unreliable. A written statement remains unchanged as long as the original, unmodified manuscript is in existence.

Even though the Alexandrian Museum and Library began to decline about 47 B.C., it did preserve many valuable scientific and literary works for long periods afterward. Despite further losses during the next five centuries, the library remained essentially intact until A.D. 391, when the Christians decided that all pagan temples, idols, and places of worship should be destroyed.

Under the leadership of Theophilus, Patriarch of Alexandria (from A.D. 385 to 412), a frenzied mob stormed the ancient Temple of Serapsis. In the process, many of the manuscripts in the library halls were destroyed. The final destruction of the mother library occurred when the Arabs conquered Alexandria in A.D. 646. Muslims and Christians presided at the wrecking of a great city and a library that had been the seat of wisdom and beauty for many centuries.

The influence of the great library is obvious when one considers that the poet Homer and other writers and doers of antiquity might have been wholly lost to us if their works had not been maintained in this library. Also, the course of cultural history might have been different if the great store of written knowledge accumulated at Alexandria had continued to be readily available beyond A.D. 391. Some of the books removed from the Alexandrian Library have been and are being found again, but many undoubtedly are gone forever.

Anatomists of Alexandria

Basic biology was not developed at Alexandria, but some practical aspects of medicine rose to a high place. Herophilus and Erasistratus were two great anatomists who worked in Alexandria. The original writings of both of these men were lost, but references to their work by contemporaries give evidence of their great accomplishments.

Herophilus

Herophilus came from Chalcedon in Bithynia to Alexandria and was at the height of his career in about 300 B.C. He was reported to have dissected some 600 human bodies, and based on this experience he wrote a general treatise, *On Anatomy*; a special study, *Of the Eyes*; and a handbook for midwives. The handbook for midwives has been described as a refreshing example of humanitarian zeal that is often apparent in the writing of Greek medical histories. In it the author maintained a contact between biological research and midwifery. Herophilus despised the traditional fear of dissecting human bodies; he may have been the first to make such dissections in public. The enlightened Ptolemic rulers placed needed facilities at his disposal and even allowed him to carry out dissections on live criminals who had been condemned to death. Thus he could study internal organs of living human specimens.

The greatest contribution of Herophilus was his description of the brain as the center of the nervous system and the seat of intelligence. Some two hundred years earlier, Alcmaeon had similarly identified the brain as the center of higher activities in man. In the intervening time, however, Aristotle, for several excellent but erroneous reasons, had transferred the center of intelligence to the heart. Herophilus returned to the views of Alcmaeon after a detailed dissection of the nervous system and the brain. During this dissection he made critical studies of the membranes and sinuses of the brain and identified the fourth ventricle as the "organ of the soul." Present-day nomenclature of the parts of the brain bears traces of his work; his own name survives in one of the venous sinuses of the brain, the *torcular Herophili* (winepress of Herophilus). This sinus is the meeting place of four great veins at the back of the head. Some early anatomists thought this sinus gave rise to a circular or whirling movement of the blood.

Besides studying the brain, Herophilus worked out the structure of the eye and closely observed its membranes and the retina. He also studied the digestive tract and named the duodenum. He found the shape of the liver to be different in different persons. (It is now known that the liver responds readily to body conditions and infections, undergoing many modifications.) According to M. Cary (*A History of the Greek World*), Herophilus came close to another discovery when he investigated the blood system and the lungs. He traced back the motions of the pulse through the arteries to the heart, which he recognized as the cause of pulsations. He did not follow out the circuit of arteries and veins, and he failed to explain the connection between the blood system and the lungs. But his study of pulse and lung rhythms was the most important step towards discovering the circulation of the blood.

Herophilus distinguished tendons from nerves (Aristotle had confused these two structures), and he worked out the anatomy of the genital organs. He defined the science of medicine as perfect knowledge of (1) what takes place during health, (2) the changes caused by disease, and (3) precautions for preserving health and curing disease. Even though Herophilus was of the school of Cos and essentially a follower of Hippocrates, he used drugs much more than did members of the original Hippocratic school. He did support the doctrine of the four humors. His Hippocratic thinking is revealed in one of his aphorisms: "The best physician is the one who is able to differentiate the possible from the impossible."

Erasistratus

The other great student of the human body during the Alexandrian period was Erasistratus, a younger contemporary and rival of Herophilus. Erasistratus was born about 304 B.C. at Cheos, a small island in the Aegean Sea. According to tradition, he was trained in Athens and for a time acted as a physician in the court of Seleucus I of Syria. His competence was recognized, and he was invited to Alexandria where he founded a school of medicine. Because Herophilus was known as an anatomist, Erasistratus called himself a physiologist. He scorned Hippocrates and turned away from the bleeding methods that were associated with the doctrine of the four humors. Erasistratus prescribed simple remedies and advocated a hygienic mode of living. Erasistratus was a follower of Aristotle and is said to have been his nephew.

Erasistratus observed the semilunar, tricuspid, and bicuspid valves of the heart and is credited with having given the tricuspid the name it now bears. His most spectacular observations were those on the functions of arteries and veins. He erroneously described the passage of the blood through the veins and into the arteries by means of small intercommunicating vessels. Erasistratus thought that when the blood reached the arteries, it changed into air or "pneuma" which was pumped throughout the body, every organ receiving the life-giving pneuma. His concept of circulation was incomplete and in error because (1) like the ancients, he did not believe the arteries ordinarily contained blood, and (2) his circulation scheme started with the veins rather than the arteries and therefore the whole system was backwards.

Erasistratus also studied wounds, found lymphatic glands in the digestive tract, and observed the lymph carrying fat (i.e., chyle). He observed the activity of the epiglottis and noted that it guided food and drink past the windpipe. The route taken by food to the stomach was followed and a grinding process of some sort was considered to be a necessary part of digestion. He approached the riddle of metabolism by measuring the intake and excrement of fowls. Sensory and motor nerves were distinguished from each other in the anatomical studies of Erasistratus.

Following the reign of Ptolemy II, the atmosphere in Alexandria was not as favorable for scholarly pursuits as during earlier periods. Later Ptolemys were not intellectually inclined. They withdrew support from research. In some cases learned men were persecuted by tyrants and politicians. The freedom and incentive necessary for science to thrive were lacking, and the Alexandrian period, which had constituted an extension of Greek culture, entered its waning years. Physicians, however, who were primarily concerned with practical applications, took an interest in nature. Some anatomical knowledge was perpetuated, and some plants were studied for their medicinal values. After the death of Cleopatra, the last of the Ptolemys, in 30 B.C., Alexandria became a Roman city. Later, it was made a center of the Christian church.

During the dispersal of Greek scholarship and culture following the death of Alexander the Great, when some scholars went to Egypt, others went to what is now Italy or to Asia Minor. Rome became a cultural center of importance but developed somewhat later than Alexandria.

The Roman Scene

Biology in Rome reached its high point between the first century B.C. and the end of the second century A.D. In large measure, Greek culture was incorporated in Roman culture as Rome gained power and influence. Most Roman biologists were medical men and many were of Greek origin. Political leaders in Rome were more interested in building an empire than in developing culture and science, and they rewarded soldiers rather than thinkers. Romans made significant contributions in law and government, but there was no parallel advance in science. Practical applications were made of Greek ideas in agriculture and medicine, but little progress was achieved in creative science. Rome's greatest contribution to biology came through her development of hygienic measures.

Because much of Italy was swampy and malaria was prevalent, cities were built on hills where the disease was less serious. Sewers were constructed to carry wastes into the Tiber River and people were warned against drinking the river water. Culinary water was brought in great aqueducts to Rome where settling tanks were constructed to purify and aerate the water. Great fountains supplied some 85 gallons per day for each person. Baths were built and they were operated on a high plane at first. Later, however, they degenerated into loafing places where people were more interested in sensual pleasure than in wholesome exercises and bodily cleanliness.

Lucretius

Lucretius (Titus Lucretius Carus, 96-55 B.C.), a cultured Roman who had been educated in Athens, spent most of his productive life writing a single poem entitled *The Nature of Things (De rerum natura)* which was unfinished at the time of his death. It is a long poem, divided into six volumes, dealing with the author's thoughts concerning nature. The poem begins with an invocation to Venus, goddess of creation. Following the invocation is a statement of objectives in which the author outlines his interest in three aspects of the nature of things: (1) beginning, (2) evolution, and (3) dissolution. He attempted to explain the universe in common terms. He believed the earth to be mortal, operating without divine intervention and, therefore, destined eventually to perish. He attempted to explain in natural terms what caused the various courses of the sun, the journeys of the moon, the position of the earth in the center of the universe, night and day, and eclipses. He also discussed the origin of animal and plant life, and such topics as mental defects, demon superstitions, and plagues.

Lucretius was a clear thinker who tried to analyze nature and to discover underlying causes. Intelligent men, he contended, should be able to comprehend nature; their mission should be to banish superstition. In developing the atomic theory that had been pioneered by Democritus, Lucretius believed that atoms formed physical objects, and were infinite in number and space. This constitutes the beginning of the modern theory developed many centuries later by Dalton and Berzilius. Lucretius adapted the atomic theory to living as well as to nonliving entities. The soul was described in terms of atoms, and dreams were believed to depend on particles floating in the air. It was mostly because of Lucretius that the atomic theory survived the Middle Ages (though in obscurity because of the disapproval of the Church). During the Renaissance Lucretius was held in high esteem. Lucretius also developed a theory of heredity which anticipated the work of Mendel. He suggested a hereditary mechanism for plants, animals, and finally man.

Pliny

Pliny the Elder (Gaius Plinius Secundus, A.D. 23-79), a naturalist and literary man in Rome, held important positions in governmental and military units. He compiled facts and fables drawn from some five hundred Greek and Latin authors into an extensive encyclopedia entitled *Natural History (Historia Naturalis)*. This compilation represented an immense register in which the author noted the discoveries, the arts, and the errors of mankind. Its great influence persisted through the Renaissance period and beyond. Pliny's *Natural History* was widely read but had little scientific value. It was an uncritical and indiscriminate mixture of animal stories, folklore, and summaries of earlier observations in nature. It tended to make the sciences subservient to practical ends. For example, Pliny included discussions of about 1,000 species of plants, all considered from the medicinal or economic point of view.

The *Natural History* was so comprehensive that it contained, perhaps largely by chance, some anticipations of modern science. Pliny stated that the earth hovered in the heavens upheld by the air, that its sphericity was proved by the fact that the mast of a ship approaching the land is visible before the hull comes in sight. He also indicated that there were inhabitants on the other side of the earth, and that the moon played a part in producing the tides.

As the popularity of the *Natural History* increased and more copies were required, it was changed by copyists and its inaccuracies multiplied. It was, nevertheless, the best work on biology available, and it did much to stimulate people's interest in nature. In sales, Pliny's book was for centuries second only to the Bible.

Pliny's deep interest in nature and natural phenomena was responsible for his death. Curiosity led him to investigate the eruption of Vesuvius which occurred on August 23 and 24 in the year A.D. 79, overwhelming the cities of Herculaneum and Pompeii. At the fateful time of the eruption, Pliny was commander of the Roman fleet that was anchored at Misenum on the Bay of Naples. When Vesuvius on the other side of the bay showed signs of activity, Pliny sailed across to get a closer view of the eruption. Landing at Stabiae he walked inland with a small group of his men, ventured too close to the active volcano, and met the same fate as the inhabitants of Pompeii. He was killed by poisonous fumes. The circumstances of his death were later recorded in a letter written by his nephew who went to investigate.

Celsus

Aulus Cornelius Celsus, who lived in Rome in the first century A.D., is credited with the most famous Latin compilation of medical works, which included information about the Alexandrian physicians. Hippocratic thought influenced him greatly, even though he lived some four hundred years after Hippocrates. His eight books, entitled *De re Medicine,* were written about A.D. 30. Celsus was a compiler, not an original observer. His name was not recorded by ancient physicians, but in 1478 his work was brought to light and found to be among the best on medicine available at that time. The first two books dealt with the effects of many kinds of food and drink on the body. Along with the dietetic information, the technique of blood-letting was described as ". . . very easy to one who has experience, yet very difficult to one that is ignorant. For the vein lies close to the arteries, and to these the nerves."

The third book treated such subjects as fevers, madness, dropsy, consumption, jaundice, and palsy. Four cardinal signs of inflammation familiar to the medical student today, calor, rubor, tumor, dolor (i.e., heat, redness, swelling, and pain), were identified. The fourth book dealt with internal diseases, the fifth was an account of drugs and their uses, and the sixth treated special subjects such as skin, eyes, ears, and teeth. An example of the type of information included in this book is a quaint means of removing a foreign body from the ear. The patient, lying on the affected side, was bound to a board and the board was struck with a hammer; "thus, by shaking the ear, what is within it drops out."

The two remaining books were on surgery and dealt with such subjects as removal of arrowheads, and operations for goiter and hernia. A cataract operation was described as follows: "A needle is inserted through the two coats of the eye until it meets resistance, and then the cataract is pressed so that it may settle in the lower part." "Tonsils," he writes, "that are indurated after an inflammation, since they are enclosed in a thin tunic, should be disengaged all round by the finger and pulled out."

In this book the attributes of the surgeon are set forth: "He should be youthful or in early middle age, with a strong and steady hand, an expert with the left hand as with the right, with vision sharp and clear, and spirit undaunted; so far void of pity that while he wishes only to cure his patient, yet is not moved by his cries to go too fast, or cut less than is necessary." The surgical instruments described in these books correspond to those found in the house of the surgeon at Pompeii (destroyed A.D. 79) that are now displayed in the Naples museum.

Applied Botany in Rome

Applied botany was one of the few areas of biology that made some advances under the Roman Empire. Medical men were the only people then taking any real interest in nature, and their reasons for studying plants were practical. They needed drugs for various ailments, drugs were obtained from plants, and therefore different plants had to be identified. Hardly any scientific terminology had been developed, and plants were most easily identified by means of pictures. Thus originated the art of botanical drawing, near the end of the first century B.C.

Crateuas

Crateuas (50 B.C.), the physician of King Mithridates VI of Pontos, is called "the father of botanical illustration." It is difficult to be sure, however, that he was actually the first to make botanical illustrations. The manuscripts of ancient times were copied by hand and earlier illustrations may not have been perpetuated. Copyists not schooled in botany may not have been able to reproduce original drawings correctly, and earlier illustrations may have been lost. Crateuas is credited with a *Materia Medica* wherein he recorded the effects of some metals upon the human body. The king is known to have been interested in toxicological investigations, and experiments may have been carried on while Crateuas was in his service. Crateuas also composed a treatise on roots *(Rhizotomicon)* divided into at least five illustrated books. He and his predecessor, Theophrastus, paved the way for Dioscorides.

Dioscorides

Dioscorides was a Roman military surgeon in the army of Nero (A.D. 54-68). He was interested in improving medical services in the complex Roman Empire. Drugs obtained from plants represented a critical item in medical practice, and Dioscorides performed an invaluable service by identifying and describing about 600 plants with medicinal value. Some names that he attached to plants have persisted and are now used in plant identification. The work of Dioscorides was compiled in a book about the medicinal values of plants called *De Materia Medica*. Later, Dioscorides' *Materia Medica* was illustrated with pictures of the plants he had described. This increased its usefulness, and the work had great influence for centuries after its original preparation. Dioscorides has been properly recognized as one of the founders of botany.

Anatomy and Medicine in Rome

Galen

Born of Greek parentage at Pergamum in Asia Minor, Galen (A.D. 131-200; Fig. 4.2), received an excellent education and became a distinguished physician and scholar. At the age of 18 he was familiar with Platonic, Aristotelian, Stoic, and Epicurean philosophies, and had already spent two years in the study of medicine. He then traveled and studied medicine and philosophy in the cultural centers of the time: Greece, Phoenicia, Palestine, Crete, Cyprus, and finally in Alexandria. When his education was completed in A.D. 158, he returned to his native city, Pergamum. Here he practiced medicine and served as physician and surgeon to the gladiators. He supervised the diet of men who fought in the arena and treated their wounds. During this period he wrote the first of his many medical treatises.

Six years later he moved to Rome, established a medical practice, and lectured on medical subjects. He was successful in his profession and won recognition when he healed the great philosopher, Eudemus, and treated other distinguished persons. Because of his success he was called a "wonder worker." Galen was confident of his own skill and critical of contemporary medical sects and theories. Enmity between Galen and his fellow physicians forced him to leave Rome. He was later brought back to be the physician to the emperor, Marcus Aurelius. From then on, much of his time was devoted to writing and to practicing medicine among the dignitaries in the court of the emperor.

At the time Galen was establishing his philosophy, Christianity was making itself felt in the Roman Empire. Galen never joined the Christian church, but he was sympathetic to Christianity and much of his philosophy was a part of Christian theology. Galen accepted the

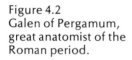

Figure 4.2
Galen of Pergamum,
great anatomist of the
Roman period.

principle of a divine intelligence as the originator and ruler of the world. He referred to the Biblical story of creation but was critical of the implication that something was created from nothing.

Galen wrote 256 treatises on subjects dealing with medicine, philosophy, mathematics, grammar, and law. Fifteen concerned anatomy and were based on the dissection of apes. Most of the treatises were medical in nature, but Galen was interested in many fields. His great book entitled *On Anatomical Preparations*, was the standard medical text for some 1,400 years. This is perhaps the longest "run" a textbook has ever enjoyed. Much of his medical knowledge was based on Hippocrates, but Plato and Aristotle were also important to his background philosophy. Because of prejudice and superstition prevalent at the time, Galen was not permitted to dissect human bodies. Most of his firsthand information was obtained from careful and detailed dissections of such animals as sheep, oxen, dogs, bears, and apes, which he presumed to be essentially similar to human beings. Although not based on dissections of humans, Galen's work remained the undisputed, authoritative reference on "human" anatomy for some 1,400 years, until the time of Vesalius.

In describing the circulatory system (Fig. 4.3), Galen followed the blood from the liver to the heart, where it was supposed to pass through pores in the septum from the right to left heart. He imagined that the blood became mixed here with air from the lungs, thus forming "vital spirits" or pneuma, the life-giving property of the organism. Pneuma was presumed to be carried with the blood to the various parts of the body. Galen believed that a "boiling up" or "fermenting" process heated the body and provided the motion to carry the blood into the body. He supposed that waste from combustion or "soot" went from the right heart to the lungs and out of the body. Thus the blood had a one-way passage through the body. He supposed new blood to be produced, presumably in the liver, to maintain the blood volume. The blood pumped out from the heart, he thought, was used in the body and never returned to the heart. The "vital spirits," however, he believed were circulated back and forth in the blood vessels, carrying the life-giving properties to the different parts of the body. This description of circulation was regarded as infallible from the second century A.D. until Harvey demonstrated the continuous circulation of the blood in the seventeenth century.

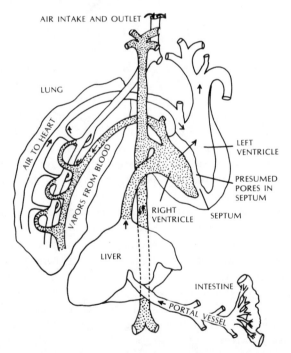

Figure 4.3. Diagram illustrating Galen's description of the circulatory system.

Reasons for the Decline of Greek Science

By the end of the second century, Greek science was virtually dead. The Alexandrian and Roman periods prolonged its influence, but did not supply new impetus. The decline actually extended over centuries before the end came. Several factors were involved in the disintegration of the high intellectual pattern established by the Greeks. Epidemic diseases, particularly malaria, were associated with the prevailing war-and-conquest way of life. The people were insecure and often demoralized during those years, and an intellectual life is difficult to maintain when there is constant fear of illness and death. Dictators, unsympathetic to culture, replaced the scholar-leaders of an earlier age. Philosophies that developed in the Roman world, particularly those of the Stoics and Epicureans, were not conducive to the development of science. Both the Stoics and the Epicureans were subjective rather than objective in approach and neither encouraged observation in nature.

It must be remembered that there was no printing at this time; each new copy of a manuscript required tedious copying by hand. Copyists were not always accurate and sometimes would "improve" manuscripts by omitting parts with which they did not agree, or by inserting their own ideas when the opportunity was presented. They were probably the world's first editors. Information was not widely disseminated and could easily be lost when a local library was disrupted. People in other locations, or at a later period of time, who were interested in the same topics had to start from the beginning and reaccumulate the same information. Thus there was little chance for the survival and accretion of a culture.

The objectivity necessary for science was nonexistent among the cultural schools of the Romans. Philosophical ideas developed by "pure logic" took precedence over those obtained from observation. A philosopher trained in logic could tear down another's work by unsupported arguments because there was no established base in classification, terminology, or irrefutable experimentation. Perhaps even more important, there had been no developments in mathematics that would permit measurements to be made and quantitative data to be used for evaluations. In other words, too much philosophy hindered the development of science.

It has also been suggested by some historians that the institution of slavery tended to keep science in the theoretical and "toy" stage. Advances in mechanics, for example, were not taken seriously and were never applied as laborsaving devices. Masters found their slaves to be "ego-boosters," wanted to keep them around, and were not interested in devising mechanical systems to replace them. History shows that laborsaving devices were invented only when labor was expensive, necessity was pressing, or survival was threatened.

The historian Gibbon has provided a summary of the decline and fall of the Roman Empire which he calls the greatest and perhaps the most awful chapter in the history of mankind.

> ... The various causes and progressive effects are connected with many of the events most interesting in human annals: the artful policy of the Caesars, who long maintained the name and image of a free republic; the disorders of military despotism; the rise, establishment, and sects of Christianity; the foundation of Constantinople; the division of the monarchy; the invasion and settlements of the barbarians of Germany and Scythia; the institutions of the civil law; the character and religion of Mohammed; the temporal sovereignty of the popes; the restoration and decay of the Western empire of Charlemagne; the crusades of the Latins in the East; the conquests of the Saracens and Turks; the ruin of the Greek empire; the state and revolutions of Rome in the middle ages... (E. Gibbon, *Decline and Fall of the Roman Empire*, vol. 6, p. 569.)

Preservers of Greek Thought in the Middle Ages

Little was accomplished in biology from about A.D. 500 to about 1500. Virtually no original contributions were recorded in any science during this long period. The cultural heritage of the past was maintained, though little used, and eventually contributed to the rise of western universities and artistic Renaissance.

Greek manuscripts such as the works of Aristotle that had been maintained in libraries of Alexandria, Athens, Rome, and other cultural centers found their way into Christian monasteries of the Middle Ages, but many were lost or destroyed. The Greek works preserved in the Christian monasteries had little influence until Latin translations were made many centuries later. By this time, many more had been lost or destroyed, resulting in decisive loss to science.

Some manuscripts that had been preserved in earlier centers of learning were taken to Persia by Persian scholars who had studied in these centers and by Greek scholars who moved there seeking a favorable place to live and study. Manuscripts that were maintained in Persia fell into the hands of the Arabs when Persia was conquered in the seventh century. The Arabs used the Greek texts and extended some to make them more useful for practical applications, but conducted few if any original investigations.

As Hellenistic learning was dominated by philosophy, the culture of the Middle Ages was dominated by theology. Christianity greatly influenced the thinking and actions of the people in the Western World, centered in Rome. The Venerable Bede (672-735), English cleric, recorded the ideas of the Church fathers, St. Ambrose, St. Augustine, St. Basil, and St. Gregory, as well as those of the encyclopedists Pliny and Isidore of Seville (560-636). His comments indicate that learning in Western Christendom was predominantly theological and moral. Little or no attempt was made during this era to understand natural phenomena or to improve the material conditions of the people. This is in marked contrast to Greek science, that held as its objective the understanding of the physical world.

Pagan philosophers of the Middle Ages asked, "What is worth knowing and doing?" to which Christian theologians answered, "That which brings about the love of God." There was no expression of natural curiosity. St. Augustine was one of the few Christian fathers to seek a rational basis for faith. He attempted to explain the biblical story of creation in terms of natural processes, and he cautioned his followers against using the scriptures as a source of scientific truth.

In the Western World of this period, facts about nature were used primarily as illustrations for the premises of religion. The moon was the image of the church, reflecting divine light. Wind was the image of the spirit. The number 11, which transgressed 10 (Ten Commandments), was sinful. Magical and astrological properties of natural objects were used as moral symbols and medical associations. The course of disease, for example, was related to the phases of the moon. Medicine and astronomy became closely associated.

Some writers have correlated the rise of Christianity with the fall of ancient science. Actually, the interest in science began to decline well before the time of Christ, and the Christians were obscure and unimportant in political and social developments for several centuries after Christ. In the Middle Ages, when Christianity became widespread, it did little to promote objective inquiry and in some instances discouraged the revival of classical learning. Christian leaders were indifferent to nature, or saw it simply as something to be "used." The church constituted a great social force during the Middle Ages, however, and was the one institution devoted to maintaining order and preserving society and the heritage from the past.

Medicine in Medieval Europe

Greek medicine was perpetuated, on a limited scale through the early Middle Ages, but little progress was made after the time of Galen. It was customary among those who studied classical medicine to content themselves with learning what was available in the Greek texts, which repeatedly were copied, compiled, and interpreted. No original investigations or new observations were recorded, nor were attempts made to expand the store of facts and ideas by recovering information that had been lost or forgotten over the years. Furthermore, the materials which were in existence were written in a language not understood by most people and not available to more than a very few scholars connected with monasteries. The common practitioner had at his disposal only the inferior but usually more practical Latin treatises on drugs and surgery.

Since most of the physicians and compilers of medical texts during the Middle Ages were monks, Christian monasteries became the centers of medical learning. Medicine was again in the hands of religious organizations, as it had been when the priests of Imhotep and Asclepius controlled the theory and practice. Monks were primarily concerned with their religious duties; to them and their superiors, medical work was of secondary importance. Church leaders tended to emphasize the importance of the soul while minimizing the body and bodily ills. To many early Christians, disease was a punishment for sin, or the result of possession by devils or the influence of witches. Treatments required prayer, penitence, and supplication, making every cure a miracle. Some of the less spiritually minded people of the time argued that the body should be strengthened physically to enable it to withstand the attacks of the devil. But such arguments had little influence in church institutions.

Isidore of Seville, a seventh century Spanish bishop and scholar, exemplifies the loss of systematic observation and the consuming concentration upon the supposed superior realities of the spiritual world. Spanish culture in the early Middle Ages was relatively well developed. Spain had been thoroughly Romanized and as part of the Roman world was considered barely inferior to Italy itself.

The greatest of Isidore's voluminous works was an encyclopedia, the *Etymologies*. It consolidated all those remnants of learning that had not been submerged by the superstitions of the pre- and early medieval world. Considered in retrospect, it probably represents the best and most complete résumé of the intellectual outlook of the Middle Ages. In its own time, it was primarily important in keeping the existing fragments of scientific knowledge in circulation, shallow and blundering though it was. Physicians were low on the social scale during these years. Oriental superstitions and magical practices were fashionable. The section of the encyclopedia, "On Medicine," was written mainly for priests and monks upon whose shoulders the art of medicine rested. It was mostly a wild disarray of definitions and origins of words because Isidore believed that knowledge should be based on words and their derivation. In his section on the four humors of the body, for example, he gave no reasons or explanations for the presumed relations of the humors to physical conditions, but included only his conception of how the terms should be used. Isidore recorded no scientific information of lasting value, but he did illustrate the character of the medieval intellect.

At the Council of Clermont in 1130, the practice of medicine by the monks throughout Christendom was forbidden because it interfered too much with religious activities. Medicine then fell largely into the hands of the secular clergy. During the Middle Ages medical schools developed in the East rather than the West and became known collectively as schools of "Arabic science."

The Persian and Arabic Scene

The Arabs in Eastern Europe and Asia created little in science but they preserved the Greek contributions more successfully than did the Christians in the Western World. When Plato's Academy at Athens was finally closed in 529, most of the remaining Greek scholars were exiled to Persia. With them went Greek manuscripts. In the years that followed, Greek thought became fused with Persian thought. When the Mohammedans conquered Persia in 631, the Arabs learned about Greek science from the captive Persians. Greek works were translated into Arabic, which became the universal language of the Arab world. The translations carried the facts and some of the spirit of Greek science.

The common language and religion of the Arabs helped foster their political union. Ambition and religious zeal led to extensive military conquest. Mohammed considered conquest by the sword as a religious exercise and duty. Arabic forces swept into North Africa, Egypt, Syria, Mesopotamia, Armenia, Persia, Spain, Afghanistan, Baluchistan, a large portion of Turkestan, a small portion of India, and the islands of Crete and Cyprus. As the Arabic empire grew and gained strength, translation projects were undertaken to make more of the Greek knowledge of medicine and philosophy available to Arabic scholars. The ninth and tenth centuries have been designated "The Golden Age of Arabic Medicine." The spread of the Arabic language had a wholesome influence on science and particularly medicine. The conciseness of the Arabic language proved valuable in all scientific writing and especially in the translation and application of technical Greek works.

Strict observance of the religious code might have prevented all scientific effort among the Arabs because the Koran, according to Mohammed's dictum, contained all necessary knowledge. Fortunately, this position was not enforced strictly and could be evaded. The religious influence, however, did restrain Arabic research to some extent. To avoid conflict, scholars adopted the practice of presenting their works as commentaries on writings of established scientists of antiquity rather than as original contributions. In philosophy and natural science it was usually Aristotle, whereas in medicine it was Galen that generally was cited as the basis for comment. Brilliantly original thought could overcome the shroud of commentaries, but usually new thoughts became obscured and the older work was reemphasized.

The Arabs were less active in the biological sciences than in the physical sciences such as astronomy, mathematics, and chemistry. Some experimentation, especially in chemistry, was accomplished in the Islamic world. All acids and alkalis identified before the nineteenth century were known in the laboratories of the Arabs. Many practical items such as the magnifying glass, the windmill, and linen paper (necessary for printing) were invented. Words commonly used in modern physical science such as alchemy, nadir, azimuth, aldurism, cipher, alcohol, and algebra have come to us from the Arabs. Arabic numbers (actually originated in India), which are much more practical than roman numerals for scientific usage, were introduced. Any who doubt the advantages of arabic numbers should try multiplying with roman numerals.

Rhazes

Rhazes, a great physician and author of the Islamic world, was born in Rayy, Persia, in 852. As a youth his main interests were centered around music, physics, and alchemy. At the age of 13 he went to Baghdad and began to study medicine under a disciple of Honian who was acquainted with Greek, Persian, and Indian medicine. His later written works indicate that he had become conversant with these cultures and that he had applied his knowledge of alchemy to medicine. Following his education at Baghdad, he returned to his home in Rayy and became head of a hospital. Later he returned to Baghdad as head physician in a hospital in that city.

After Rhazes' reputation as a physician was made, he devoted himself to traveling from court to court and practicing his art. He became famous as a court physician and amassed great wealth, but his money was distributed among the poor and disease-ridden people and he died poor and blind.

Rhazes is famous not only for his practice of medicine, but also for his literary contributions in medicine. He is credited with 237 books of which 36 are now available. His most famous work was a medical encyclopedia, *Continens,* which was assembled from his notes after his death. His most highly regarded (by present-day opinion) work is a small book on smallpox and measles. This is one of the few of his works which has been translated into English. It offers perhaps the first medical description of these two diseases with clear and concise clinical descriptions. The writings of Rhazes contain ingenious and penetrating observations on hiccoughs and results of spinal injury. He introduced mercury compounds as purgatives for human beings after first trying them on monkeys. Unfortunately most of his works have been lost or destroyed. They are known mainly from references made by contemporary and later writers.

Avicenna

One of the most illustrious of the Persian scholars was the physician-philosopher, ibn-Sina, known in the Western World by the Latin name, Avicenna. He was born in the year 980 in the province of Balkh, now Afghanistan. He had mastered the Koran by the age of seven. About that time his father moved the family to the city of Bokhara which was the capital of Turkestan. Here the boy continued his education with the study of arithmetic and logic. He became interested in medicine, which he studied from the works of Aristotle and Galen.

At the age of 17, Avicenna was a practicing physician and successfully cured the Samanid Sultan of Bokhara of an ailment. After this accomplishment, he was appointed court physician and invited to study in the Sultan's library where translations of Greek manuscripts were kept. At the age of 22, Avicenna left Bokhara and traveled from court to court healing rulers, practicing medicine among the common people, and philosophizing. His great success and increasing popularity led to his imprisonment through the influence of jealous rivals. While in prison he began writing his famous book, *Canon of Medicine,* which was later completed in the city of Ispahan (Isfahan) at the court of a friendly prince. Here Avicenna found the peace and security that he needed to continue his writing and his practice of medicine.

Avicenna's writings were in two general fields, medicine and philosophy. They demonstrated his ability to observe, retain ideas, and use logic. Avicenna's philosophical contributions have been classified in five basic areas: logic, physics, psychology, metaphysics, and ethics. He has been credited with over a hundred medical works.

Avicenna's writings were in two general fields, medicine and philosophy. They demonstrated his ability to observe, retain ideas, and to correlate known facts with the systems of Aristotle and Galen. The *Canon* was composed of five volumes. The first two were devoted to physiology, pathology, and hygiene; the third and fourth described methods of treating disease; and the fifth was a materia medica dealing with the nature and properties of substances used to treat disease. Although voluminous, the *Canon* was clear and easy to read, exemplifying the simplicity that could be achieved with the Arabic language.

Although most of Avicenna's written work was theoretical, he was a practical physician. His abilities in using the medical arts to heal the sick were attested to by the fact that he was a court physician for many rulers. Three of Avicenna's more practical medical writings were as follows: (1) *On Vision, Haemolytic Jaundice, and Meningitis.* Avicenna anticipated the modern explanation of vision. He understood the difference between obstructive and haemolytic jaundice and was the first to provide a good description of meningitis. (2) *Malignant Cancer.* Avicenna had experience with tumorous growths and cancer, and he described cases in considerable detail. His descriptions and methods of treatment are quite acceptable by modern standards. The only hope of cure in a malignant disease, he said, is surgical treatment in the early stages. Excisions must be wide and bold; all veins running to the tumor must be included in the amputation. This is not sufficient, however, and the affected area should also be cauterized. Even then, Avicenna added, cure is not certain. (3) *Use of Anesthetics.* It is clear from this discussion that the drugs Avicenna effectively used as anesthetics had been known to the Chinese centuries before.

Avicenna's translated work had great influence in the West. It became available in 1000 when the West was looking into the theoretical as well as the practical aspects of medicine. For some five centuries the *Canon of Medicine* was accepted as a guide in European universities.

Other Scholars

Ibn-Rushd (Latin, Averroes) was a more original thinker than Avicenna, but he was mostly a philosopher rather than a physician. He was born in Cordova, Spain in 1126. As Arabic science was declining at the close of the twelfth century, the fanatical religious reaction in Spain caused him to be banished for heretical opinions. The banishment was terminated a few years before his death, however, and he spent his latter years in Spain where he died in 1198.

Averroes' works were disguised as commentaries on Aristotle, but his views of natural evolution and other natural phenomena were nearer to the present-day concepts than were those of Aristotle. During the Middle Ages the Christian school considered him as merely an interpreter of Aristotle. Later writers, however, particularly opponents of the ecclesiastical philosophy, lauded him and preserved his name and his writings. Some of his ideas are recognizable in present-day concepts of natural science.

When the crusaders from Western Europe reached the Holy Land in the eleventh and twelfth centuries and met the Saracens (Arabs), they had an opportunity to compare medical knowledge in the East with that in the West. The comparisons favored the Arabs and the crusaders learned a great deal. Although the medical literature from the Arabs was much more accurate than that obtainable directly from the West, it did have weaknesses.

As the Arabic literature moved west, many misconceptions were perpetuated. The literature was characterized by an over-adherence to the classical Greek authorities, with a magnification of errors and imperfections introduced by previous copyists. Its emphasis on astrology, aversion to anatomical studies, degradation of surgery (Avicenna ranked this as an inferior branch of medicine), and predilection toward cautery and laudable pus in surgery, all had an adverse influence on the development of medicine in the West. Constantinus Africanus (1020-1087) of Salerno, and Gerard of Cremona (1140-1187), who worked in Toledo, were chiefly responsible for the early translations of the major medical works of the Arabic culture into Latin.

The Western World obtained most of the available works of the Greeks through roundabout courses, often involving several successive translations. For example, much of Aristotle was translated from the Greek into Persian, then into Arabic, and in the eleventh and twelfth centuries into Latin. Early in the thirteenth century Michael Scot translated more of Aristotle's work from Arabic to Latin. From Latin, the Greek works were translated into German and other modern languages.

Medical School of Salerno

The first important Western European medical school of the Middle Ages developed at Salerno in southern Italy. In 846 this institution was already cited as ancient, though it was not officially founded until 1150. Salerno was in a strategic position between the eastern and western cultures. It was close to Sicily, where the Greek-Arabic influence had become concentrated, and yet it was also an integral part of the Greco-Latin civilization. This city not only had access to both Eastern and Western sources of information but constituted a crossroads where many cultures came in contact and a cross-fertilization of ideas was continually in progress. Because of the diversity of thought, the school at Salerno was said to have been founded by a Latin, a Greek, an Arab, and a Jew. Constantine the African (Constantinus Africanus), a Carthaginian monk, was especially influential in the growth of the school. Salerno's period of greatest influence as a medical center began in the eleventh century about the time of Constantine the African and the Norman conquest and lasted until the beginning of the thirteenth century, when its decline began.

Salerno was not a church institution but an organization of men interested in medicine. Medical papers written by members of the school contained, among other things, clinical descriptions of such diseases as dysentery and urogenital irregularities. Therapeutic agents such as mercury ointments for skin diseases and iodine for goiter were described. Excellent surgical work was done by members of the Salerno school. Students came from all parts of the world and, upon obtaining their training, were prepared to practice in any country.

Salerno was famous not only among the medical people but also among the populace. Many crusaders stopped there on their way home from the Crusades to have their wounds treated. Often they would ask for a rule that could help them maintain their health when they were no longer near men of medicine. Terse statements, or aphorisms, were frequently given in answer to such requests. These spead and became a series of prescriptions sometimes more philosophical than medical; but who can deny the truth of a statement such as the following: "If you lack medical men, let these three things be your medicine: good humor, rest, and sobriety."

Summary

When Alexander died in 323 B.C., his Greek empire fell apart. The Greek scholarly influence, although somewhat diluted, spread to a wider geographical area and gave rise to the Hellenistic culture. Principal centers of learning were Alexandria in the East and Rome in the West. At Alexandria a "museum" was established where scholars worked and manuscripts were accumulated. In Rome much energy was devoted to the building of an empire. Science was applied to agriculture, medicine, and sanitation but was not cultivated.

During the Middle Ages some of the Greek manuscripts were available in Christian monasteries and in Persian and Arabic political and religious centers but only limited applications were made of Greek science. Those manuscripts which were used were mostly in the field of medicine. Virtually no original contributions were made in science until the end of the period, when the Renaissance awakening began. At Salerno, Italy, the cultures of the East and West came together and a medical community of considerable importance emerged over a long period of time and was formally founded as a school in the twelfth century.

Landmarks in biology from third century B.C. to twelfth century A.D.

Contributor	Century	Location	Field or contribution
	B.C.		
Herophilus	3rd	Alexandria	Human anatomy.
Erasistratus	3rd	Alexandria	Human anatomy ("physiologist").
Lucretius	1st	Rome	The Nature of Things (one can think out nature).
Crateuas	1st	Rome	Illustrations of plants for medicinal use.
	A.D.		
Pliny	1st	Rome	Encyclopedia of natural history.
Celsus	1st	Rome	Books for medical practitioners.
Dioscorides	1st	Rome	Herbal for medicinal use.
Galen	2nd	Rome	Anatomy and practical medicine.
Isidore	7th	Seville, Spain	Encyclopedia with section on medieval medicine.
Rhazes	8th	Persia	Encyclopedia of practical medicine.
Avicenna	10th	Persia	Codification of medical knowledge.
Constantinus	11th	Salerno, Italy Toledo, Spain	Translated Arabic (Greek) medical works into Latin.
Averroes	12th	Cordova, Spain	Philosophy and medicine.
Gerard	12th	Cremona, Italy	Translated Arabic (Greek) medical works into Latin.

References and Readings

Cary M. 1932. *A History of the Greek World 323-146 B.C.* London: Methuen and Co.

Clagett, M. 1955. *Greek Science in Antiquity.* Part II. New York: Abelard-Schuman.

Coonen, L. P. 1957. "Biologists of Alexandria." *Biologist* 39:13-18.

Dioscorides, P. *De Materia Medica.* 1478. Colle di Valdésia: Johannes de Medemblick.

——————. *Codices Selecti.* 1970. Graz, Austria: Akademische Druck-u. Verlagsanstalt.

Duckworth, W. L. H., trans. M. C. Lyons and B. Towers, eds. 1962. *On Anatomical Procedures.* Cambridge: At the University Press.

Farrington, B. 1947. *Greek Science.* London: Oxford University Press. (Also Penguin Books, 1953).

Forster, E. M. 1961. *Alexandria, A History and Guide.* Garden City, N. Y.: Anchor Books.

Galeni. *Prima Classis Humani.* 1541. Venetiis: Lunte Florentini.

Gibbon, E. 1910. *The Decline and Fall of the Roman Empire.* 6 vol., Everyman's Library. New York: E. P. Dutton and Co.

Graubard, M. 1964. *Circulation and Respiration.* New York: Harcourt, Brace and World.

Green, R. M., trans. 1951. *Galen's Hygiene.* Springfield, Ill.: Charles C. Thomas.

Greene, W. C. 1933. *The Achievement of Rome.* Cambridge, Mass.: Harvard University Press.

Hadzsits, G. D. 1963. *Lucretius and his Influence.* New York: Cooper Square Publishers.

Hamilton, E. 1957. *The Echo of Greece.* New York: W. W. Norton Co.

Kilgour, F. G. 1957. "Galen." *Sci. Amer.* 196(3):105-114.

Latham, R. E., trans. 1951. *Lucretius, On the Nature of the Universe.* London: Penguin Books.

Lucretius. *On the Nature of Things.* Translated by H. A. J. Munro, 1929. London: G. Bell and Sons.

Needham, J. 1934. *A History of Embryology.* Cambridge: At the University Press.

Parsons, E. A. 1952. *The Alexandrian Library: Glory of the Hellenic World.* Amsterdam: Elsevier Press.

Plinii, C. *Liber Secundus de Mundi Historia.* 1553. Francofurti: Petrum Brubachium.

Plinius Secundus, C. *Natural History.* English translation by H. Rackham, 1938. Cambridge, Mass.: Harvard University Press.

Sarton, G. 1954. *Galen of Pergamon.* Lawrence, Kansas: University of Kansas Press.

Sarton, G. 1959. "Forgotten men of science." *Saturday Review* 42:50-55 (March 7, 1959).

Singer, C. 1957. *A Short History of Anatomy from the Greeks to Harvey.* New York: Dover Publications.

——————. 1958. *From Magic to Science.* London: E. Benn, 1928; New York: Dover Publications.

Wethered, H. N. 1937. *The Mind of the Ancient World, A Consideration of Pliny's Natural History.* New York: Longmans, Green and Co.

Wilson, G. 1929. *Great Men of Science*. (Galen pp. 53-58). Garden City, N. Y.: Garden City Publishing Co.
Wright, F. A. 1932. *A History of Later Greek Literature*. London: G. Routledge and Sons.

CHAPTER 5

FOUNDING OF UNIVERSITIES AND ARTISTIC RENAISSANCE

THE REDISCOVERY of Greek culture stimulated cooperative group activities for the purpose of accumulating and using the Greek knowledge in the areas of present-day Italy, France, Spain, and other Western countries. Western countries designed to accumulate and use the Greek knowledge. This movement, promoted by scholars who had already come together in a few centers, became a sustained cultural awakening and resulted in the founding of universities (Fig. 5.1). The first universities reflected a sort of rebellion against the narrow educations offered by monasteries. The school at Salerno has been considered by some scholars to be the first university in the West. It did provide training in philosophy, theology, and law as well as medicine and perhaps could qualify as a university; but its reputation came in the special field of medicine. The word "university" in those days

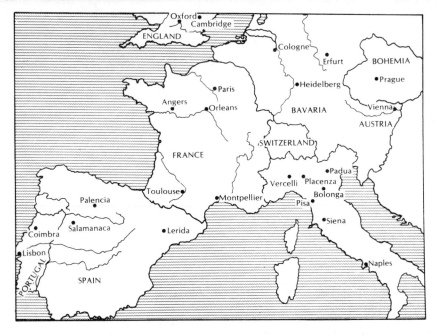

Figure 5.1. Locations of some leading universities in Europe founded in the twelfth to fifteenth centuries.

merely connoted a corporate status under which religious leaders and political rulers recognized and aided schools. European universities developed as collections of more or less independent colleges held together by the legal advantages afforded by a university organization.

Men with special "faculties" for particular subjects were chosen to teach, and groups of such men with appropriate qualifications eventually became known as "faculties." The "Universitas" thus developed as a group of units (colleges) made up of teachers and scholars. When a teacher or master moved to a new place, his students went with him. The trend developed that universities were a law unto themselves and were not subject to the laws of the individual community in which they happened to be established. A conflict arose frequently between the town and gown factions of the university community.

Medical students attended lectures on practical and theoretical medicine and on natural and moral philosophy. A standard requirement was attendance at lectures and at discussions of the works of Hippocrates, Galen, and Avicenna. Two other required subjects were surgery and astrology. Training in anatomy during the early thirteenth century mainly involved studying the work of Galen with no actual dissections. Autopsies became established procedures connected with law enforcement before they were acceptable in medicine.

Surgery was generally subordinate to medicine and was considered a manual craft. Licenses were granted to those who passed appropriate examinations, but licensed surgeons were not classed as medical "doctors." For the degree in medicine, the candidate was required to be twenty-four years of age, to have five years' standing in the study of medicine, and to have lectured on some medical book. The surgery license required considerably less training.

University of Bologna

Bologna became a center of learning during the period of the Roman Empire, and eventually developed into one of the first European universities. No real organization occurred, however, until the thirteenth century when a school of civil law was officially founded. This law school became a model of university organization for Italy, Southern France, and Spain. It is difficult to determine an exact date when Bologna might be considered to have attained university status. An early reliable reference is associated with the election of Pope Gelasius II, in 1118. At that time it was described as a university.

A more or less spontaneous association developed at Bologna between teachers and scholars because of the eagerness of young men to acquire knowledge. Students controlled the institution and, at first, operated it quite independently, without any connection with political or religious authorities. Interested students sought qualified teachers. Lectures at Bologna were at first held in the private apartments of teachers. Not until the beginning of the fifteenth century did the groups meet in places especially designed for instruction.

In Bologna, the scholars set up a functional university and elected a rector. They formed a student guild which was legally recognized in the city. The professors of the law college, however, refused to recognize the students' legal rights. The students, representing as they did one of the strongest intellectual groups of the period, established Bologna as a university and claimed complete control over the professors. As a form of student body organization, the students were divided into "nations." Organizations of French, Spanish, English and other national groups developed at the university.

Some professors gathered their students around them and drew away from the main university. Thus, Placentinus had left Bologna to establish schools first at Mantua and afterwards at Montpellier. This occurred in the third and fourth quarters of the twelfth century. Many migrations from Bologna to other communities occurred in succeeding years. One notable migration of students and professors occurred in 1321 when a faction of the University of Bologna moved to Siena. This followed an incident when a student was executed in Bologna because he tried to abduct a notary's daughter. In retaliation, his friends and some professors left Bologna and went to Siena. Later a reconciliation took place between the students and the city. The city of Bologna built a church in honor and recognition of the students of the university. Students became an increasingly important factor in the university city of Bologna as well as in other Italian university cities, and laws were passed forbidding students and professors to move away from the community. These laws, like others established by the community, were frequently challenged by the scholars.

The Bologna School of Medicine was formally organized in the thirteenth century, although medicine had been taught at Bologna for many years before that time. Teachers of medicine were not paid a fixed salary. Like other professors, they received fees from their students. A popular teacher received large sums and could amass considerable wealth. Later, the municipality assumed the responsibility of paying a fixed salary to three teachers who occupied particular chairs in medicine: the practice of medicine, the philosophy of medicine, and astrology.

At the beginning of the thirteenth century, Ugo of Lucco, a famous surgeon who had developed simple methods for the treatment of fractures, was given a high salary to come to Bologna as a lecturer. Friar Theodoric of Lucco (1205-98), son of Ugo, also became a famous physician and was the first to advocate simple treatment of wounds. He maintained that the formation of pus was not necessary for healing, and that drugs applied to wounds sometimes actually hindered the healing process. In one of Theodoric's books, directions were given for the use of anesthesia during operations: sponges were drenched in narcotics, such as opium or mandrake, after which they were dried and stored. Before the operation they were soaked in hot water for an hour and then applied to the nose of the patient, who was instructed to breathe deeply. The operation was not begun until the patient was asleep. Anesthesia had been used before the time of Theodoric by the Chinese in the third century B.C. , by Avicenna in the tenth century A.D. , and by several others. Practical, widespread use of anesthesia, however, was not developed for some six centuries after Theodoric.

Another great surgeon of the School of Bologna was William of Saliceto (1209-77). His book on surgery became popular and was used for several centuries. The name of Taddeo Alderotti is also associated with the School of Medicine at Bologna. This famous physician, who began teaching about 1260, was immortalized in the verses of Dante. Dante had studied at Bologna and may have attended Alderotti's lectures.

Henrie de Mondeville (1260-1320), a pupil of William of Saliceto, became a famous teacher of anatomy. He prepared drawings at Bologna and later taught at Montpellier. Mondino (1270-1326), also a great anatomist at Bologna, wrote a textbook that was used in virtually all European universities for three centuries after its preparation. No attempt was made to discover new truths, but Mondino set out to verify the work of the Greeks by actual observation. It is doubtful that he dissected human bodies, but he studied numerous domestic animals and advised his students to learn anatomy by actual dissection.

The most distinguished occupant of the chair of astrology at Bologna was the poet and philosopher, Cecco D'Ascoli, who was burned at the stake in 1327, a victim of the Inquisition.

During the fifteenth century the School of Medicine, or School of Artists (medicine belonging to the Liberal Arts), was provided a special building at Bologna. To ensure quietness for lectures and study, noisy shops and industries were not permitted in the vicinity of the university buildings.

University of Paris

The University of Paris is said to have been founded in 1110, but state and papal charters were granted later. Its early beginnings centered around a group of Dominican and Franciscan monks who were located in Paris. First established as a school of theology, the university grew and became a model for those that developed later in northern Europe. Free-lance teachers gathered around them groups of students who wanted to learn. Instruction at first was informal and personal. Greek classics formed the basis of the curriculum. Four faculties, law, medicine, theology, and art, made up the first academic framework of the University of Paris. In the twelfth century there was no appreciation for science nor the scientific method. By modern standards the classwork would be considered superficial indeed. It was book learning and became known as the scholastic method in education. The teacher's task was mainly to pass on information that was available to him. For the most part he took no interest in helping his students to think independently, and he took little responsibility himself in adding to the store of knowledge. Anyone who could read the works of Aristotle and other classical authors could be a teacher.

As early as 1127, or perhaps 1150, students at Paris became influenced by developments at the University of Bologna. A guild of students was formed. At this period the students began to have more influence on the activities of the university. Students were divided into nations as at the University of Bologna—the French, Normans, Picards, and English nations were among the first to develop. The institution soon became popular and large numbers of students were enrolled. There were not enough teachers to work individually with the increasing numbers of students. Classes were then organized and the lecture method was employed for instruction. Some great teachers acted independently and challenged authority, and some good students were sincerely interested in learning.

The general tendency, however, among students in a system in which the scholastic method was employed was, then as now, to memorize facts and to recite back the information they were given by the teacher. They were inclined to sharpen their wits and their senses rather than to develop their reasoning power. Town and gown disturbances occurred in Paris as early as 1200. Many tavern brawls are noted in the writings of this period. In 1229, because of a student disturbance, the university was dissolved and groups went to Toulouse, Orleans, Reims, and Angers, each group starting a small studium in the new location. In 1231 some students returned to Paris, but the newly initiated, smaller universities remained intact.

At this time no regular school buildings were available in Paris. Classes met in places that could be provided for this purpose. The Church of St. Julien-le-Pauvre, which still stands in Paris, was a meeting place for the early university.

Two men of special interest to biologists, who rose above the level of their colleagues, were at the University of Paris during the thirteenth century. They were the German, Albertus Magnus (1206-80), and the Englishman, Roger Bacon (1214-94).

Albertus, a member of the religious order of Dominicans, was a follower of Aristotle. He considered it his special mission to harmonize the teaching of Aristotle with church doctrine. Saint Thomas Aquinas (1225-74) was carrying out a similar mission in Italy. When Albertus found statements in Aristotle's writings that did not agree with his own observations, he was courageous enough to state his disagreement. Aristotle, for example, had said that the heart was the seat of the soul, but Albertus attributed this function to the brain. Unfortunately, however, he followed Aristotle in some mistakes that had been corrected in other places during the intervening years.

Albertus was interested in physics and geology as well as biology. Without the aid of modern instruments, he described and classified mineral elements into a system that was good although limited in scope. Free arsenic was discovered by Albertus, and he made important chemical combinations using this element. His main contribution in biology was a scholarly beginning in describing and classifying plants. His descriptions of plants, like those of the elements, were good as far as they went, but he had no natural basis for classification and knew nothing about minute structures. In his later life, he went back to his native Germany and located in Cologne, where he became a bishop in the church.

Roger Bacon, an Englishman who had studied at Oxford and later at Paris, was a Franciscan monk. His interests were directed toward the physical sciences rather than biology, but he indirectly contributed to biology. He was against the scholastic method of education which was prevalent at his time, and he demonstrated the values of observation and experimentation. He was so far ahead of his time in this respect that he had little influence on educational processes. So many strange things happened as a result of his experiments that he was accused of witchcraft. When Pope Clement III was suffering from a stone in his bladder, Bacon was summoned as a consultant. The Pope was so impressed with Bacon's scientific attitude and wide knowledge that he commissioned him to write two volumes, some 900 pages, on science. In the course of his work, Bacon studied and wrote on optics, lenses, and theories of light. He advocated the use of lenses for correction of eye irregularities and traced the function of sight to the brain. His greatest contribution to science as a whole was the introduction and demonstration of the experimental method.

University of Padua

The University of Padua (Fig. 5.2) had its beginning as a law school that transferred from the University of Bologna in 1222. Padua and Bologna were both members of the Lombard League (a group of communes that united to gain strength in conflicts between forces of the Emperor and those of the Pope) and had much in common during

Figure 5.2.   University of Padua building as it appears today.

this period. Emperor Frederick II attempted to crush the school at
Bologna in 1225 because of the school he created at Naples. In
1228 a contract was drawn up providing for the move of the school
from Padua to Vercelli. Vercelli agreed as a part of the contract to
construct more than 500 housing units for students. Some students
and their professors went to Vercelli, but the university proper remained at Padua.

Continuing its stormy existence, the University of Padua nearly
perished in the middle of the thirteenth century but was revived in
1260. During this period the University of Bologna was colonizing
in every direction, with some units having as many as 2500 students.
In 1332 a group of Bologna students seceded to Imola, and Padua
invited them to come to Padua. Until this time, Padua had specialized
in law, but it now expanded to include theology and other liberal
subjects. As the law students at Padua increased in number, they were
divided into their national groups, each group having a number of
votes in proportion to the number of students. Most of the groups
had 10 votes on jurisdictional matters; the Germans, however, since
they were few in number, had only two votes. When curricula in

medicine and art were established, they were subordinated to the law school, which included canon and civil law. All students and professors at this time were required to swear obedience to the statutes of the jurists.

The university of medicine and arts was later established as an independent part of the university and divided into seven nations. Attempts were made to raise the standards of the university. By law all who did not have sufficient support for their education and all those under 15 years of age were disqualified as students. In order to be sure that the best possible faculty members would be employed for the university, citizens of the community were excluded from salaried chairs by the constitution of the University of Padua. The salaried doctors were nominated by a city committee that provided funds and formally elected by the students. At first the college offering doctor's degrees was limited to a few students, but the numerical restrictions were removed when facilities were available for a larger enrollment.

In the latter part of the fourteenth century, Padua was again struggling for survival and almost closed its doors. In 1361, Visconti, ruler of Milan, revived the university, found good teachers, and paid them substantial salaries from the common treasury. He ordered his subjects to attend the University of Padua. It flourished at this time despite another school founded in Milan (Plaisance 1412). Humanities were taught besides law. In 1387 natural sciences were taught along with astronomy. Libraries were established and the school was truly a university. The college organization was not established in the University of Padua until 1363, when a *collegium tornacense* was founded by six students of law. Later many small colleges were established in the university.

During the rule of the Carrara family (a medical family that owned land in and around Padua, took its name from the city of Carrara, near Padua, and ruled as feudal lords and despots), one of the Dukes, Francis Carrara, provided for Padua's first university building, which was occupied in about 1400. This was paid for from the ox tax and wagon tax proceeds. Faculty members were also paid from these tax funds. The neighboring city of Venice gave strong support to the University of Padua. The city of Padua conferred on the rector of the university the right to wear a robe of purple and gold and to use the title "Doctor" throughout his life. He was presented with a golden collar of the Order of St. Mark. It was under Venetian tutelage that Padua reached the height of her glory in the sixteenth century. Venetian citizens during this period were forbidden to study in any other place except Padua. The University of Padua was influenced by France after the conquest of Milan by Louis XII in the early sixteenth century. Many French professors then became associated with the University of Padua.

University of Pisa

Pisa was incorporated as a law school and gave its first formal diplomas in 1342. At this time the law school was quite well established and beginnings had been made in medicine and various humanities. During the latter part of the fourteenth century the University of Pisa was in trouble because of lack of funds. The results of the plague epidemic included a low intellectual level in many parts of Europe. At this time the University nearly perished. In 1406 Florence conquered Pisa and injected new blood into the University of Pisa. Parts of the University of Florence were transferred to Pisa. Florence had supported a medical school in earlier periods of time and had made provisions for human bodies to be used in the studies of anatomy. These bodies were not from the natives of Florence but from conquered people of a captive race. In 1472 much of the University of Florence was transferred to Pisa, and with this new infusion Pisa again was prosperous. Chairs were established in poetry, speech, mathematics, and astronomy. The faculty of law was reorganized and medicine was restored. At this time professors were listed in two categories—ordinary and extraordinary, according to age, seniority and the significance of their program. This distinction remains today in the universities of Italy and Germany. Pisa became a center for students of the classical humanities and has persisted as such to the present time.

University of Montpellier

The documents and writings of the professors from Montpellier in the period around 1272 express great concern over the charlatans and quacks, mostly women, who were practicing some aspect of medicine. King James I of Aragon said in a pronouncement made at Montpellier in 1272, "Such a practice not only denounces the good name of the university but is also a dangerous practice causing deaths and bad consequences." At this time a license was required to practice medicine, and the university was given the authority to license those who were well prepared to practice. This became one of the first state laws concerned with university affairs. At a later date surgery also came under the control of the medical schools and surgeons were required to be licensed before they could practice. Jewish professors from Spain came to Montpellier and were teachers in the medical school.

University of Salamanaca

The University of Salamanaca was the greatest Spanish university of the revival age. It was founded by Alfonso X of Léon, King of Léon and Castile (1252-84), in 1258. The law school became the most famous, but course work was given in a number of other fields including medicine. In the latter part of the thirteenth century this university had chairs of civil law, canon law, logic, grammar, music, astronomy, pharmacy, and medicine. It became one of the great universities of Europe, attracting most of the students from Castile and other areas of Spain. This was a progressive university that kept up with the times. There is evidence that women students were enrolled in the fourteenth century. Salamanaca encouraged the unorthodox designs of Christopher Columbus and the Copernican system found early acceptance in its lecture rooms. Professors of oriental languages were appointed, and students were taught this material.

The main university building was erected in 1413 through the influence of Ferdinand and Isabella. In the period around 1543, modern texts were being used at the university. In astronomy the text of Copernicus that was only 17 years old was in use. The drawings and anatomical writings of Vesalius were used for the courses in anatomy. Pupils were required to make six complete dissections during the year and 12 partial dissections. A building was constructed for the study and teaching of anatomy. Also during this period a great university library was established and no less than 40 branches were placed in the city. About 6,000 students and 60 professional chairs made up the university.

University of Lerida

The kingdom of Aragon erected its first universities at Lerida and Huesca in the fourteenth century. Professors were nominated by the faculty and representatives of the city at large voted on the nominees. In 1391 Lerida had a law school and a medical school along with other disciplines taught at the university. To encourage medicine, the city of Lerida granted the university the privilege of a corpse a year for dissection in the medical school. The corpse was supposed to be that of a condemned person. It was decreed that the criminal assigned for this purpose must be drowned. In the wake of the great plague, professors from Lerida joined with those from Montpellier in discussions concerning the causes of the plague. They became interested in public sanitation. A tribunal of police was developed for enforcing sanitation. They were also concerned with quackery in medicine. Many court cases are on record against quacks. These include some surgeons, barbers, witches, and fortunetellers. Sometimes university faculty members split the incomes of these quacks and they received illegal licenses for practice through this kind of bribery.

The University of Lerida was prosperous in the sixteenth century and was later transferred to Cervera.

University of Palencia

The University of Palencia was founded by King Alfonso VIII of Castile. There was already an old Episcopal school in this locality. King Alfonso invited masters from famous schools in Paris and Bologna to come to Palencia and teach for competitive salaries. Unlike other universities, Palencia did not find it necessary to have the Pope grant a charter. This became a studium generale with a curriculum to attract students from other cities. Courses taught were theology, canon law, logic, grammar; some other subjects were also developed. Its history was one of failures and rivalry. By 1263 it had all but faded away and by the end of the century it ceased to exist.

University of Lisbon

The University of Lisbon was founded by decree by King Deniz. A jurisdictional dispute developed between the king and the pope. In establishing the university, the king obtained the pope's approval but certain strings were attached to this approval. The students were expected to be under the religious jurisdiction of the Church. They were expected to confess and to attend mass. They did not care for this restriction and refused to perform some of the activities required by the Church. When pressure was put upon them, riots developed. This same situation had occurred already at the University of Paris. In order to quiet the riots at Lisbon, the king transferred the university to Coimbra, where he lived, in 1308. It was transferred back to Lisbon in 1338, back to Coimbra in 1355, to Lisbon again in 1377, and to Coimbra in 1537. By the end of the sixteenth century it was located outside the city of Lisbon. It was reestablished in Lisbon in 1911. This university contained a theological faculty and gave course work in the arts and grammar. Although the study of medicine was brought to Spain by the Moors, the development of a medical school in connection with the University of Lisbon was under the Christian influence.

The School at Toledo

This was a center of learning but was not a university in the strict sense. When the Moors invaded Spain, the rabbis from Lucena, near Toledo, moved to Toledo. During this period Narbonne and Montpellier were intellectual centers. The many Jews living in Toledo practiced and taught the medicine they had learned from the Moors in Spain. Jewish professors from Spain became the teachers in the medical school at Montpellier.

Toledo is noted for translations of the Arabic medical texts into Latin. The Jews had translated the medical knowledge of the Arabs, which had come mostly from the Greeks, into the Hebrew language; now they made a further translation into Latin. This was one of the main contributions of the School at Toledo. Even though Toledo was famous as a center of learning with many well-known scholars, it never was a true university. A college known as St. Ildephonse was established there in 1508.

Other Universities

Other leading western universities established during the same period and the two centuries that followed are: Oxford (1167), Cambridge (1209), Naples (1224), Prague (1348), Vienna (1365), Erfurt (1379), Heidelberg (1385), and Cologne (1388).

Effect of Black Death Epidemic on Culture

A major cause of the fourteenth century lapse or decline in culture was the disease, bubonic plague, called "Black Death." Epidemics of this disease occurred repeatedly during the century, but the great disaster that influenced the history of the time and left its mark on centuries that followed was in the years 1348-1350. During this period one-fourth of the people in Europe were killed by the plague. The overall population of Europe dropped from some 85 million to some 60 million in just two years. The disease was first reported in the ports of Italy and it spread rapidly through the European countries. Death came to infected people almost immediately after the first symptoms appeared.

Recently it has been shown that bubonic plague is associated with rats, fleas, bacteria, and unsanitary living conditions. The cause was not known then, and superstitious people found many answers to their questions. Some thought the disease was punishment for sin and they crowded into churches pleading for protection. Others approached the threat of death differently. They relaxed their moral restraints and enjoyed the pleasures of life in the face of impending death. Listlessness and immorality rose to new heights and cultural pursuits were given little attention.

Emotional and ill-considered wrath on the part of people looking for a scapegoat fell upon the Jews and the physicians. The degree of stability and progress gained in the twelfth and thirteenth centuries was lost and the cultural level reverted to that of the Middle Ages. The whole of Europe was characterized by economic as well as cultural stagnation. In the latter part of the fourteenth century the European continent was in need of cultural and artistic renaissance. This came only sporadically and extended over the next two centuries.

Artistic Renaissance

Some cultural centers had seen a rebirth of the spirit of inquiry and a new awareness of nature during the twelfth and early thirteenth centuries, but the plague which had swept across Europe, together with other factors, had halted this trend. Late in the fourteenth century and continuing through the mid-sixteenth century, a new revival in culture was directed toward humanistic values. Accomplishments of this transitional period between the Middle Ages and the sixteenth- and seventeenth-century cultural developments were more literary and artistic than those of the former periods of history. The period of artistic rebirth is called the Renaissance. It was a flowering that could be attributed to artistic and intellectual roots that extended back to the Greeks and Romans but had remained almost completely dormant during the Middle Ages.

It must be emphasized that the Renaissance was not a sudden break with the past, but rather a period of increased vitality in physical and intellectual discovery. America was discovered and partially explored during the latter part of the Renaissance and sea travel became common. New coastlines, continental areas, and oceanic islands were discovered that had strange plant, animal, and human inhabitants. Gunpowder, which had been discovered in the twelfth century, came into practical use and markedly changed the mode of warfare and conquest. The religious reformation that occurred in the latter part of the Renaissance stimulated a critical attitude toward general culture as well as toward religion. This movement was associated with a wholesome development in mathematics and the basic science areas, particularly astronomy.

Science is accomplished by human beings; therefore, it is a humanity. The method of science requires objectivity but the basis of science must be human experience. No single road has led to the development of science but an awareness, on the part of individuals, of natural objects and phenomena is indispensable to scientific progress. The Renaissance brought, at least to some observers, artists, and philosophers, the excitement of discovery and the satisfaction of accomplishments. Modern science grew out of the humanistic philosophy of the Renaissance.

The famous Renaissance artist Sandro Botticelli (1444-1510), for example, was a keen observer of nature and a lover of plant life. Several of his paintings portray plants with artistic beauty as well as accuracy. His picture entitled "Spring" (now at Florence), painted in 1478, shows Venus in a grove of orange and myrtle with cupids hanging over her. Venus welcomes the approach of Spring who enters with Flora and Zephyr. Flowers, accurately and beautifully rendered, flow from the mouth of Flora. The entire painting includes some thirty kinds of plants, at least twelve of which can be distinguished with sufficient detail to be identified now as separate species.

Accurate observation of the world of nature that came with the Renaissance was a preconditon for the emergence of modern biology. Observer-artists of the period demonstrated that works of art can bring the artist and his audience to a critical awareness of nature. The influence of the Renaissance artists on biology depended on their accurate observations, their knowledge of structure and function of objects on which they worked, and their skill in representing the objects beautifully and effectively.

Invention of Printing

The invention of a way to print from movable type was an important physical development of the Renaissance period. In addition to making knowledge more accessible, the new printing process encouraged efforts to discover and present ancient Greek and Latin manuscripts. When only single copies of a manuscript could be made from an old text, wide dispersal was impossible, but printing made it seem worth whatever time and effort were required to find and reproduce old texts. By the end of the fifteenth century, many of the works of Aristotle, Theophrastus, Dioscorides, and Pliny were available to the educated public. This does not mean that all these manuscripts had been discovered for the first time since their original production. Some were never actually lost but only neglected. Some had been discovered and translated during previous revivals, particularly in the ninth and twelfth centuries, but the first sustained interest on the part of large groups of people came in the Renaissance centuries. Not only did the Greek texts appear, but Latin translations were made which could be read by more people.

The Renaissance awakening eventually promoted original observations and experiments, and printing assumed even greater importance. For the first time an investigator could record his results and interpretations on permanent printed pages. These pages could be widely distributed and provide other investigators, even at a distance, with access to his work. Investigators located in areas where no previous work had been done on their subjects no longer had to start all over. With access to worldwide literature they now could start where predecessors had stopped. Printing also permitted critical review of original investigations by contemporaries, and provided opportunities to compare experimental results and interpretations.

Renaissance Anatomy

The human mind can resolve tangible problems more easily and effectively than intangible relations that may require indirect experiments as a basis for analysis. Because anatomy is tangible and readily observable, it was developed more extensively in the ancient and medieval periods than were other areas of biology such as physiology.

Aristotle and his Greek contemporaries studied anatomy. Extensive anatomical work, including numerous human dissections, was done at Alexandria before the time of Christ. Galen dissected many different animals, including apes, but no human bodies. His writings, considered for many centuries to be infallible documents on human anatomy, actually were based on observations of lower animals. The few human dissections that were made during the Middle Ages had no lasting influence. At the Salerno school, pigs provided the main source of anatomical information. Anatomists at Salerno were interested in the structure of the human body, as was Galen, but human dissections were not considered proper and were so rare, therefore, that most of the human anatomical knowledge had to come from speculation based on other animals. As the history of anatomy is considered in perspective, it is surprising that studies based on pigs and apes were as useful as they were to anatomists who were primarily concerned with human anatomy.

Public dissections called "anatomies" were organized by the new medical schools of the Renaissance to provide opportunities for medical students to see the internal parts of the human body and to make the best possible use of the few human bodies available for dissection. Most professors of anatomy had been trained in the scholastic pattern of the day and had never actually made a dissection. Because they considered the task of dissection below their dignity and did not care to soil their hands, technicians or prosectors were employed to make the dissections. It was much more pleasant and less confusing for professors to handle and observe the clean pages of Galen than the bloody body structures being dissected.

The professor usually sat some distance above the operating table in a room built like an arena and read Galen aloud. If the reader and the dissector were well synchronized, the students in the lower part of the arena were supposed to see the anatomical parts as they were being described by the reader. Actually, neither the professor who was some distance away nor the students who were seated around the operating table could see clearly the parts being dissected. The only person who really saw what went on was the prosector and he was supposed only to work, not to talk. If a discrepancy occurred between the description and the dissection, the written word of Galen was invariably accepted rather than the mute evidence of the body. Curiously, in such circumstances, the body itself was considered wrong; Galen, who had never seen a human body dissected, was considered right.

Anatomy, particularly human anatomy, developed even further under the skillful hands of Renaissance artists who made dissections in order to be able to render body structures accurately in painting and sculpture. Artists of the Renaissance apparently had more motivation to really learn anatomy than the professional anatomists and thus they became better acquainted with the bodies of animals and man. They had to acquire first-hand knowledge as to how the bodies of animals and man were organized and they made dissections to find out. The greatest of the Renaissance artists who influenced biological

Figure 5.3
Leonardo da Vinci,
Renaissance artist
and anatomist.

thought were Leonardo da Vinci and Michelangelo Buonarroti. They were not satisfied with the kind of dissection they observed in the anatomies and preferred to use the knife themselves. These two and many other Renaissance artist-anatomists did at least part of their studying and most of their work in Italy, where cadavers were available and sometimes live specimens could be dissected.

Leonardo da Vinci

Leonardo da Vinci (1452-1519; Fig. 5.3) was one of the great anatomists of all time. He became famous not only for his anatomical work but in almost every phase of Renaissance art and science. Leonardo was born in Vinci, near Florence, Italy, and spent his early years in Florence and the surrounding country, where he watched birds and insects in the meadows and nurtured a profound interest in nature. He observed trees and flowers on the hillsides and spontaneously recorded the things he saw by sketching on whatever material he had at hand. His father recognized his artistic potential, and when Leonardo was 14 years of age he was apprenticed to Verrocchio, one of the leading artists of the Renaissance. Art in this period was characterized by reality. External structure of human beings, as well as that of other animals, was extensively studied, especially for the preparation of statues and other pieces of sculpture that were developed to perfection during the Renaissance. In order to make them lifelike, the artists scrutinized living examples externally and internally, with the greatest of care.

Leonardo attended anatomies whenever they were held in his vicinity. He performed dissections on animals and probably on human bodies while he was still a young man. Later he was granted permission to carry out anatomical studies at the hospital of Santa Maria Nuova in Florence. Leonardo's artistic approach to anatomical science was new and refreshing. His contemporaries and most of his predecessors followed precisely the descriptive work of Galen and that of earlier Greek and Alexandrian anatomists. Leonardo dissected actual, unmodified human bodies for the purpose of studying them in all of their beauty and to facilitate representing them properly in art. He showed scientific interest and curiosity in his early undertakings. He also showed originality and skill as a technician in working out devices to permit easier observation and experimentation.

The Renaissance was an exciting time for its leaders. New information was being obtained in many areas of culture, yet it did not seem impossible for a person to be interested in, and even to master, several branches of the knowledge accumulated at that time. Leonardo was among those interested in various artistic and scientific developments. He was attracted to anatomy from the standpoint of art but soon he became so absorbed in it that he carried on anatomical studies for their own sake. In 1510 he wrote that he hoped to master the whole field of anatomy during that year. Furthermore, it was his intention to write an encyclopedia on man. He never attained either goal. This was the fate of many of his sweeping objectives.

Besides being enthusiastic and ambitious, Leonardo was a perfectionist who was never satisfied with his accomplishments in art. Nothing was ever considered finished, and he was always experimenting to make his productions better. Some sixteen years were devoted to a bronze statue of a horse which was never really completed. He worked twelve years on the famous painting of the Last Supper which he executed on a monastery wall in Milan, Italy. Three years were spent in getting the twist of the lip of Peter exactly right. Although the Last Supper is an acclaimed masterpiece in art and anatomy, its creator was never satisfied with the painting. The painting of the Mona Lisa, now at the Louvre in Paris, is a famous masterpiece, but Leonardo was not entirely satisfied with it. Of course, Leonardo engaged in many other activities and made other accomplishments while adding refinements to his major works of art.

During the course of his anatomical studies, Leonardo was reported to have dissected some 30 human bodies. His findings were recorded in some 129 notebooks, which were well illustrated with detailed sketches. Successful results of his experiments showed that he had anticipated some modern techniques in studying the soft parts of the body. The eye, for example, which was delicate and difficult to dissect, he embedded in coagulated albumen. To accomplish this feat, the whole eye was placed in the fluid white of an egg. The mass was heated until the albumen became solid. Soft and flexible parts of the eye were then held firmly enough to enable him to cut transversely with a sharp blade without the loss or distortion of even the finer structures. Leonardo also developed methods of injection to facilitate his study of the hollow ventricles of the brain. These ventricles are surrounded by very thin membranes which are easily distorted and lost. By using a syringe, Leonardo filled the ventricles with melted wax. After the wax solidified he was able to dissect with precision and to follow three ventricles of the brain. Leonardo also did some vivisection experiments on animals. In one he observed the heartbeat of a pig. This was accomplished by drilling through the wall of the thorax, making an incision, and inserting pins to hold the heart cavity open.

Leonardo, the artist, was one of the first since the ancient Greeks to appreciate the Aristotelian method of acquiring knowledge through observation and experimentation. He had no use for the alternative Platonic method of introspection, and he had little respect for established authorities when dealing with observable structures. If observation failed, he insisted, the fault was with the observer, not the method. In one of his notebooks he said, "Wrongly do men cry out against innocent experience, accusing her often of deceit and lying demonstrations!" Leonardo believed that only through complete knowledge of the function of muscles could he understand the structure and relations with other body parts. To merely observe a muscle was not enough; he had to discover its connections and find out how it functioned.

Leonardo did not make any lasting contribution to science as such, and, strictly speaking, he was not a scientist; but he did demonstrate the productivity of a scientific approach. A scientist must have persistence to carry his experiments to completion. Leonardo had the curiosity and the persistence to continue his pursuit of perfection in art, but he did not design correctly nor complete his physical experiments. His treatises on anatomy, however, and the illustrations in his notebooks show that he not only had great skill as an artist but also great ability as an anatomist. By using illustrations to portray structural characteristics, he presented a large volume of information with only a few words.

Some 20 notebooks and bound volumes of Leonardo's work, amounting to approximately 4,000 pages, are now available. Many have only sketches and drawings; others are covered with minute writings running from right to left on the paper rather than from left to right. Had the notebooks been known in the sixteenth and seventeenth centuries, Leonardo probably would have been recognized as the father of anatomy. Unfortunately, the notebooks were lost until the eighteenth century.

Michelangelo

Michelangelo (1475-1564; Fig. 5.4) was much like Leonardo in his approach to anatomy. Interested in the subject in connection with his painting and sculpture, he made actual dissections of internal structures in order to represent the features of the human body accurately and in proper relation with each other. As an amateur scientist and observer, Michelangelo did not approach the great activity of Leonardo, but as a master artist his work was much more voluminous. Three of his most famous statues portraying his knowledge of the human musculature are: (1) *David*, a large statue that stood in the out-of-doors at Florence for some 400 years. It was becoming weather worn and has been moved to the Museum of the Academy of Fine Arts at the University of Florence; (2) *Moses*, now at the Basilica St. Peter's in Vincoli in Rome; (3) *Pieta* Christ being taken from the cross, which was done when Michelangelo was 22 years old, now at St. Peter's Basilica in the Vatican near Rome.

Michelangelo was an individualist. He not only carried out the preparatory dissections himself and formulated the plans for statues, but he selected the blocks of marble and did all the creative part of the sculpture himself. Only the cleanup work was left to students and assistants. Artists generally tend to be individualistic, but Michelangelo was unusually independent. This may be the reason why none of his students became great artists. An apprentice to Michelangelo was only allowed to do menial tasks. Though Michelangelo did not leave notebooks giving detailed procedures for learning about external and internal anatomy as did Leonardo, his masterpieces in art give ample evidence of his profound knowledge of the human body.

Michelangelo also contributed in other forms of art and architecture. One of his projects was the dome on St. Peter's Basilica in the Vatican. In planning this structure, he learned much from the Pantheon which was designed by Agrippa before the time of Christ.

Figure 5.4
Michelangelo,
Renaissance painter,
sculptor, and anatomist.

This building was so perfect that Michelangelo said it was "built by nonhuman hands." It was a dome 144 feet in diameter that was prefabricated with clay cement and raised to its finished position. In respect to Agrippa and the Pantheon, Michelangelo reduced St. Peter's dome by 3 feet, making it 141 feet in diameter.

Renaissance Surgery

The development of surgery was handicapped through the ages by three main factors. The first was the pain that was caused by the incision and by the manipulation of the internal organs. Few persons had the physical and emotional stamina to submit to surgery without anesthesia; and of those who did, most succumbed to shock. The second handicap was the hemorrhaging that attended cutting into the body. Though ligaturing was tried, most attempts to stop severe bleeding were unsuccessful. The lack of knowledge about infection was the third deterrent to surgical progress. Virtually nothing was known about the cause of infection, and cleanliness rarely was considered important. Consequently, almost all surgical procedures resulted in infection, and most postsurgical infections were fatal.

The history of surgery is interwoven with that of anatomy and that of medicine, but it has a pattern of its own. Because of the lack of communication, each civilization essentially did its own pioneering in surgery. At certain periods of history it has been set apart arbitrarily as a lesser application of medicine or a menial trade, and not a respectable one at that. Through long periods of history, however, medicine was interwoven with superstition, while surgery remained a more realistic area in which the operator could see what he was doing. After the time of Galen, surgery as a science underwent a long period of neglect which was virtually unbroken until the time of Leonardo da Vinci and Vesalius.

De Chauliac

Guy De Chauliac (1300-70) from Avignon, France, physician to Pope Clement VI and his two successors, was one of the exceptions to this general rule. Much more advanced than his contemporaries or most of his successors over a period of several hundred years, he had a sound background of learning and commanded respect in his time. Through his influence, surgery found a respectable place, and it was recognized that this art should be in the hands of men of good character and adequate training. He advocated cutting to remove cancerous growths in the early stages of their development and he treated ulcers surgically. He successfully used traction as a treatment for fractures. He was one of the early surgeons to develop the crude use of anesthesia. De Chauliac synthesized the fields of medicine, anatomy, and surgery, and he pioneered in demonstrating that these areas have much in common.

Paré

The status of surgery in Paris during the Renaissance was fairly typical of that in other centers of culture. Three medical hierarchies ruled Paris during this period: (1) the Faculty of Physicians at the University, who fancied themselves vastly superior to ordinary practitioners and consequently did not work with their hands; (2) the surgeons of the College of St. Come, who confined themselves to surface applications and minor surgery and were in a higher social plane than the barber surgeons; and (3) the barber surgeons, who

Figure 5.5
Ambroise Paré, famous
French surgeon of the
sixteenth century.

did most of the actual cutting. It was largely through the skill of the
last-mentioned group that advances of any sort were made. Through
application and practice, barber surgeons learned to perform vene-
sections and others of the more complicated surgical operations.
Many of the barber surgeons became so adept at specific operations that sometimes they had to leave the area in which they were
practicing to escape the jealousy of the less capable physicians and
surgeons of higher social class. Fortunately the barber surgeons were
more numerous than other classes of surgeons. Many of them became
recognized by their patients as being superior in skill, and this made
their plight not entirely intolerable.

The most noted surgeon of the post-Renaissance was the Frenchman, Ambroise Paré (1517-90; Fig. 5.5). He was trained as a barber
surgeon and obtained vast experience. Through his surgical ability
he managed to become the adviser and friend of several kings of
France. Being an ambitious student he became acquainted with
ancient and medieval as well as current knowledge of surgery. Not

trained in Greek, he spoke and wrote in his native vernacular. His formally trained contemporaries snubbed him, but his personality was strong enough to maintain his innovations against tradition and dogmatic opposition. Paré's power of observation and his common sense were exceptionally good and brought him remarkable success. Humble in his practice and with a profound faith in the healing powers of nature, he was often quoted as saying, "I dressed the wound and God healed the patient."

The accepted treatment for gunshot wounds at the time was boiling oil poured over the wound, which caused great suffering and many complications. One day on the battlefield the supply of oil ran out. Patients not treated, to the surprise of Paré, got well more quickly than usual when other treatment was administered. This led Paré to experiment by using less oil, and finally he gave it up altogether. He found that a clean wound would heal satisfactorily with no treatment. Paré revived the ancient dictum of Hippocrates, "first do no harm," and capitalized upon it. He discovered that it was untrue that copious pus formation on a wound was a good sign indicating a discharge of unhealthy humors from the body and the operation of healing processes. Other physicians and surgeons before Paré had opposed the idea of "laudable pus," but Paré was the first to bring substantial evidence into the argument.

Paré's next contribution was a method of stopping bleeding following amputations, which were common in the Middle Ages and Renaissance because of the lack of aseptic surgery. Infection, particularly gangrene, was usually associated with surgery, and limbs frequently had to be removed to save the life of the patient. The common method at that time for stopping bleeding was cauterization. Paré found that better results could be accomplished by ligaturing the blood vessels. This was less painful and more effective. Artificial limbs were also introduced by Paré to replace losses through amputation.

Though midwives dominated obstetrics even through the Renaissance age of enlightenment, Paré obviously had some knowledge of the process of childbirth. His revival of the podalic version, or the turning of a breech baby so that it arrived head first, made the procedure popular again after it had fallen into disuse. In cases of excessive bleeding from the womb, Paré had the courage to try to induce labor to stop the hemorrhaging.

Before his death, Paré made many minor as well as major contributions to surgery and medicine. His less famous contributions were generally simplifications of complex treatments and moved in the direction of greater cleanliness. Paré performed autopsies in addition to his varied activities and wrote a treatise on the legal aspects of surgery. He published eight major works. At the time of his death he was recognized as one of the most enlightened men of surgery and medicine.

Summary

Western universities heralded the cultural awakening after the Middle Ages. The University of Paris, where Albertus Magnus and Roger Bacon taught, developed as a teacher-centered, scholastic institution dealing with speculative as well as applied subjects, whereas the University of Bologna was student-controlled and more professional. Theodoric of Lucco and Mondino were prominent members of the medical faculty at Bologna. Biology in these and other western universities of the thirteenth and fourteenth centuries existed almost solely insofar as it was associated with medicine.

The artistic Renaissance that developed in the fourteenth to sixteenth centuries brought to some observers, artists, and philosophers the excitement of discovery and the satisfaction of accomplishment. Roots of modern science found nourishment in the humanistic philosophy of the Renaissance. Printing with movable type and illustrating with wood blocks gave greater circulation and significance to observations and interpretations. Some French and Italian leaders of the Renaissance are listed as follows:

Contributor	Dates	Field or Contribution
De Chauliac	1300-70	Physician and surgeon; improved status of surgery.
Botticelli	1444-1510	Art showing critical observation and rendering of plants.
Leonardo	1452-1519	Anatomy and art on human and animal subjects.
Michelangelo	1475-1564	Sculpture and painting of living subjects.
Vesalius	1514-65	Anatomy (Chapter 6).
Paré	1517-90	Devised more effective and humane methods of medicine and surgery.

References and Readings

Belt, E. 1955. *Leonardo the Anatomist.* Lawrence, Kansas: University of Kansas Press.

Brehaut, E. 1912. *An Encyclopedia of the Dark Ages: Isidore of Seville.* New York: Longmans, Green and Co.

Chambers, M. M., ed. 1950. *Universities of the World Outside.* Menasha, Wisconsin: George Banta Publishing Co.

Corner, G. W. 1927. *Anatomical Texts of the Earlier Middle Ages.* Washington, D.C.: Carnegie Institute of Wash.

Daly, L. J. 1961. *The Medieval University.* New York: Sheed and Ward.

Elgood, C. 1951. *A Medical History of Persia.* Cambridge: At the University Press.

Hart, I. B. 1962. *The World of Leonardo da Vinci.* New York: Viking Press.

Haskins, C. H. 1933. *The Renaissance of the Twelfth Century.* Cambridge, Mass.: Harvard University Press.

————. 1940. *The Rise of the Universities.* New York: Peter Smith.

————. 1927. *Studies in the History of Medical Science,* 2nd ed. Cambridge, Mass.: Harvard University Press.

Holmyard, E. J. 1957. *Alchemy.* Harmondsworth, England: Penguin Books.

Irsay, S. d'. 1933. *Histoire des Universités.* Paris: A. Picard.

Langer, W. L. 1964. "The Black Death." *Sci. Amer.* 210:114-121.

Lassek, A. M. 1958. *Human Dissection, Its Drama and Struggle.* Springfield, Ill.: Charles C. Thomas Co.

Laurie, S. S. 1912. *The Rise and Early Constitution of Universities.* New York: D. Appleton and Co.

MacCurdy, E., trans. 1958. *The Notebooks of Leonardo da Vinci.* New York: George Braziller.

McMurrich, J. P. 1930. *Leonardo da Vinci, The Anatomist.* Baltimore: Williams and Wilkins Co.

Mizwa, S. P. 1943. *Nicolaus Copernicus.* New York: Kosciuszko Foundation.

Rait, R. S. 1912. *Life in the Medieval University.* London: Cambridge University Press.

Rashdall, H. 1936. *The Universities of Europe in the Middle Ages.* London: Oxford University Press.

Sarton, G. 1937. *The History of Science and the New Humanism.* Cambridge, Mass.: Harvard University Press.

————. 1957. *Six Wings, Men of Science in the Renaissance.* Bloomington: Indiana University Press.

Schachner, N. 1938. *The Medieval Universities.* New York: Frederick A. Stokes Co.

Singer, C. J. 1957. *A Short History of Anatomy from the Greeks to Harvey.* New York: Dover Publishers.

Taylor, G. R. 1963. *Science of Life.* New York: McGraw-Hill Book Co.

Thomas, B. 1937. *The Arabs.* Garden City, New York: Doubleday, Doran and Co.

Vallentin, A., trans. E. W. Dickes. 1938. *Leonardo da Vinci, The Tragic Pursuit of Perfection.* New York: Viking Press.

Wieruszowski, H. 1966. *The Medieval University.* New York: D. Van Nostrand Co.

CHAPTER 6

SCIENTIFIC OBSERVATIONS AND EXPERIMENTAL METHODS

MODERN BIOLOGY is dependent on observation and experimentation; it is an experimental science. The optimum procedure, however, for approaching experimentally the study of vital functions in living organisms has only recently become established. Many early biologists struggled to make their subject objective and experimental, but it was not until the latter part of the nineteenth century that the beginning made in earlier history came to fruition and biology became essentially experimental.

Direct Observation Applied to Human and Comparative Anatomy

Some sixteenth century universities provided courses and demonstrations for medical students based on Galen's work supplemented by

Figure 6.1
Andreas Vesalius,
Renaissance anatomist.

references to Aristotle's biological books and to the Hippocratic
texts. The procedure of quoting earlier authorities rather than observing actual objects was followed in other aspects of medical training
besides anatomy. In many cases the students did little more than
memorize lists of symptoms of disease compiled by Arabic commen-
tators. Dissections of the human body that were made were not
performed by scholars or doctors but by demonstrators or barbers who
had no knowledge of the medical texts.

A plea for a study based entirely upon observed fact came from
Andreas Vesalius (1515-1564; Fig. 6.1), a young professor of anatomy
at the University of Padua. He agreed that medical practice should be
based on a knowledge of anatomy, but he disapproved of the way
anatomy was taught. It was improper, he insisted, to learn anatomy
from ancient texts when it was possible to make direct observations
of the human body. Vesalius went far beyond all of his predecessors
in wholly denying the validity of an earlier authority and declaring

that knowledge was only to be acquired by completely independent research. Furthermore, new discoveries could be accepted only after multiple experiments; in the case of anatomy, by repeated dissections and observations. In his lectures, Vesalius did his own dissection and commented upon each system as he exposed it to view. His book *De Humani Corporis Fabrica* (The Fabric of the Human Body) (1543) emphasized the empirical approach. Instead of following the order of presentation used by Galen, Vesalius followed the sequence of an actual dissection.

Background and Early Training of Vesalius

Even before Vesalius became professionally qualified as an anatomist he had many experiences in actual dissection of animals. While a young boy at his home in Brussels, Belgium, he dissected mice, moles, cats, dogs, and weasels on his mother's kitchen table. The son of an apothecary physician, he had opportunities to observe the activities of medical practitioners. Bolstered by a keen intellect and a strong motivation, he began in 1530, at the age of 16, to train for medicine at the University of Louvain, a Catholic university near Brussels. Not finding sufficient challenge at Louvain, he transferred to the University of Paris in 1533, when he was 18 years old. Here he was engaged for three years in a study of medicine and as part-time assistant in anatomy. In Paris he studied under Jean Guinter of Andernach, Professor of Anatomy, and Jean Fernel, Professor of Medicine. A third teacher, Jacobus Sylvius (1478-1555), under whom Vesalius studied for a short time, taught anatomy outside the medical school in the College de Treguier. He was a sophisticated anatomist who had studied and taught classical languages, Greek, Roman, and Hebrew, until he was 50 years old. Sylvius had become so interested in treatises on anatomy by Greek and Roman authors that as a side activity he developed a series of classical lectures on the subject. He considered the works of Galen to be infallible, the final word that could not be improved upon. Sylvius did no original research but merely organized and rephrased Galen's work. His lectures were exercises in classical oratory.

As he worked as an assistant, Vesalius expressed doubts concerning some of Galen's work and insisted on describing body structures as he saw them. He quarreled with his professors on many points, but when he learned that Galen had dissected animals but had never actually dissected a human body, he lost all confidence in Galen's work. In the summer of 1536, when war broke out between France and the Empire, Vesalius was an enemy alien and was compelled to leave the University of Paris without graduating. He enrolled in the medical school of the University of Louvain from which he obtained the degree bachelor of medicine in 1537. Pursuit of practical anatomy had been difficult and dangerous in France, and conditions for anatomists were only slightly better in Belgium, but Vesalius took advantage of every opportunity. On one occasion, while walking with his mathematician friend, Gemma Frisius, he came upon a gibbeted body that had been picked so clean by birds that it presented an almost intact skeleton. With his friend, he succeeded in separating the bones and smuggling them into his quarters where the bones were later assembled into his first articulated human skeleton.

Vesalius at Padua

Back at home in Brussels, Vesalius continued his study of anatomy, securing specimens from graveyards and execution chambers. In his own living quarters, he assembled the bones of human skeletons, handling each bone so frequently that he came to know bones by touch and could recognize them with his eyes closed. After a short period of independent study, he joined the Imperial Army of the Emperor of Germany and became a military surgeon. In this capacity his knowledge of human anatomy was greatly increased by numerous opportunities to operate on soldiers and make direct observations on human structural parts.

On his release from military service he went to Venice and engaged in further study that developed into anatomical practice and teaching. In public dissections, he surprised his observers by dissecting and lecturing at the same time. He enrolled at the University of Padua and completed the examinations for the doctor of medicine degree in December 1537. On the following day he accepted a professorship at Padua in surgery and, at the age of 23, became a member of the most famous medical faculty in Europe. At this time the University of Padua possessed no building of its own and it is not known where the classes and particularly the dissections were held. Students flocked to his lectures, and he soon became a popular professor. He developed a series of demonstrations that attracted literally hundreds of students and physicians. His popularity increased because he lectured in Latin rather than Greek. Dispensing with the barber assistant, he dissected, demonstrated, and lectured (Fig. 6.2). As a further pedagogical device he drew large diagrams illustrating structure and function of the human body. Practitioners as well as scholars could understand him. Information about insights on Vesalius's first anatomy course comes from the notebook of one of his students, named Vitus Tritonius. This notebook contains not only notes but also the student's crude copies of some of Vesalius's diagrams. In 1540, to settle a point he had labored to prove for many years, Vesalius assembled the bones of an ape and those of a man and demonstrated some 200 differences before a large audience. It was further demonstrated that in all instances of differences Galen's description of "human" anatomy was in error.

Another student of Vesalius, John Caius, who entered the Padua medical school in 1539, lived with Vesalius for eight months (October 1539 to June 1540), but did not share his critical attitude toward Galen. Caius's chief literary production was the recovery, study, and translation of the Greek text of Galen on which he published a number of books between 1544 and 1549. After receiving his medical degree in

Figure 6.2.   Vesalius dissecting and lecturing. (Courtesy of
Parke, Davis and Co., Copyright© 1958.)

1541 and serving for two years as lecturer on Greek scholars, Caius toured Europe and went to Cambridge where (in 1557) he became the re-founder of Gonville Hall. Caius later became president of the College of Physicians and Surgeons in London.

Fabrica

Vesalius was anxious to spread the word of his new methods and his new anatomy, which he did by describing both in more detail than had ever been attempted before. A result of this work was the great book of Vesalius, Fabrica (1543). The text is not only extensive but is so arranged that an elaborate system of cross-referencing permits description of the interrelationships of structures required by Vesalius's further concern with anatomical systems and what he considered the dynamics of anatomy, that is, physiology. The illustrations, beautifully prepared by Vesalius himself and by a student of Titan from Venice, are essential to the text. References are made to the most minute details and to comparisons of such details between illustrations. For the first time illustrations had a genuine pedagogical value for the student. With some idea of the importance of his work, Vesalius spared no expense to make its appearance as fine as possible. Johannes Oporinus, an excellent and meticulous printer of Basel, was selected to print the book. Fabrica, a rare combination of Renaissance art, correctly delineated and presented the anatomy of bones, muscles, blood vessels, nerves, internal organs and the brain all in proper perspective.

Illustrations were not mere diagrams but accurate representations of the structure of a living body as Vesalius had observed it. Having no new knowledge of the heart, and finding Galen's description of the general aspects to conform with his observations, Vesalius followed Galen in his description of this organ. When it came to the septum between the two sides of the heart he was puzzled. He postulated invisible pores through which the blood could pass from one side to the other. Among the curious errors sustained by Vesalius for many years as a Galenic legacy was the belief that the blood vessels possessed in their coats three kinds of fibers—vertical, horizontal, and diagonal—which either propelled, retained, or drew the blood. The portal vein is indicated as terminating in five branches, a holdover from the old tradition of the five-lobed liver. This drawing must have been produced very early because Vesalius in the text denies any such termination of the portal vein in five branches. In spite of his attempt to base his account of the human body entirely upon his own observations, Vesalius's descriptions contain many of the old errors.

Vesalius, like Aristotle, believed that a complete understanding of the essentials of any system was impossible without comparative studies. An illustration in *Fabrica* of a human skull resting on a dog skull is the first noteworthy illustration of comparative anatomy. The purpose was to show that Galen's ascription to the human of the premaxillary bone and suture of the dog revealed the animal source of his alleged human anatomy. Vesalian skeletons, as the major representatives of osteology, surpassed by far all previous depictions in their nature of artistry and correctness. There was an element of whimsy which provided a dynamic quality to otherwise static specimens. The usefulness of comparative anatomy, reemphasized by Vesalius, helped to make possible subsequent developments in physiology and to transform applied medical study of the Middle Ages into wider biological inquiries of the seventeenth and eighteenth centuries. He had great difficulty in obtaining human bodies and those of other animals for dissection and had no method of preserving the parts. Most dissections were made during the winter when cold weather would aid in preservation, but even then a body could be used for only a few days. Vesalius's first course at Padua was restricted to the period of December 6th to 24th because the body of the 18-year-old man being used for demonstration could not be retained longer without preservatives. While Vesalius was in Basel in connection with the publication of *Fabrica*, a bigamist who had tried to dispose of his first wife, and thereby his problem, was caught and executed. The body was turned over to Vesalius for dissection and then articulation of the skeleton. The skeleton is now a valuable relic at the University of Basel. Vesalius's most accurate observations, therefore, were made upon the hard parts such as bones that are least subject to decay. His account of the skeleton and the muscles showed his greatest skill and originality.

Although Vesalius's great work, *Fabrica*, was completed, according to the colophon (inscription containing facts relative to production), in June 1543, its author was unable to obtain bound copies for presentation until the beginning of August. But whether examined as printer's sheets or as a bound book, the remarkable title page, showing the first picture of an anatomical theatre, must have attracted immediate attention. The artist of the title page is unknown, although his skill and that of the blockcutter are beyond reproach. The body subjected to dissection was that of a female criminal who sought to escape execution and dissection by declaring herself pregnant, only to be denied that escape by the vigilance of the midwives. An illustration of this dissection appears elsewhere in *Fabrica* (Bk. V, Figure 24) where, as Vesalius writes, the peritoneum and the abdominal muscles have been opened and pulled to the side. Then we have resected all the intestines from the mesentery, but we have left the rectum in the body as well as the whole of the mesentery of which we have to some extent separated the membranes so that its nature is exposed to view. However the present figure has been drawn for the special purpose of indicating the position of the uterus and bladder exactly as they occurred in this woman. We have not disturbed the uterus in any way and none of the uterine membranes has been destroyed. Everything is seen intact just as it appears to the dissector immediately upon moving the intestines to one side in a moderately fat woman.

The incorrect curvature of the spine resulted from the manner in which the skeleton was erected, with the spinal column supported by a rigid metal bar. The Vesalian representation of the sternum displays it correctly as made up of six segments rather than the seven which Galen described.

The bones of the foot shown in *Fabrica* are distinctly those of. Vesalius's own observations. Of all the Vesalian illustrations, the best known and most often reproduced are the fourteen drawings of muscles in their proper position on the body, "the musclemen." The seventh muscleman is depicted and supported by a rope in a way in which, as Vesalius wrote in *Fabrica,* a cadaver had been suspended for the depiction of all fourteen musclemen. The fifth muscleman is depicted with an exaggerated rectus-abdominus muscle for which Vesalius has frequently been criticized by those who have examined the illustrations but have not read the text. As he says in the text, this was to demonstrate the appearance of the muscle if one accepted the Galenic description. As pointed out by E. Jackschath (1903), if the musclemen are placed in a certain sequence, a hilly landscape can be envisioned. This was identified by Willy Wiegand (1952) as the Euganenean Hills some six miles southwest of Padua.

Epitome

To accompany *Fabrica*, Vesalius composed a brief work planned as an "epitome" of the larger work, very elementary in nature and for the use of beginners. The Latin text is trivial, but the illustrations, some duplications of those in *Fabrica*, are excellent. Two of them commonly known as the "Adam and Eve" figures were representative of human surface anatomy and were accompanied by a simple text describing the external aspects of the body. This popular version of *Fabrica* could be used by barbers, surgeons, and medical students who knew no Latin but could follow the sequence from the pictures.

The title page of the German edition of the *Epitome* lists the publication date as August 9, 1543. This translation was prepared with

Vesalius's blessing by Albanus Torinus (1489-1550), physician, professor, and at the time Rector of the University of Basel. It carries a second dedication to the Duke of Wurttemberg, for whom Torinus was physician. Special difficulties were encountered in the translation from Latin into German because the latter language did not then possess all the necessary technical terms and extensive circumlocutions were necessary. Nevertheless, the extreme scarcity of the book today indicates its popularity and usage. The employment of the vernacular meant that it could be read by uneducated people, and it was literally read to pieces.

Another of Vesalius's writings, the "Letter on the China Root" (1546), is actually two works under a single title. The first deals with a new drug, the China Root (*Smilax China*), thought to be specific for various ailments, especially syphilis. Vesalius's account, written in reply to a query from a colleague in Mechlin, indicates considerable dubiety concerning the value of the drug. The second work is of greater importance since it is a defense of *Fabrica* against certain attacks, notably those that stemmed from the earlier student-teacher feud with Jacobus Sylvius and the further explanation of certain subjects about which Vesalius decided more illumination was desirable.

Although Vesalius was only at the University of Padua for a few years, he made a remarkable beginning in observation and experimental procedures in anatomy. Because Padua was famous for its medical faculty and attracted many students from other lands, any reforms in medicine that were accepted at Padua, including the interest in comparative studies, soon reached all the major medical centers throughout Europe.

Vesalius, Physician to Imperial Household

With the publication of *Fabrica*, Vesalius had the curious idea that his career as an anatomist was over. He had emphasized through his book the need for physicians to study anatomy but he himself, as a physician, would not undertake regular medical practice. In accordance with his family's tradition, he sought imperial employment and accepted a position as physician at the court of the German Emperor, Charles V. Henceforth until the abdication of Charles V in 1556, Vesalius was to practice medicine in the court, from time to time serve in the army as military surgeon, and to be called upon to minister to the emperor's many complaints, which were compounded by his gluttony.

At the invitation of Casimo de' Medici, Duke of Tuscany, Vesalius went to Pisa (1544) and gave a series of anatomical demonstrations. They were presented in a building on the Via Lung'Arno, as now recorded on the marble plaque over the doorway. These were so successful and aroused such interest that the Duke made an effort through a very high salary (800 florins) to entice Vesalius to the chair of anatomy at the newly refounded University of Pisa. The emperor, however, refused to release Vesalius, who had been made more valuable in his eyes by the appreciation from elsewhere.

Adolph Occo, 1494-1572, and Achilles Pirmin Gasser, 1505-1587, were two of the leading physicians of Augsburg (Swabia; now southern Germany) with whom Vesalius became acquainted (in 1547). Close ties of both friendship and profession developed, and whenever in Augsburg Vesalius participated with these physicians in consultations and post-mortem examinations. It was in consequence that he was able to present his remarkable description of heart block and a notable account of a case of hydrocephalus in the revised edition of Fabrica. In 1555 Vesalius was called to Augsburg for consultation with Occo and Gasser over the condition of a member of the great banking family of Welser. Vesalius offered the remarkable diagnosis, relative to the banker's age, of fatal aneurysm of the aorta, later verified by post-mortem examination.

In 1551 Leonhart Fuchs, Professor of Medicine at the University of Tübingen, whose interest in botany (Chapter 9) is recalled by the name of the flower "fuchsia," published for use of his students a digest of Vesalius's Fabrica, thus giving his allegiance to the new anti-Galenic anatomy. Vesalius, who had suffered much from plagiarists, was not especially flattered, and his irritation was indicated by a later reference to Fuchs's book as "not a compendium but a dispendium," that is, lost effort, and its plagiarist author as having achieved "golden mediocrity."

Jacobus Sylvius, erstwhile teacher of Vesalius in Paris, and later a major defender of Galen, attacked Vesalius in two books, an edition of Galen's On the Bones (1549) and Rejection of the Calumnies of a Mad Man (1551). The object of attack was anonymous although apparent to most in the earlier book, and in the second Vesalius was clearly identified as a madman and subjected to an incredibly virulent attack which indicates the strength of conservative feeling on the subject of anatomy. Such attack, however, was the last stand of the conservatives against the new anatomy based upon independent observation rather than ancient authority.

Within two years of the publication of Fabrica and Epitome, illustrations from both works as well as the text of Epitome were used by

Thomas Geminis in England in his *Compendiosa Totius Anatomie De-lineatio* published in London (1545). Both Vesalius and his brother Franciscus were evidently soon informed of this publication, as they refer to it in their Letter on the China Root (1546), but as both described the illustrations in Geminis's work as crude, poor copies of the Vesalian plates, it is clear that neither could actually have seen the fine copper engravings produced by Geminis. The illustrations had, in fact, been copied with great care and accuracy, and a new technique of line engraving on copper which Geminis introduced in England made possible a sharpness of line unobtainable in even the finest woodcuts. The title page of the London (1553) edition of the *Compendiosa* of Thomas Geminis reproduced in facsimile edition was published in London in 1959, followed by an extensive introductory essay by C. D. O'Malley.

The English version (1553) of the *Compendiosa* was the first dissecting manual in English, and as such proved popular. The content of the second edition (1559) differed from that of the earlier edition (of 1545) only in the addition of some woodcut plates and accompanying text. The kind of illustration in which superimposed flaps were used to demonstrate the successive layers of the internal organs was a feature of publications of the period. The descriptions accompanying the Vesalian tables were translated into English and a "treatyse of anatomie" compiled by Nicholas Udall apparently largely from the writings of Thomas Vicarys replaced the Latin text of *Epitome,* which had appeared somewhat mutilated in the first edition of *Compendiosa* (1545). While the English text represented no new advance for anatomical studies, it did provide for a wide circulation of the Vesalian engravings in the English-speaking world and helped to disseminate the new anatomy of Vesalius in England.

The tradition of greatness in the study and teaching of human and comparative anatomy at Padua, begun by Alessandro Benedetti around 1490 and brought to a peak of brilliance by Vesalius, was continued by his successor, Gabriele Fallopius (1523-62) and Fallopius's student, Hieronymus Fabricius d'Aquapendente (1537-1619) (that is, Fabricius from the village of Aquapendente) whose influence on William Harvey and his epochal work was profound. With its distinguished and innovative medical faculty, Padua attracted students from England, Belgium, Holland and Germany throughout the sixteenth and the first half of the seventeenth centuries, achieving a preeminence unmatched elsewhere.

A feature of the medical program at Padua was the public anatomical dissections that were usually carried out in wooden structures temporarily erected for each event. An anatomical theatre constructed during the professorship of Fabricius and one at Bologna are the only Renaissance structures of their kind still completely preserved.

Fabrica, Second Edition

The second or revised edition of Fabrica (1555) was produced in Basel, once again by Oporinus. The new edition was superior in all ways to the first except for the frontispiece. Apparently some disaster in the course of preparation had required the rapid copying and preparing of a new wood block that was clearly inferior to that used for the first edition. Apparently the artist had never seen Vesalius, but, judging the disproportion of head to body in Vesalius's portrait to be correct, reproduced that erroneous disproportion on the new frontispiece. Many changes were made in the text, indicating increased knowledge on the part of the author. Examples are: denial of the permeability of the midwall of the heart, a recognition of the venous valves, a discussion of hydrocephalus and heart block. Some of the illustrations were revised and a few new ones added.

Chapters in the first edition of Fabrica are introduced by historiated initial letters containing various scenes representative of certain subjects discussed. They are not relevant to the main theme of the illustrations and may therefore be considered as pictorial notes to the text displaying aspects of dissection or surgery. These letters were mostly redrawn for the revised Fabrica. Most notable of the new ones is the larger figure which shows Apollo flaying Marsyas. This has sometimes been declared to represent Vesalius flaying his opponents, although there is no authority for such a view. Comparison of the two "venous man" illustrations indicates the figure in the second edition to be somewhat different in minor details from that in the first edition, probably owing to damage to the finer lines of the original wood block. A number of corrections and additions were made in the second edition as compared with the first edition.

Like the first edition of Fabrica, the revised edition was accompanied by an issue of Epitome, not a new edition, however, but merely the remaining old stock with a new final leaf bearing the date 1555 and shown on a special illustration. The only extant copy known is the one that belonged to the distinguished Swedish bibliophile, Erik Waller, and is now in the library of the University of Uppsala.

Vesalius, Royal Physician in Spain

In 1556 Charles V, a very sick man with a complicated series of ailments and a "veracious, uncontrollable appetite, especially for foods most ill-suited to him," abdicated the imperial throne and died two years later in Spain. At the time of his abdication, he bestowed the title of Count Palatine upon Vesalius, granted him a generous life pension and dismissed him from his service. Vesalius was almost immediately enrolled as one of the royal physicians of Philip II of Spain, in particular to serve the Netherlanders at the Royal Court of Madrid. Until the summer of 1559, however, he remained in Brussels where he developed a flourishing practice and built a fine home.

In 1558 and 1559 Vesalius was called upon for service to two royal houses other than that which he served officially. The first of these services was to the ailing wife of the Prince of Orange, later to be known as William the Silent of Holland. In a short letter, the only surviving one in French and now in the historical library, Yale University Medical School, the Prince was informed by Vesalius that although his wife was not suffering from a "continuous fever" she had been attacked by a "great melancholy" which might be dispelled by the Prince's return home. Such hope as Vesalius offered, however, was shattered by the death of the patient on the 24th day of March, 1558.

In 1559 the long years of warfare between France and Spain were ended by the Treaty of Cateau-Cambrésis. The peace was supposedly strengthened by several dynastic marriages, one of them between Philip II and the daughter of Henry II of France, which entailed a lavish celebration and a tournament. On the 30th day of June, the third day of jousting, Henry, running a final course against the Count Montgomery, was wounded above the right eye by a splinter from his opponent's shattered lance. Although there were many physicians available, as well as the famous French surgeon Ambroise Paré, Philip II, who had remained in Brussels and allowed the Duke of Alba to be proxy for his marriage, dispatched Vesalius to attend the injured French king, now his father-in-law. Arriving in Paris, Vesalius participated in the case which terminated fatally in a few days. Nevertheless, his presence permitted him to take part in the post-mortem examination and to write a report which provides a clear picture of cerebral compression supervening upon concussion after injury to the brain. On the 23rd day of August, 1559, Philip II sailed from Flanders, arriving in Laredo on the 8th day of October. When he traveled overland to Madrid, Vesalius, his wife Ann, and their daughter of the same name were with the royal entourage.

The strong national feeling of the Spanish physicians apparently prevented Philip from making Vesalius his personal physician. Vesalius, as a physician to the Netherlanders at the royal court, was, however, occasionally called by the king or by members of his family. Philip inherited many of Charles V's ailments, although not his lethal appetite. Vesalius was granted a large annual salary of 300 florins plus 30 sous daily for sustenance. This was increased by a flourishing private practice, especially among the foreign embassies in Madrid.

During the summer of 1561, Gabriele Fallopius, then holding Vesalius's old chair of anatomy at the University of Padua, sent his famous predecessor a copy of his just published book on anatomical observations, a respectful criticism and emendation of *Fabrica* containing some very important additions to anatomical knowledge, the best known being the description of the fallopian tubes. It was well received by Vesalius, partly perhaps because Fallopius frequently referred to him as the Divine Vesalius. In his reply (1564), which is of little consequence except for its autobiographical content, Vesalius remarked upon the obstacles to scientific work in Spain, mentioning that he was unable to obtain so little as a human skull or the recent literature of anatomy. It seems likely that he already had realized his mistake in going to Spain and presumably would seek to leave when the opportunity arose. His letter suggests that he would make a return to academic medicine, probably once again at Padua.

During the spring of 1562, the Marquis of Terranova suffered a penetrating wound of the left chest during a tournament at Palermo, Sicily and developed emphysema which seemed to endanger his life. His physician, the distinguished Sicilian Gian Filippo Ingrassia (discoverer of the third ossicle of the ear, the stapes), called upon the medical profession for advice. Vesalius wrote him a long account of his successful treatment of this condition by means of surgically inducing drainage and referred to his successful use of this procedure in a number of cases in Madrid. The technique was so impressive for its day, and Vesalius's account such a brilliant exposition, that even though the Marquis had recovered from danger in the meantime, Ingrassia published this statement of the "Mighty Vesalius" for the sake of later generations.

In 1562 an incident involving Don Carlos, eldest son of Philip II and heir to the Spanish throne, helped form Vesalius's resolve to leave Spain. Don Carlos was in Alcala in 1562 recovering from malaria contracted in Madrid. He was attracted to Mariana de Garcetas, daughter of the caretaker of the royal quarters of Alcala. On Sunday the 19th of April, catching sight of the girl in the palace garden and seeking to join her, Don Carlos, in the unsympathetic words of the English Ambassador, "in hasty following of a wench, daughter of the keeper of the house, fell down a pair of stairs and broke his head." The case attracted great attention throughout Spain. Numerous royal physicians and surgeons were ordered into attendance upon the injured prince, and at least one notorious quack was persuasive enough to be admitted for a time to the attending medical group. Vesalius, accompanying Philip II to his son's bedside, was also a participant. Don Carlos recovered after several months during which he had been in considerable danger, but it is impossible to assess credit for the recovery. Among other procedures, the hundred-year-old mummified remains of a Franciscan friar of Alcala, the Blessed Diego, were placed beside Don Carlos during a spell of delirium, and many, including the king, considered this as responsible for the miraculous recovery. Thanks to Philip, the friar was canonized in 1588, but even the most pious physicians resented such slight of their medical abilities and it seems likely that Vesalius, never a pious man, thereby became even more strongly determined to remove himself from Spain at the first opportunity.

Back to Padua

In 1564 Vesalius was finally able to leave Spain, an event which has given rise to a seemingly indestructible legend. According to the story (first recounted by Ambroise Paré), an unnamed anatomist in Spain dissected a woman pronounced dead. Upon opening her body he found evidence of life which brought the unhappy anatomist within the toils of the Inquisition. He was saved only through the intercession of the king and the promise of an exemplary pilgrimage to the Holy Land. The story was repeated in 1603 by the English physician Edward Jorden who supplied Vesalius's name as the anatomist. Still other accounts declare that he was driven from Spain by the conservative medical element. Most likely of all is the account of the Belgian botanist, Charles de L'Ecluse, who arrived in Spain shortly after Vesalius had left. According to his account, Vesalius had been ill or feigned illness which gained him permission for a leave of absence during which time he also intended to make a pilgrimage to the Holy Land.

Vesalius traveled to the Spanish border town of Pergignan with his wife and daughter and so into France, where at Cette, after a violent family quarrel, mother and daughter returned to Brussels and Vesalius moved on to Marseilles, thence to Genoa and across northern Italy to Venice. Since Fallopius had died in 1562, a victim of tuberculosis, Vesalius regained his old chair of anatomy at the University of Padua.

At some time in March 1564, Vesalius boarded a ship from Venice bound for the Holy Land. After a stop at Cyprus, the ship continued to Jaffa, the usual port for the landing of pilgrims, who continued thence overland to Jerusalem. Nothing is known of Vesalius's visit to the Holy Land except for one recorded glimpse of him on the Plain of Jericho accompanied by "a certain Franciscan Friar," with both men more concerned about local botany than with pious sightseeing. When exactly Vesalius planned to return is unknown, although it would have been necessary for him to be in Padua in mid-October for the beginning of the new academic year. His return trip was booked on one of the pilgrim ships which frequently, according to complaints, provided poorly for the passengers. On this particular return journey, a severe storm delayed the ship for more than a month, during which time food and water were exhausted. Some of the passengers succumbed and were thrown overboard, and when finally the vessel did manage to reach the island of Zante, according to the several accounts, Vesalius managed to get ashore but died soon thereafter, in October, 1564. No trustworthy information is available to indicate where the ship made land nor where Vesalius's remains were buried.

Fate of Vesalian Books and Wood Blocks

After the celebrated Vesalian wood blocks had been used for the second edition of *Fabrica,* published in 1555, they had performed their final service insofar as Vesalius and his publisher, Oporinus, were concerned. Thereafter all trace of them was lost until they were found in the possession of Andreas Maschenbauer, an Augsburg publisher who used them in an edition of an anatomical treatise (1708-23). Again lost, rediscovered, and used for another publication, this time in Ingolstadt, 1783, they passed from there to Landshut and finally into the comparative security of the library of the University of Munich. There they were locked away and forgotten until 1893, after which they were employed once more in 1934 for a luxurious edition of Vesalian plates published by the Bremer Press, a remarkable instance of lasting qualities of wood and enduring workmanship. However, man's increased destructive power eventually accomplished what time's ravages had failed to do. On the 13th of July, 1944, a block of the University of Munich was burned as a result of bombing, but the cellar behind fireproof doors preserved its contents, including the wood blocks. Three days later, on the 16th of July, bombs again fell on the now ruined building and set fire to the still glowing ashes. This time the contents of the cellar could not be saved and the blocks were destroyed.

In the library of the University of California at Los Angeles is a copy of an English translation of the bloodletting letter that was made by J. B. de C. M. Saunders and C. D. O'Malley and published in 1947. This is entitled *Andreas Vesalius Bruxellensis, The Bloodletting Letter of 1539*, an annotated translation and study of the evolution of Vesalius's scientific development, Henry Schuman, New York, 1947. An edition of a book by Guinter is in the University of California at Los Angeles. This is entitled *Guinterius Anatomicarum Institutionium Libri III*, Lyons, 1541. The first edition of this appeared in 1536 and in it the author paid a compliment to the anatomic skill of his former student in Paris, Vesalius.

The first edition of Vesalius's epochal treatise on the structure of the human body, *De Humani Corporis Fabrica*, Basel, J. Oporinus, 1543, is available in the University of California at Los Angeles library. The significance of this outstanding work has been summarized by the late C. D. O'Malley:

> When one examines the text of *The Fabrica* it is immediately apparent why this book is one of the great classics of medicine. Never before had the structure of the human body been so thoroughly discussed with such care for anatomical minutia and with such effort to integrate the various parts of the structure. In the course of the description hundreds of long-held erroneous Galenic and medieval doctrines were shattered simply because Vesalius ignored early authoritarian rights and relied upon his own researches, observation and reason. There are errors, naturally, and from time to time Vesalius failed to follow his own principles of investigation, but the correctness of descriptions, especially of bones, muscles and brain were an outstanding achievement and in fact there are contributions to anatomical knowledge in all seven books of *The Fabrica*.

Experimental Methods in Physiology

Interest gradually shifted from the skeleton and muscles to the sense organs and the internal parts; the eyes, ears, heart, lungs, arteries, veins, and kidneys were more fully described. Later, stages of fetal development were studied and compared with the equivalent stages in the formation of the chick. Studies describing and comparing the anatomy of a variety of animals appeared. Thus, by the end of the sixteenth century, descriptive anatomy had advanced well beyond its ancient attainments and was generally recognized as a study that must be based upon observation.

In the anatomical investigations of Vesalius and those who followed, many observations not only contradicted some of Galen's assumptions but rendered impossible certain aspects of his physiological theories. Nevertheless, Vesalius and his sixteenth century successors continued to accept and teach the physiological theories of the past.

An understanding of the vital functions of the body obviously presupposes a knowledge of anatomy. Aristotle had thought that the functions of the parts of animals could be inferred from the knowledge of the comparative anatomy of each system. Galen had realized that in order to understand how the parts worked it was necessary to see them in a living animal. Vesalius also recognized the close association between anatomy and physiology and had claimed that vivisection was necessary if the functions of the parts were to be adequately investigated. The works of Vesalius and his immediate successors were based on observation but were not really experiments. It is true that some so-called experiments were performed, but they did not conform to the modern definition, i.e., observations and measurements made under controlled conditions to test the validity of a hypothesis or theory. Many early "experiments" were isolated attempts to demonstrate to students how certain parts worked. They were not a part of any inquiry and certainly were not sustained attempts to confirm or test in detail any physiological theory. Indeed, the idea of using an experiment as a tool to investigate nature was only beginning to evolve during the late sixteenth and early seventeenth centuries.

Aristotle, the Alexandrian physiologists, Galen, and others had carried out some experiments and beginnings were made very early in the use of the experimental method, but these attempts were not associated with basic issues in biology. Usually they were demonstrations of small segments of information that did not represent broad theoretical considerations. Galen, for example, severed spinal cords of monkeys at different levels and showed that power of motion and sense were destroyed in the structural parts below the incisions. This illustrated a fundamental aspect of the functioning of the spinal cord, but the experiment was not designed to test any of Galen's basic assumptions or broad generalizations.

Experimental procedures depend on the formulation of hypotheses and the designing of experiments to test these hypotheses. To be truly experimental, after a problem is identified, the first step is the formulation of a hypothesis that provides a possible answer or explanation of the phenomenon in question. Experiments must then be designed to test the hypothesis. The adequacy of the testing procedure is judged not only by the actual amount of experimental evidence generated, but also by the rigor of the arguments and the precision with which the testable consequences can be formulated. If the initial hypothesis is not confirmed, others are devised and tested.

Development of Physiology as a Science

Studies of the circulation of the blood provide a good example of the early evolution of physiology. Aristotle had summarized the work of earlier Greek scholars on blood vessels and added his own refinements, probably from actual dissections of animals. He identified the heart as the central organ controlling blood flow, the seat of vital activity, and the place where the blood received its animal heat. From the lungs, according to Aristotle, came the pneuma or spirits. Pulsation was described as a boiling process in the heart that occurred when the blood and pneuma came together. Herophilus, anatomist of Alexandria, counted the pulsations and worked out a rhythm and a rate of the heart. He also described the function of the valves of the heart. Erasistratus, another physiologist in Alexandria, elaborated on the pneuma as the cause of the heartbeat and the source of body heat. He distinguished among different kinds of spirits that controlled the functions of the body. The vascular system he considered to carry vital spirits, whereas he described the nervous system as carrying animal spirits.

Galen followed the procedure of Aristotle, in that he summarized the work of his predecessors and supplemented their findings with his own observations on such animals as sheep, dogs, swine, and apes. He described the aorta and the main veins. The valves of the heart were known to him, but he regarded them as fireplaces for heating the blood, which occurred, according to Galen, in the heart. The heart was supposed to provide the motion for moving the blood out of the heart. Galen's plan of body physiology was complicated. The food entered the stomach, underwent coction or boiling and proceeded to the liver where it was changed to blood. From the liver it was carried to the right heart where a portion entered the pulmonary artery and went to the lungs. The blood was believed to be used in the lung. There was no return to the heart. The smaller portion of blood was said to pass through the imaginary pores of the septum into the left heart where it was mixed with the pneuma that had entered the heart separately from the lungs through the pulmonary arteries. The mixture of blood and pneuma was expelled through the aorta and carried out by the blood vessels to nourish the body. Galen, like Aristotle, believed that the arterial system was concerned with the distribution of the vital spirits. Air, or the special part of the air called pneuma, was taken in by way of the trachea and the lungs and passed through the pulmonary vein to the left side of the heart. In spite of its veinlike structure, the pulmonary vessel was regarded as part of the arterial system and was called "the veinlike artery." In the left ventricle of the heart the pneuma was supposed to mix with a certain amount of blood to form the vital spirits, which were distributed to the rest of the body by the aorta and the arterial system. The lungs were believed to exist solely for the purpose of cooling and ventilating the heart and the movement of the chest and heart were regarded as one and the same.

The actual mechanism of the heart was completely misunderstood. It was supposed that the heart sucked in air and blood during the active stage called systole and allowed them to drain off during diastole. Galen's description of circulation, like his anatomical work, was considered infallible for some fourteen centuries. Vesalius, who had corrected many of Galen's errors in human anatomy, had not become aware of the erroneous description of blood movement in the body. Vesalius was puzzled when he dissected the heart and found no pores from the right to the left heart, but he had no microscope and assumed that the pores were invisible to the unaided eye.

Colombo, successor to Vesalius at Padua, had hinted that the blood might flow from the right to the left side of the heart through the lungs. This so-called "lesser circulation" was in fact mentioned by Michael Servetius (1511-53), a student of Vesalius, who knew about pulmonary arteries and veins and was familiar with the notion of the movement of vital spirits from right to left ventricle. But Servetius says nothing about how the blood was propelled into the pulmonary artery or the aorta. Servetius might have resolved the problem experimentally, but vivisection was banned and he did not have an opportunity to do the necessary experiments. Furthermore, Servetius was burned at the stake along with most of his writings and whatever he may have said about circulation is not likely to have had much influence.

Background and Training of Harvey

William Harvey was born at Folkestone, England, in 1578. When he was a small boy, local butchers are said to have given him hearts of animals from which he might watch the flow of fluids through the heart. He undoubtedly received some knowledge of anatomy and the impression of circulation from these early observations. In a later period, when he was a mature physician, the king ordered huntsmen to give him materials that might be useful for his observations.

He attended Canterbury Grammar School and at the age of 16 entered Gonville and Caius College, Cambridge, from which he obtained his B. A. degree in 1597. Late in the year 1599 he went to the University of Padua where he studied with Hieronymus Fabricius of Aquapendente, a student of Bartholomaeus Eustachius (1520-74), and Gabriele Fallopius, who had carried on the tradition of Vesalius at Padua.

Fabricius ranked second only to Vesalius as an influential teacher and scholar of comparative anatomy. He taught and carried out research projects at the University of Padua for more than 60 years. When Harvey arrived at Padua, Fabricius had already suggested the presence of valves in the veins and had made more general observations of circulation of the blood.

Undoubtedly Fabricius had a view of circulation following his discovery of valves in the veins. The logical necessity of one-way traffic of blood in the vessels must have followed this discovery of valves in the veins. For a demonstration of these valves, press the blood vessel at the wrist and back of the hand downward toward the fingers. The vessel becomes empty of blood and blood does not follow the pressing of the fingers. The vein remains empty at least for a short time. On removing the pressure, the vessel is instantly filled from the fingers upward. This is due to the valves in the veins. Harvey in his book *On the Movement of the Heart and Blood in Animals* gives diagrams to illustrate the experiments stopping blood in the forearm and hand from which he inferred valves, as Fabricius before him had described. From this beginning he ascertained a mode of action of several parts of the heart mechanism and of the correlated channels of the blood.

Why did not the contemporaries or predecessors of Harvey discover continuous circulation of the blood? Perhaps the one general answer would be that they did not follow the experimental method completely to its conclusion but instead held reverently to the dogma perpetuated from Galen. Cesalpino of Arezzo used the word "circulatio" and made references to movement of the heart and action of the blood but could not extricate himself from the accepted doctrine that the heart and vessels served to distribute to the body its vital spirits and native heat rather than blood as found in the veins.

When Harvey obtained his doctorate in 1602, he returned to London and established a private practice of medicine. In the same year (1602) he took a doctor of medicine degree to become eligible for a fellowship in the College of Physicians, to which he was admitted in 1604. In 1608, at age 30, he became assistant surgeon at St. Barthole-mew's Hospital, and in 1609 he became physician.

Harvey's Contribution on Circulation of Blood

When William Harvey (1578-1657; Fig. 6.3) introduced into biological inquiries the method of investigation that combined dissection and vivisection with simple experimentation, he made the first decisive break with ancient physiology. Building upon previous anatomical dis-coveries, he opened the way not only to a new viewpoint about the vital functions but also to a new methodological approach. Harvey's notes give complete proof of his discovery of continuous circulation some 12 years before the accepted date (1628) of publication of his paper. Certainly he owed many insights to his immediate predecessors, particularly Fabricius. He probably drew upon all the ideas that had come to him from various sources, but the discovery of continuous circulation undoubtedly resulted from his own experimental work.

Harvey's experience at Padua certainly strengthened his view on continuous circulation. In a brawl with a student group there, a friend of Harvey received a dagger slash across the arm that severed an artery. Blood came gushing from the wound in spurts. It looked as if a pump were at work. Harvey knew that blood from a vein runs smoothly and his professors had said there were two different kinds of blood in the body—the first being blood from the liver to supply nourishment, that is, animal spirits, and blood from the heart which pulsated rapidly and gave heat and energy to the body (vital spirits). The blood appeared the same to Harvey. He tasted it and it tasted the same. Harvey wondered (hypothesized) if it was the same and was being pumped through the entire body. Harvey's hypothesis was that the blood flows continuously through the body. His method of testing the hypothesis was a series of observations and experiments.

With the aid of a hand lens, Harvey observed the action of the heart in cold-blooded vertebrates. He had been taught that all parts of the heart moved in unison, but on observation the upper part seemed to beat first. Harvey had observed from his studies of comparative anat-omy that the hearts of all vertebrate animals had much in common. Apparently the same process of pumping blood through the body was necessary for all animals and it was possible to see the flow through the various parts of the heart.

As a part of his medical studies, Harvey attended two "anatomies" where bodies were completely dissected by barber surgeons. Later, in

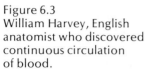

Figure 6.3
William Harvey, English
anatomist who discovered
continuous circulation
of blood.

London, six anatomies were held per year on executed criminals. From these Harvey received some impressions as to how the process of blood circulation might work. He was able to note the pressure points on the wrists, temples, and sides of the neck where the pulse can be felt. These also suggested the pumping action from a central pump, the heart. From his own surgical and medical practice, Harvey observed the blood flow and checked his theories on circulation from observations of people.

In August 1615, Harvey was appointed Lumleian Lecturer or Professor of Anatomy at the College of Physicians and Surgeons. This professorship had been set up by Lord John Lumley with a fund to pay for the lectures. On April 16, 1616, Harvey began his first Lumleian Lecture by doing in the hall of the college a dissection that lasted three days. On the second day he came to the chest and described the heart as a pump, with blood circulating in a closed system of vessels through the body. This was his first public discussion of his circulation theory. He said in his lecture:

> It is proved by the structure of the heart that the blood is continuously transferred through the lungs into the aorta as by two clacks of a water bellows to raise water. It is proved by a ligature that there is a passage of blood from the

arteries to the veins. It is therefore demonstrated that the continuous movement of the blood in a circle is brought about by the beat of the heart.

After this lecture, Harvey set out to further test his theory by precisely measuring the pressure and the amount of blood at various points in
the body. First he pressed his finger along the vein between the valves
and showed a section that was empty. If points 2, 3, and 1 are identified
as on the accompanying diagram (Fig. 6.4), Harvey demonstrated by
moving his fingers between points 1 and 2 that the vein was empty.
He moved his finger up the arm from point 1 to point 3. That part
filled with blood. This demonstrated the activity of the valves in controlling the one-way traffic of blood through the body.

Next Harvey demonstrated that the body could not produce as
much blood as Galen's explanation would require. He accomplished
this experimentally by measuring the actual blood passing a given
point. Two measurements were required in this experiment; first, how
frequently does the heart beat, and second, how much blood is
ejected into the aorta with each beat. Harvey estimated the pulse rate
at 33 beats per minute, partly because of his own desire to be conservative and safe. Harvey estimated further that the left ventricle ejects
three ounces of blood per beat based on the observation that the left
ventricle holds more than two ounces of blood. In his notes he says:

> If the left ventricle ejects only one fourth of its contents with each beat and
> the heart beats a thousand times in a half-hour, at least 500 ounces will have
> been transferred, more than 30 pounds.

This is more than that contained in the entire body.

In 1628 when Harvey was 50 years old, he published, in Latin, *Exercitatio anatomica de motu cordis et sanguinis in animalibus* (On the
Movement of the Heart and Blood in Animals), a 72-page book outlining the views that he had solidified some 12 years before. In the first
chapter in his book, Harvey pointed out the difficulties inherent in
the theory of Galen. In particular he argued against the assumption

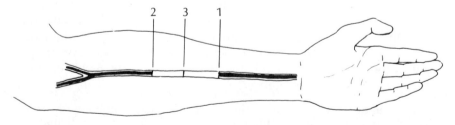

Figure 6.4.   Harvey's demonstration of valves in veins of forearm.

that the movement of the lungs and the heart were the same and that the veinlike artery could possibly perform the three tasks assigned to it, that is, the expulsion of fumes from the heart to the lungs, the transport of air from the lungs to the heart, and the provision of vital spirits from the heart to the lungs for their own sustenance. Nor could he accept the idea that all the blood present in the arterial system was conveyed from the right to the left side of the heart by way of pores in the dense septum, pores of which no trace could be found.

Harvey next reported on his investigations into the action of the heart. He had performed vivisections on cold-blooded (poikilothermic) animals whose heartbeat in cold weather is relatively slow and on warm-blooded animals whose heart motion slows as death approaches. As a result he was able to describe the various stages of the heart's action. He pointed out that it is during the quietest period when it is relaxed that the heart fills with blood. The blood first enters the auricles, which then contract and send the blood into the ventricles. The main action of the heart is the contraction of the ventricles, during which operation the tip of the heart is raised. The orifices become smaller and the whole heart becomes larger and narrower. As a result, the blood in the ventricles is ejected into the pulmonary vein, which is like an artery, and into the aorta.

He demonstrated with the aid of experiments the action of the various valves and showed that liquid was unable to pass back from the auricles into the vena cava or into the veinlike pulmonary artery, or from the ventricles to the auricles. Having established how the heart worked, Harvey went on to consider the pulmonary circulation. He showed that in the fetus there is a connection between the ventricles but that in mature animals with lungs the septum is impermeable. No blood or liquid would pass from the right side of the excised heart to the left ventricle, although liquid would pass through the vessels into the lungs and back into the left auricle. If Harvey was not the first to suggest the existence of the pulmonary circulation, he was certainly the first to back his beliefs with a vast array of observations and experiments.

When Harvey had established to his own satisfaction the action of the heart and the existence of the pulmonary circulation, he considered what is now called the systemic circulation. He counted the number of heartbeats per minute and measured the amount of liquid an excised heart would hold. Even if the heart were very inefficient and ejected only a very small portion of this liquid at each beat, a simple calculation showed that the total amount of blood entering the aorta every hour must far exceed the fluid within the body.

Therefore, Harvey concluded the blood leaving the heart by way of the aorta must be returned to it by the vena cava. He pointed out that it was possible to exhaust the body of blood very quickly by severing an artery. He showed that if a living animal were opened and a ligature applied to the vena cava, the area below the obstruction and away from the heart became full of blood whereas the heart itself became pale and soon slowed down. By a series of similar experiments he investigated the action of the valves in the veins and showed that they prevented the back-flow of blood. He concluded that the blood in the veins flowed toward the heart. If, however, ligatures were applied to the arteries of an animal, the region away from the heart became pale and bloodless while the blood itself backed up between the heart and ligature. In the arteries, therefore, the blood flowed from the heart toward the various parts of the body.

This new understanding of the motion of the blood through both the pulmonary and systemic circulation systems removed the confusion concerning the pulmonary vessels. According to Harvey's theory, the vessel connecting the right ventricle of the heart and the lungs was a true artery in both structure and function for in it the blood flowed from the heart to the lungs. Harvey, therefore, renamed it the pulmonary artery, the term now in use. Similarly he called the vessel connecting the lungs and the left auricle the pulmonary vein since, in it, the blood flowed from the lungs to the heart. As a result of his experiments and observations, Harvey felt he had established his main thesis and showed that the blood flowed from the right ventricle through the lungs to the left auricle and from the left auricle through the arteries and back by way of the veins to the vena cava to the right auricle. However, without a microscope Harvey could not see the capillaries and was never able to demonstrate the existence of connections between the arteries and veins. They were believed to possess some incorporeal essence that accounted for their ability to confer and maintain life.

Harvey, who insisted on the importance of the observable, was naturally opposed to any idea of incorporeal spirits. He felt that the explanations based upon the assumptions of such spirits were completely unsatisfactory and that they served as a common disguise for ignorance. But there was a manifest difference between the living organism and the dead body. The warm blood circulating through the body was different from the lump of coagulated gore on the floor of the butcher's shop. Harvey could not explain the difference nor could he even describe it in any clear way. In order to mark off this difference he was forced back into the language of spirits and always referred to the blood within the body as "spiritous." However, he tried to make it plain that in retaining this word he did not wish to imply that he believed the blood in the arteries or veins was in a different state because it was intermixed with a mysterious incorporeal principle. The blood was spiritous in the same way that good wine or brandy might be said to be spiritous. Encountering the criticisms of his theory on blood circulation, Harvey said, "A path is open for others starting here to progress more fortunately and more correctly under the more propitious genius."

Harvey's work on blood circulation was not only a starting point for the modern science of mammalian physiology but it was also the first milestone on the road to the modern experimental approach to the biological problem. In announcing his great discovery, Harvey modestly suggested that it was merely a return to Aristotle.

Harvey's Later Years

When Charles I came to the throne, Harvey was even more closely connected with the court, and he attended the king as his physician. During this period Harvey had the opportunity to travel frequently to the Continent of Europe and is believed to have met a number of European doctors and to have discussed his views and demonstrated his discoveries to them. The king, Charles I, was apparently interested in Harvey's anatomical research and ordered the royal huntsmen to provide Harvey with specimens. Harvey noted that the king had observed certain of his dissections and had provided facilities for some rather elaborate experiments. Harvey remained with Charles I throughout the civil war and went to Oxford when the court transferred there from London. After the Royalist defeat, Harvey did not follow the king's associates into exile but retired to live with his brothers in the country near London. There he remained until his death in 1657, at the age of 79.

Harvey's most important research was concerned with the movement of the heart and blood vessels and with animal reproduction. His work on these subjects had a considerable effect on the development of physiology. Harvey published three books: the famous *Anatomical Exercise on the Movement of the Heart and Blood in Animals* (1628), a small book entitled *An Anatomical Disquisition on the Circulation of the Blood to Jean Riolan the Younger, of Paris,* containing replies to his critics addressed to the French doctor and published in 1649, and a large book, *Exercises on the Generation of Animals* (1651). In addition, Harvey collected material for a book on morbid anatomy and prepared a series of notes on the action of the muscles.

Neither of these was ever published although the notes on the latter have recently been deciphered, edited, and translated into English by a modern scholar, Dr. Gweneth Whitteredge.

Chronology of Events

A.D.

130-200   Galen, Roman physician who wrote many treatises on medicine and philosophy including works on anatomy and physiology.

1543-1555   A. Vesalius, *The Fabric of the Human Body* (1543); *Epitome* (1543); "Letter on the China Root" (1546); *Fabrica* Second Ed. (1555).

1545   T. Geminis, *Compendiosa*.

1551   L. Fuchs, digest of *Fabrica*.

1553   M. Servatius mentioned pulmonary circulation.

1559   R. Columbo suggested circulation from right heart to lungs and back to left heart.

1561   G. Fallopius, anatomist at Padua who studied nervous system and generative organs.

1570   B. Eustacius, anatomist at Padua who studied voice organs, sympathetic nervous system, blood, and nerves.

1599   H. Fabricius, teacher of Harvey at Padua who had discovered valves in veins.

1628   W. Harvey, *On the Movement of the Heart and Blood in Animals.*

1934   Vesalian wood-block plates used for publication.

References and Readings

Ashley-Montague, M. F. 1955. "Vesalius and the Galenists." *Sci. Monthly* 80:230-239.

Bridges, J. H. 1892. *Harvey and His Successors*. London: Macmillan and Co.

Chauvois, L. 1957. *William Harvey*. New York: Philosophical Library.

Cushing, H. W. 1943. *A Bio-Bibliography of Andreas Vesalius*. New York: Henry Schuman.

Fermi, L. and G. Bernardini. 1961. *Galileo and the Scientific Revolution*. New York: Basic Books.

Franklin, K. J. 1961. *William Harvey, Englishman, 1578-1657*. London: Macgibbon and Kee.

Galilei, Galileo. *The Dialogues Concerning Two New Sciences*. Translated by H. Crew and A. de Salvo, 1963. New York: McGraw-Hill Book Co.

Gasking, E. 1970. *The Rise of Experimental Biology*. New York: Random House.

Gumpert, M. 1948. "Vesalius: Discoverer of the human body." *Sci. Amer.* 178:24.

Harvey, W. 1628. *Exercitatio Anatomica de Moto Cordis et Sanguinis in Animalibus*. Translated by C. D. Leake, 1930. Springfield, Ill.: Charles C. Thomas.

_____. *Lectures on the Whole of Anatomy*. Annotated translation by C. D. O'Malley, F. N. L. Poynter, and K. F. Russell. 1961. Berkeley: University of California Press.

Keele, K. D. 1965. *William Harvey.* London: Thomas Nelson and Sons.

Keynes, G. L. 1949. *The Personality of William Harvey.* Cambridge: At the University Press.

Lind, L. and C. Asling. 1949. *The Epitome of Andreas Vesalius.* New York: Macmillan Co.

Marcus, R. B. 1965. *William Harvey, Trailblazer of Scientific Medicine.* London: Chatto and Windus.

O'Malley, C. D. 1964. *Andreas Vesalius of Brussels, 1514-1564.* Berkeley: University of California Press.

Payne, J. F. 1897. *Harvey and Galen.* London: Oxford University Press.

Vesalius, A. 1543. *De Humani Corporis Fabrica.* Basel, Switzerland. (Translated in Rook, *The Origins and Growth in Biology,* Pelican Books, London, 1964.)

Whitteridge, G. 1971. *William Harvey and the Circulation of the Blood.* London: Macdonald.

CHAPTER 7

SCIENTIFIC SOCIETIES AND EXPERIMENTAL SCIENCE

THE EXPERIMENTAL work of Harvey on blood circulation and that of other scientists in the physical sciences such as Copernicus, Kepler, Galileo, and Newton had demonstrated that questions concerning natural phenomena could be settled by observation and experimentation. The contributions of the physical scientists promoted a revolution in thought. An earth-centered universe changed to a system of planets centered around the sun. Other changes in thought followed, about motion, time, space, and relations among bodies. In fact, it gradually began to seem possible that knowledge gained in this way could surpass and replace that acquired by studies of ancient authorities. Despite the tremendous success of investigators using experiments, however, the question of how best to obtain new knowledge remained unsettled.

The small scientific academies that sprang up all over Europe during the late sixteenth century and early seventeenth century gave much of the necessary stimulus to the experimental method of approaching scientific subjects. Some of these academies were promoted by princes and other prominent people who became patrons or honorary members, thus lending state or church support to the movement while indulging some personal interest in the study of nature. Small laboratories were established by individual investigators on problems that were exciting to them. Collections of plant and animal specimens were maintained in private and family museums which encouraged continued scientific activity. These individuals talked with others about their work and groups developed sporadically, depending entirely on the interest of members for their sustenance. Some of these informal groups gave rise to informal societies that became academies. Although sometimes supported by wealthy and influential people, shared and vigorous interest in science rather than available material support was the incentive for their origin and continuation. When interest of members lagged, activities were postponed or abandoned. When new stimuli or new ideas came to the group, a new enthusiasm and a wave of accomplishment followed. Some small academies of the period explicitly claimed to be devoted to experimental philosophy. This enthusiasm for experimentation indicated the growing realization that a true knowledge of nature required more than generalizations from a few facts. Persistent and sustained experimental work was necessary to develop science.

The first real contributors to science after the Middle Ages were thus curious amateurs who came together spontaneously to discuss problems in which the participants were interested. Historians of science have called these organizations "curiosity cabinets." They were loose societies whose members gathered together for interest and amusement, but which made significant and lasting contributions in science.

Freedom of thought and the incentive to develop new ideas are necessary for wholesome growth in science. It is not surprising, therefore, that the real advances in science were promoted by independent and curious men who did original work for the love of learning and who met together of their own free will to discuss learned subjects. This spontaneous movement resulted in the development of societies devoted to science, so-called "academies." The word "academy" had been used some 2,000 years before to identify Plato's school where discussions were held in a grove of trees. In the sixteenth century the term was revived to describe a gathering of people who united for the purpose of learning. Individuals who met primarily to discuss subjects in which they were interested were often actually engaged in original research.

Although the early universities did much to awaken an intellectual activity and promote a general recovery from the low cultural level of the Middle Ages, they did little to foster experimental science. It is true that some university men such as Albertus Magnus and Roger Bacon made observations and devised scientific experiments. But in general, the early universities often seemed to hinder rather than to aid science. The University of Paris, for example, one of the most influential of the early universities, provided courses in such subjects as art, music, theology, law, and medicine, but none in basic science. The universities did in a way provide an incentive to master "tool" subjects, especially language, which are necessary to perpetuate any intellectual activity.

Universities of the sixteenth century were usually authoritarian. Little interest was shown in objective study or in natural objects and phenomena. It must be remembered that elementary and secondary schools were poorly developed and their curriculums were not standardized. Then, as today, some students were better prepared than others by previous training to profit from university courses. Wide variations existed among the universities in terms of the age at which students could enter. The early University of Paris, for example, would be comparable in terms of age of students to present-day secondary schools. Most of the students were only 12 to 16 years of age when they attended. At Bologna and other Italian universities that were somewhat more sophisticated and where stricter admission requirements were enforced, the average student was older— more comparable to a graduate student in present-day universities. The main objective of some early universities was to preserve the knowledge of the past, and little or no concern was shown for creative activity. In later periods in time the universities contributed greatly to science.

Progenitors of Academy Movement

The curiosity and enthusiasm that stimulated the academy movement were exemplified by that versatile experimenter of the Renaissance, Leonardo da Vinci (described in Chapter 5). Leonardo entered into widely ranging discussions with many different groups that were devoted to science and art.

Francis Bacon

Francis Bacon (1561-1639; Fig. 7.1) was more a philosopher than a scientist. He became the father of the English school of philosophical thought and originated much of the philosophy perpetuated by John Locke and others. Bacon's free-thinking philosophy somewhat resembled Aristotle's, in that he visualized a great plan for the origin and governing of the earth and its inhabitants. Bacon was an effective writer and popular lecturer, but he lacked the objectivity characteristic of modern scientists. He gathered many facts but did not sift, organize, and coordinate them effectively. Several indirect contributions to science, however, came from his work. He fought against the scholastic methods of teaching and advocated free criticism, but he did not recognize a major precept of modern science, that is, the fact finder must be self-critical. According to today's thinking, the investigator must present to other scientists and the public at large only facts and conclusions that he can support from objective data. Bacon considered science to be important only as a tool to fill in the details of a plan which was already established and functioning. In spite of his inadequacies by modern scientific standards, Francis Bacon started a movement for free discussion which led to one of the most important events in the history of science, the formal organization of academies, and in particular of the one that became the Royal Society of London.

Peiresc

One of the first men to be actively identified with functioning scientific academies was the versatile Fabri de Peiresc (1580-1637). This wealthy and influential Frenchman served for many years as a voluntary agent for the exchange of knowledge among individuals and organizations in France and other countries. He wrote letters, visited scientists in different countries, and did everything possible to support science. His acquaintance with other prominent men and investigators in different countries of Europe kept him aware of virtually everything that was being done in science in Western Europe over a period of years. He followed the work of Galileo with great interest and repeated some of Galileo's observations with the telescope.

Italian Academies

The academy movement developed earlier and more effectively in Italy than elsewhere in Europe. One of the first academies met at

Figure 7.1
Francis Bacon, British
philosopher and statesman
who prepared the way for
the academy movement.

Naples in the home of Giambattista della Porta (1543-1617) in the latter part of the sixteenth century (1560). To gain membership in this group, which became known as *Academia Secretorum Naturae*, a candidate was required to discover a new fact in natural science. Experiments were performed at academy meetings and some 20 books of results and discussions were compiled.

Academy of the Lynx

A better-known early academy was the Academy of the Lynx (*Academia dei Lincei*) founded in Rome in 1603 or before. The word "lynx" was selected to suggest the motto, "sharp eyesight and keen observation." A symbolic picture of a lynx with upturned eyes holding in its paws the powers of darkness became the standard of the academy. It was intended to symbolize the struggle of scientific truth against ignorance. The society was founded by Duke Federigo Cesi (1585-1630), a rich young Italian and skillful experimenter. At first only three men met with Cesi at his home to discuss studies in which they were interested. In 1609, the membership was 32, including three distinguished academy leaders, della Porta, Peiresc, and Galileo.

Galileo made an instrument for observing small objects and another member, Johannes Faber (1574-1629), an entomologist, gave the instrument the name "microscope." A charter member, Francesco Stelluti (1577-1653), used the microscope in preparing a zoological study on bees which became the first (1625) scientific report to be published by the society. Other important publications from the society carried botanical and astronomical observations. One in particular included a description of plants and animals of Mexico. Following the death of Cesi in 1630, the society became inactive, but it was revived many years later (1870).

Academy of Experiments

The Academy of Experiments (*Academi del Cimento*) was established at Florence, Italy, in 1657, under the patronage of the Grand Duke Ferdinand II (Medici). Although it had an active life of only ten years (1657-67), it substantially influenced the development of science. Its nine members included Castellio and Torricelli, both disciples of Galileo; Giovanni Alphonso Borelli (1608-79), a mathematician with interests in biological mechanisms; and Francesco Redi (1621-97), who disproved spontaneous generation of flies.

This was the first institution devoted extensively to the publication of scientific materials. All members contributed their findings without signing their individual names; the actual writing was done by the secretary. The *Report of Experiments,* printed in 1666, was devoted mostly to physics. It contained original descriptions of the first barometer, which had been invented by Torricelli in 1643, the first true thermometer, and the hydrometer. Reports dealing with the expansion of water on freezing, universal gravity, and electrical properties of matter also were included in this volume. Science began as an interdisciplinary effort and now in the latter part of the twentieth century is turning back to that pattern.

The researches of the Academy of Experiments and, to a large extent, those of other early academies, were conducted along well-determined scientific lines. Speculative presentations were usually avoided, and conclusions were restrained. The balance and self-restraint was due in large measure to a mutual cooperation among several scientists in joint researches.

French Academy of Science and Others on the Continent

The French Academy of Science originated, like each of the Italian academies, as a small group of curious men who initially met in their own homes to discuss things in which they were interested. Descartes and Peiresc had done much to lay the foundation for the French Academy. Marin Mersenne (1588-1648), a skillful writer, who popularized the work of Descartes and Galileo in France, was the leader of the Academy movement in France. Melchisedic Thevenot (1620-92), who became the patron of the microscopist, Swammerdam (Chapter 8), was one of the most influential members.

Formal meetings were held in public places as early as 1630. At that time, Cardinal Richelieu agreed to accept the organization and not to prosecute the members. The French Academy of Science was officially organized in Paris in 1666 by Louis XIV. Regular sessions were then held in the Royal Library, which was also used as a laboratory for demonstrations. Members of the French Academy repeated many of the experiments performed by the investigators in the Academy of Experiments and other societies. They also conducted some original research. In these investigations, they dissected a considerable number of animals and plants for the purpose of discovering peculiarities of individuals, but they did not consider similarities and differences between species. Although limited in scope, their studies on the natural history of animals did correct certain errors recorded by earlier investigators.

A similar academy movement was promoted in Germany. A group of scientists met at Leipzig as early as 1651, but formal academies were not organized until later. The Berlin Academy of Science was founded in 1700. Similar developments occurred in Denmark, Holland, Switzerland, Belgium, Portugal, and other European countries.

Royal Society of London

In the British Isles, small, informal "curiosity cabinets" developed in London, Edinburgh, and Dublin during the latter part of the sixteenth and early part of the seventeenth centuries. An attempt to establish a formal organization was made in 1616 by Edmund Bolton, an eminent English scholar, who obtained hearings among British government leaders during the reign of James I (1603-25). The king and his advisors were favorably impressed, and the proposal was discussed in parliament, where the matter was favorably received. Plans were drawn in considerable detail and members were tentatively chosen, but the king died before the organization became a reality. In 1645, during the reign of Charles I, a group of scholars including John Wilkins, Jonathan Goddard, and George Ent began holding weekly meetings in London, sometimes at the homes of the members and sometimes at

Gresham College, "to discourse and consider philosophical inquiries, and such as related thereunto; as physics, anatomy, geometry, astronomy, navigation, staticks, magneticks, chymicks, mechanics, and natural experiments."

In about 1648, some members of the London group moved to Oxford where they formed a similar scientific organization. John Willis, the anatomist (whose name is perpetuated by the circle of Willis in the brain), along with other residents of Oxford, joined the society and broadened it to discuss diverse aspects of biology (i.e., physiology) as well as the physical sciences. The group at London continued to meet at Gresham College or elsewhere in London and called their society the "Invisible College." When a member was at Oxford he attended Oxford meetings. A connecting link was thus maintained between the two scientific groups.

Civil war and political strife in London in the 1650s made it more difficult for the society to hold meetings. Gresham College was used for a time as a military garrison and was not available for scientific meetings. During this period, the Oxford group carried on the tradition of the society. When conditions became more favorable in 1660, the London meetings were revived and the society grew rapidly in strength and prestige. In 1662, it incorporated as the Royal Society of London under a charter signed by Charles II.

The establishment of the Royal Society is among the most significant events related to scientific development in England. Meetings have been held continuously since the official founding, and numerous journals and books have been published. Most of the early officers and members of the Royal Society were not connected with universities but were independent experimenters and free thinkers. The lack of scientific background and diverse interests of the early members in unfamiliar natural phenomena of every kind deprived them of the advantage of united effort upon a limited set of problems. However, the vitalizing influence maintained by the Society more than made up for any lack of coordinated effort. Medical men in the Society gave biological problems much attention. Dissections were performed by the anatomists in the group, and observations on living processes were recorded.

The first official secretary of the Society was Henry Oldenburg (1615-77; Fig. 7.2), a native of Bremen who had come to England as a diplomatic agent. At Oxford he became acquainted with the physicist Robert Boyle and other men interested in science. Oldenburg was not an experimenter, but he was a versatile writer with great enthusiasm for scientific subjects. He carried on a massive correspondence with scientists and societies all over Europe and spared no time or energy in promoting the Royal Society.

Curator of Experiments, Robert Hooke

Another early leader in the Royal Society was Robert Hooke (1635-1703). Throughout his professional life he was closely identified with the Society, and he did much to shape its destiny. His particular background and interests had prepared him well for the responsibilities which he held with the Society.

At the age of 18, Hooke entered Christ's Church College at Oxford University, where he attracted the attention of Robert Boyle and was soon engaged as an assistant to Boyle. Hooke developed a profound respect for Boyle, who was only eight years his senior. This respect soon was coupled with admiration, and a warm friendship developed that continued throughout their lives. As an assistant to Boyle, Hooke worked on the construction of an air pump, a mechanical device of inestimable usefulness. Hooke was interested in air and air pressure and undoubtedly contributed much to the development of "Boyle's Law," which states that the volume of a gas is inversely proportional to the pressure.

Although primarily a mechanic and a physical experimenter, Hooke made many contributions that had biological implications. His interest in the microscope, which fostered its further development and a more widespread recognition of its usefulness, is his main connection with biology. Hooke's first scientific publication, a small tract that appeared in 1661 and dealt with the surface tension phenomenon, won attention at the "Invisible College."

At the time, the Royal Society consisted of about 20 distinguished experimental scientists. In 1662, when the loose, spontaneous organization received royal patronage and a charter, it became fashionable to join, and the society was greatly enlarged by an influx of gentlemen not trained in science. Real scientists were soon outnumbered and outvoted by nonscientists. This new trend in membership made necessary some revision in organization and administration if the society was to maintain its original objectives and keep functioning on a high plane. Some enlightenment in science had to be provided for the new members. To fill this need, the position of Curator of Experiments was created, and Hooke, then 26 years of age, became the first curator and the only scientist to be paid by the Society. His duties were to devise experiments as demonstrations for each weekly meeting and to carry out other research officially recommended to him. In the next year (1663), after the society received a charter, Hooke was elected a fellow. For many years Hooke arranged three or four experiments for each weekly meeting.

Since England was a sea power, much of the Society's experimental work was related to navigation. Hooke invented a wheel barometer and designed a method for keeping weather records. The need for a portable timepiece in navigation inspired him to improve the watch by inventing a more efficient spring. Because of his work in navigation, he became interested in astronomical observations. These, in turn, prompted him to improve the telescope. The work with lenses took him into the broader field of optics, and he became concerned with other applications of lenses, particularly in the microscope (see Chapter 8).

Following the death of Henry Oldenburg, in 1677, Hooke and Nehemiah Grew became joint secretaries of the Royal Society. Hooke held his position for five years and edited seven numbers of *Philosophical Transactions*. He was offered the position of permanent librarian in 1679, but declined and the post was given to William Perry. When Hooke died in 1703, the Fellows of the Royal Society attended his funeral in a group, paying a tribute to his part in founding and maintaining the Society.

The Royal Society of London has been important in promoting biology in England and throughout the world ever since its founding. Illustrious men such as Sir Joseph Banks have taken positions of leadership and successfully maintained the high scientific level envisioned by the founders. Later accomplishments of the Society and its leaders will be cited in succeeding chapters.

English Experimenters

Formulation of any scientific hypothesis is a creative act of imagination, but it is not the result of undirected imagination. In a developed science, hypotheses are formed with due regard to known facts and in the light of the overall view of nature presented both by the accepted theories and by the assumptions that underlie them. Very few facts were established in any field in the middle of the seventeenth century. The whole outlook on nature was in a state of change because the views of the ancient authorities had been shown to be wrong in many important ways. Metaphysical assumptions on which these views depended were also in disrepute. The only completely new system was the mechanistic one suggested by Descartes, and most scientists turned to this system for inspiration.

Boyle

Robert Boyle (1627-1691) was one professor of a university who also took an active part in scientific societies. He was well established at Oxford University and was also an active member of the Royal Society. He was mostly a physical experimenter, widely known for formulating Boyle's Law and other advances in physical sciences, but he also made some substantial contributions both directly and indirectly to biology. In 1660 Boyle published his *New Experiments Physiomechanico, Touching the Spring of the Air and its Effects.* Boyle had obtained a new type of vacuum pump that enabled him to partially evacuate a large glass cylinder within which he could perform experiments. He was chiefly concerned with physical problems arising from previous investigations of pneumatics. But among his many experiments he tested the effects of very low air pressure upon a lighted candle and upon a number of small animals. When placed within this partially evacuated chamber, the candle went out within a few seconds and the animals soon died in convulsions. Subsequent attempts to decide which lasted longer when placed within the chamber proved inconclusive.

In one sense these experiments merely underlined a well-known comparison. Life had always been likened to a flame, and many ancient writers had suggested that heat was in some way connected with vitality. Moreover, the fact that the air in the mine or enclosed space would not support life if it failed to keep a candle burning was well known. Boyle's experiment, however, did raise a new issue. If in combustion and respiration the air merely received sooty vapors, as was usually believed, then the candle would be expected to burn and the animal to live longer in an evacuated vessel than in an enclosed space of similar size. Boyle, impressed by the speed with which the animals died, went on to show that the reverse was true. Animals lived longer and candles burned longer in a closed vessel of comparable size than in the evacuated chamber, and in this case there was no change in the air pressure when the animal died or when the candle was extinguished. Finally he showed that animals lived a little longer if the air within the enclosed vessel was lightly compressed. Boyle therefore suggested that there were two possibilities. Either the air itself was necessary to sweep up the sooty vapors or more probably the air must play some essential role in both respiration and combustion.

Hooke

Robert Hooke (1635-1703), as curator of experiments of the newly formed Royal Society, was required to perform "experiments" at Society meetings. On one occasion in 1667 he repeated an experiment previously described by Vesalius. Removing the chest, ribs, and diaphragm from a living dog, he kept the animal alive by blowing air through the trachea into the lungs. The experiment proved conclusively that motion in the chest itself was not essential for life but served merely to bring about the alternate expansion and collapse of the lungs. The procedure was then repeated and the lungs were pricked so that the air could escape. In this case, the animal was kept alive by passing a stream of air through the lungs, thus showing the movement of the lungs was necessary only to ensure an interchange of air and disproving the widely held belief that the motion of the lungs aided circulation by agitating the blood.

Lower

Meanwhile at Oxford University Richard Lower (1631-1691), a young doctor who had been continuing Harvey's work on the blood, showed that if the nerve supply to the heart was cut the heart ceased to pulsate. He reaffirmed that the heart was merely a pump and measured more accurately than Harvey its output and the speed of the blood as it moved through the main arteries. In the course of his work Lower attempted to inject various drugs and liquid food into the veins of living animals. His partial success led him to try experiments on blood transfusion. In 1665 Lower reported that he had revived a dog that had almost bled to death by giving it a transfusion of blood from another dog. In November 1667 before the members of the Royal Society in London, he demonstrated his technique by giving a man who was slightly "cracked in the head" a transfusion of blood from a sheep. The man received very little blood but he survived without any change in his condition. Lower, however, was not the first to try this experiment.

Doctors had repeatedly suggested that diseases were due to impurities in the blood. The discovery of the circulatory system and the new outlook that emphasized the similarity between the body and a machine made the idea of transfusion as a cure for disease almost inevitable. Lower was greatly impressed with Hooke's experiments on artificial respiration, and he repeated them in order to study the changes in the blood as it passed through the lungs. He tested the blood before and after it left the lungs and demonstrated that so long as the supply of fresh air was maintained, blood entered the lungs dark colored and left them bright red. If air was withheld, blood leaving the lungs was dark in color. He concluded that the bright red blood associated with arterial blood was not required in the left side of the heart, as many believed, but in the lungs and that it resulted from exposure of the blood to the air. Lower pointed out that this change was not due to the vital heat or to any specifically vital process because it could occur in vitro. If a vessel of blood were left in the air, the surface of the blood became bright red although the blood beneath was dark and venous. If the bright surface layer were removed, the layer below would take on the bright red color immediately after being exposed to the air.

Physiologists had really gone as far as they could with their investigations until more was known about the composition of the air. During this period most people regarded air as a simple element that could absorb other substances as a sponge takes up water. One vague theory, however, held that air contained an inert part and a more active part which was nitrous. The active part was assumed to be of nitrous nature because nitre was the active ingredient of gunpowder and because the air was thought to be essential both for combustion and for the production of nitre from animal manure. As early as 1641 Dr. George Ent (1604-1689), a friend of Harvey, had suggested that the nitrous part of the air was important for the maintenance of life. In the light of Lower's experiments, members of the Royal Society, including Lower and Hooke, immediately assumed that the color change in blood was due to the intake of nitrous particles that were ultimately lost in the organs as the blood circulated. At this period, during the seventeenth century, the chemists were not interested in the chemical composition of air and ignored the results of these experiments. Not until a hundred years later was this aspect of respiration clarified.

Societies in the United States

During the seventeenth century the American colonists were busy settling a new land, and they had little time for scientific and cultural activities. The academy movement, however, did spread in a limited way to the New World even during the early years of colonization. The Royal Society of London was especially influential in the colonies. In fact, the Royal Society almost moved to Connecticut at the invitation of Governor John Winthrop during the time of the Civil Wars in London. Robert Boyle, John Wilkins, and several other members of the Royal Society were corresponding with Governor Winthrop about the possibility of moving to America when the Civil Wars ended. The migration was not accomplished, but Secretary Oldenburg wrote by order of the Royal Society inviting Governor Winthrop to be the chief correspondent of the Royal Society in the West. A number of prominent colonists subsequently were admitted to the Royal Society in recognition of their scholarly accomplishments. The membership eventually included Cotton Mather, the three John Winthrops, James Bowdoin, Paul Dudley, and Roger Williams of New England; Benjamin Franklin and David Rittenhouse of Pennsylvania; William Byrd II and John Mitchell of Virginia; and Alexander Garden of South Carolina. These men had caught the vision of the academy movement, and they helped to establish scientific societies in the colonies.

Scientific Societies in Boston and Philadelphia

One of the first (perhaps the first) of the scientific societies in the colonies was established in 1683 at Boston under the leadership of Increase Mather. It was patterned after the academies of Europe, particularly the Royal Society, and became the Boston Philosophical Society. Newtonian philosophy, based on rigid natural laws which could be expressed in mathematical terms, found its way to America

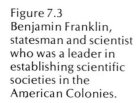
Figure 7.3
Benjamin Franklin,
statesman and scientist
who was a leader in
establishing scientific
societies in the
American Colonies.

and influenced not only the Society but the country itself during the time it was taking form as a free nation. Arguments over politics and theology killed the Boston Society after it had functioned for only a few years.

Stimulus and encouragement for scientific work in the colonies came from French scholars through Benjamin Franklin (1706-90; Fig. 7.3), Thomas Jefferson (1743-1826), and others who established a close cultural alliance between France and America. In 1727, Franklin gathered a small group of curious men around him in Philadelphia. Meetings of this group were held each week on Friday evenings. A mutual improvement society, called the Junto, was thus established. Natural history and philosophical subjects were discussed. In 1744, the membership included Thomas Bond, physician; John Bartram, botanist; Thomas Godfrey, mathematician; William Parsons, geographer; Phineas Bond, natural philosopher; Thomas Hopkinson, statesman and author; William Coleman, Associate Justice of Pennsylvania; and Benjamin Franklin, printer and philosopher. Despite long intermittent periods of inactivity, the Junto eventually developed into the American Philosophical Society. Similar societies were established in other principal cities in the United States as circumstances became more favorable to cultural pursuits. The first state academy was the Connecticut Academy of Arts and Science founded in 1799.

Specialized Societies in Subject Areas

Later trends favored specialization according to content or practical applications of areas of science. Medical societies were the first to develop independent status in the United States. As early as 1735, a medical society was organized in Boston. It functioned for about six years. One of the oldest medical societies is the New Jersey Medical Society at New Brunswick, dating from 1766. In 1771, a group of Harvard students formed an Anatonomical Society. The Massachusetts Medical Society was incorporated in 1781 in response to a petition signed by a group of young physicians. During this same period, other specialists came together and formed agricultural societies, chemical societies, geological societies, and others. Most of these were loosely organized groups, meeting and working enthusiastically when interest was high and disbanding when interest lagged.

Out of these early beginnings came substantial and influential societies of entomologists, botanists, and other biological specialists which are now organized on national and international levels. More recently, specialized societies made up of individuals that have common interests have tended to form broad associations or unions for mutual benefits. The American Association for the Advancement of Science, for example, has a membership of more than 100,000. The weekly periodical, *Science*, first published in 1880, reaches the entire membership, and annual meetings are attended by thousands of scientists. Sections are organized according to specialty in such a way that individuals interested in a particular area of science meet together at the annual meetings. A recent addition to the major organizations in America is the American Institute of Biological Sciences, which includes most of the country's biological societies. The first large meeting of this society was held at the University of Minnesota in 1951. A periodical, *BioScience*, reaches the entire membership of A.I.B.S.

Effect of Societies

The general effect of the early scientific societies was to encourage organized observations and experiments and to stimulate a new spirit of investigative curiosity. This effectively brought scientists together and provided an opportunity for them to discuss their interests with each other. The societies also provided mediums for the preservation and publication of scientific treatises, thus making such work more readily available to wider circles of people. New tools such as the microscope, telescope, thermometer, hydrometer, and barometer were invented and were vitally linked with the development of more precise experimental procedures. Thus, improved laboratory facilities and laboratory methods could be traced to the activities of the early societies. Public lectures were also initiated under society auspices and helped inform more people about science.

Awards for Scientific Achievement

Among the noteworthy activities of the Royal Society of London was the establishment of awards or prizes for meritorious work in science. This did much to stimulate activity among people who otherwise might not have used their talents in scientific pursuits. The oldest and most prized of the medals awarded for outstanding achievement by the Royal Society is the Copley Medal, established from a legacy of Sir Godfrey Copley, F. R. S. (i.e., Fellow of the Royal Society). Copley's will, dated 1704, left the sum of 100 pounds in trust to the Royal Society to stimulate scientific research. In the immediately following years, the interest from the fund was used by the curator of the society to provide demonstration experiments at the meetings. Beginning in 1736, however, the directions for use of the interest accruing to the Copley funds were changed in order to exert greater influence on the development of science. It was decided that a prize should be given each year to the author of a valuable scientific discovery or experiment. The annual award first consisted of a gold medal valued at five pounds prepared with a portrait of Copley on one side and the coat of arms of the Royal Society on the other. In 1881, the Copley family increased the fund sufficiently to provide a gift of 35 to 50 pounds, depending on the interest rate of the endowment funds, along with the medal. More than 200 scientists have received the honor up to the present time. Among the American scientists who received the Copley medal were Theobald Smith (1859-1934), bacteriologist, and Thomas Hunt Morgan (1866-1945), geneticist.

An award of particular interest to biologists is the Darwin Medal, established for the promotion of biological study and research. This medal was made possible by the Darwin Memorial Fund, created to commemorate the work of Charles Darwin, F. R. S. (1809-82). A silver medal and a grant of 100 pounds is awarded biennially by the Council of the Royal Society. The American biologists who received this honor include Henry Fairfield Osborn (1857-1935) and Thomas Hunt Morgan.

Scientific Publications

Perhaps the most important factor in perpetuating and developing science is a readily accessible medium for disseminating scientific accomplishments. Greek science, and that of other ages before the advent of printing, suffered for want of an adequate means of transmitting and preserving ideas and observations. Specialized periodicals that record and transmit scientific findings now cover all fields of science and are available in all major languages.

Beginnings in publication were made by the early academies and societies, but the first scientific periodical as·such was published in France by Denys de Sallo (1626-69), beginning in 1665. Sallo was a patron of science, and he took it upon himself, with the help of his assistants, to review scientific work, to prepare abstracts or summaries of accomplishments, and to circulate these among scientists. Through the influence of his friend Jean Baptiste Colbert, who was Minister of State under Louis XIV, the necessary support was obtained to continue to circulate such material at regular intervals. The project was acceptable to the government, and the publications were well received among scientists. The title *Journal des Savants* (Journal for those learned in the literature of science) was chosen for the periodical. It began as a review journal, but original accounts of observations and experiments were included later. As time went on, it carried more original than review articles. Similar publications were soon established in England, France, Italy, and Germany.

Philosophical Transactions

One of the most important serial publications of all time was the *Philosophical Transactions* of the Royal Society of London, which originated three months after the *Journal des Savants*. It began as a personal project of the first secretary, Henry Oldenburg, who maintained a correspondence with other societies and with corresponding members of the Royal Society. This publication, like the French journal, originally was devoted to reviews of work completed and in progress. Gradually, however, the emphasis in *Philosophical Transactions* also changed from reviews to accounts of original investigations. Monographs by the microscopists Malpighi, Hooke, and

Grew, and letters from Leeuwenhoek were published in the *Transactions*. Through the great influence and tireless effort of Secretary Oldenburg, a precedent for high quality in scientific publication was established.

In the foreword of each volume of *Philosophical Transactions,* a statement was and is made in which the Society disclaims responsibility for the material in the articles included. No conclusions are considered final and absolutely authoritative. The responsibility for accuracy rests with the authors, who are specialists in their fields. When consulted by a government or another agency, the Society never gives an opinion as a Society, but qualified fellows are asked to reply and give their opinions. No scientific body has better claim to authoritativeness, but this society denies the legitimacy of such a claim. Science can grow to maturity only when there is a general understanding that the last word cannot be said on any subject. There must be no dogma to hinder inquiring minds in their continuing search for truth.

Journals Devoted to Particular Fields of Science

The next major development in the publication of scientific periodicals occurred in the following century. Some publications began to reflect the move towards more narrowed specialization in particular fields of science while others were devoted to broad unifying areas. The Englishman, W. J. Hooker (1785-1865), began publication of *The Botanical Magazine*. His son, Sir J. D. Hooker (1817-1911), continued the publication for many years. Another important contributor in botanical publications was the Swiss botanist, Augustin de Candolle (1778-1841). His studies included extensive morphological and physiological investigations in which he became a world authority. He contributed to the French journal, *Annales du Museum d'Histoire Naturelle* (Annals of the Museum of Natural History) and edited the book entitled *French Flora*.

Another contributor to the *Annals of the Museum of Natural History* was the French comparative anatomist Georges Cuvier (1769-1832). The German physiologist Johannes Müller (1801-58) studied medicine at the University of Bonn and became professor of physiology. He edited the German periodical *Archiv für die Physiologie* (Archives of Physiology) and wrote the book entitled *Handbook of Physiology*, which was responsible for introducing experimental physiology in Germany.

The Proceedings of the Zoological Society of London were first published in 1830. *Entomological Magazine* began in 1833 but lasted only five years, to be replaced by *The Entomologist,* which, except for a period between 1840 and 1864, has continued publication to date. Numerous other magazines devoted to specialized areas of biology have sprung up, some to perish but many to continue and expand.

The *Linnean Society Transactions* were published first in 1791 and have been continued until the present time. In 1856 the *Journal of the Linnean Society of London* began publication. A wide variety of articles dealing with both plants and animals has been published by this journal.

Museums

The word "museum" originated from the Greek Temples of Muses and came to be applied to places devoted to culture and contemplation. The great museum at Alexandria in the third century B.C. was mainly a library or a research institute where written manuscripts were collected, preserved, and used by scholars. Some 400,000 volumes were collected in the four principal departments: literature, mathematics, astronomy, and medicine. It is interesting to speculate how the history of science might have differed if this great library had been preserved. The Serapeiana (Chapter 4) was destroyed by the Christians at the time of Theophilus about A.D. 391, and the Alexandriana was completely destroyed by the Mohammedans in A.D. 646. It was a most serious loss to the world's intellectual achievement.

The word museum was not used during the Middle Ages and was revived in the seventeenth century with a slightly different connotation. It was then applied to collections of rocks, animal and plant specimens, or other natural curiosities of various kinds. The first museum on record to have a serious and scientific purpose was that of the geologist Georg Agricola (1490-1555), who collected a useful series of rocks and minerals.

Human beings have an inherent tendency to collect and to hoard objects of interest. Coins and curios have been collected by individuals throughout the ages. In the sixteenth century, systematic collections of natural objects were brought together and museums with scientific value were established. They were naturally concomitant to the academy movement and were settings for the display and preservation of specimens that were collected or prepared by investigators.

Museum collections of dried and preserved specimens served a valuable purpose in facilitating the comparison and classifications of individuals and groups of plants and animals. Many early naturalists maintained home museums. Among the biologists, Vesalius made use of an anatomical museum, and Cesalpino prepared dried specimens of plants in a herbarium. These men and many others thus began to collect and preserve specimens more critically and compared one with another. A scientific purpose became their motivation for collecting, and a first logical step was to work out a way to systematize their specimens. Practical demonstrations with actual specimens and dissections soon became a recognized function of the museum. Foundations were thus laid for the development of the modern museum as a research and teaching institution. John Hunter's museum in London (Chapter 10) and the British Museum in the same city were two of the earliest and best museums of this kind. In the United States the American Museum of Natural History in New York, the Smithsonian Institution in Washington, and the Chicago Natural History Museum have followed the same tradition.

Chronology of Events

B.C.	
300	Alexandrian Museum.
A.D.	
1560	G. della Porta, leader of early academy at Naples.
1597-1625	F. Bacon, essays with observations about life.
1603	F. Cesi, founded Academy of the Lynx in Rome.
1609	F. Peiresc, leader of academy movement.
1630	Early meetings of French academy.
1651	Meetings of scientists at Leipzig.
1652	Meetings of "Invisible College."
1657	Ferdinand II (Medici), patron of the Academy of Experiments at Florence.
1660	R. Boyle, *New Experiments Physiomechanico.*
1662	Royal Society of London founded with Henry Oldenburg, Secretary, and Robert Hooke as Curator of Experiments.
1665	R. Hooke, *Micrographia.*
	D. de Sallo, first publication of *Journal des Savants.*
1666	Report of Experiments from Academy of Experiments. French Academy of Science officially organized.
1667	R. Lower, blood transfusion from sheep to man.
1683	Boston Philosophical Society founded.
1700	Berlin Academy of Science founded.
1704	Copley Medal fund established.
1727	B. Franklin, Junto organized at Philadelphia.
1735	Medical Society of Boston organized.
1799	Connecticut Academy of Arts and Science founded.
1830	*The Proceedings of the Zoological Society of London* first published.
1856	*Journal of the Linnean Society* first published.

1880   *Science* first published by American Association for the Advancement of Science.
1951   American Institute of Biological Science founded.

References and Readings

Abbot, E. A. 1885. *The Life and Works of Francis Bacon.* London: Macmillan and Co.

Andrade, E. N. da C. 1954. "Robert Hooke." *Sci. Amer.* 191:94-98.

Bacon, F. *Advancement of Learning,* ed. J. Devey. 1901. New York: Collier.

Bates, R. S. 1945. *Scientific Societies in the United States.* New York: John Wiley and Sons.

Brasch, F. E. 1931. "The Royal Society of London and its influence upon scientific thought in the American Colonies." *Sci. Monthly* 33:337-355, 448-469.

Cameron, H. C. 1952. *Sir Joseph Banks.* London: Batchworth Press.

Drake, S. 1966. "The Accademia dei Lincei." *Science* 151:1194-1200.

Espinasse, M. 1956. *Robert Hooke.* Berkeley: University of California Press.

Fay, B. 1932. "Learned societies in Europe and America in the eighteenth century." *Amer. Historical Rev.* 37:255-266.

Hall, A. R. and M. B. Hall, eds. 1967. *The Correspondence of Henry Oldenburg.* Madison, Wis.: University of Wisconsin Press.

Hartley, H., ed. 1960. *The Royal Society: Its Origins and Founders.* London: Royal Society.

Hartley, H. 1960. "The tercentenary of the Royal Society." *Amer. Sci.* 48:279-299.

Keynes, G. L. 1960. *A Bibliography of Dr. Robert Hooke.* Oxford: Clarendon Press.

Kraus, M. 1942. "Scientific relations between Europe and America in the eighteenth century." *Sci. Monthly* 55:259-272.

Kronick, D. A. 1962. *A History of Scientific and Technical Periodicals.* New York: Scarecrow Press.

Lange, E. F. and R. F. Buyers. 1955. "Medals of the Royal Society of London." *Sci. Monthly* 81:85-90.

Martin, D. C. 1960. "The tercentenary of the Royal Society." *Science* 131:1785-1790.

Matzke, E. B. 1943. "Concept of cells held by Hooke and Grew." *Science* 98:13-14.

Montagu, B. 1844. *The Works of Francis Bacon.* Philadelphia: Carey and Hart.

More, L. T. 1944. *The Life and Works of the Honourable Robert Boyle.* London: Oxford University Press.

Ornstein, M. 1938. *The Role of Scientific Societies in the Seventeenth Century.* Chicago: University of Chicago Press.

Porta, della, G. B. 1957. *Natural Magick.* New York: Basic Books.

Power, H. 1664. *Experimental Philosophy.* London: John Martin and James Alestry. (Microscopical observations of the flea, field spider, locust, mites, nettles, leaves and many other things.) Reprinted, New York: Johnson Reprint Corporation, 1966.

Redi, F. 1668. *Experiments on the Generation of Insects.* Trans. M. Bigelow. Reprinted, Chicago: Open Court Publishing Co., 1909.

Roth, L. 1937. *Descartes' Discourse on Method.* Oxford: Clarendon Press.

Sprat, T. 1958. *History of the Royal Society.* eds. J. I. Cope and H. W. Jones. Saint Louis, Mo.: Washington University Studies.

Stimson, D. 1948. *Scientists and Amateurs.* New York: Henry Schuman.

Thornton, J. L. and R. I. J. Tully. 1954. *Scientific Books, Libraries, and Collectors.* London: Library Association.

Wolf, A. 1939. *A History of Science, Technology and Philosophy in the 16th, 17th, and 18th Centuries.* New York: Macmillan Co.

CHAPTER 8

MAKERS AND EARLY USERS OF MICROSCOPES

Through the influence of the academies and the independent activities of members and patrons, many advances were made and objective observation and experimentation on natural phenomena found a place in the seventeenth century culture. Publications transmitted and preserved scientific accomplishments, and thereby greatly facilitated progress in diverse fields. Galileo, the leading spirit of the period in physical science, wrote and circulated reports of his observations and deductions, thus setting them forth for others to criticize and evaluate. Substantiated facts became highly valued, whether they fit accepted dogmas of the time or not. Galileo's example of objectivity and his moral victory in the conflict with authoritarianism were vital milestones in the transition from the old order of authority to the new order of freedom of thought.

As the main themes of modern biology began to take shape, instruments such as the telescope, pendulum, thermometer, barometer, hydrometer, air pump, and watch spring were adapted to scientific usage. The microscope, most important of all instruments to biology, was developed to practical usefulness in the seventeenth century. This tool opened up new investigative worlds in cytology, histology, and microbiology. Its history is of great significance to biologists.

Early Use of Lenses

Beginnings in the use of lenses can be traced to the Assyrians before the time of Christ. Seneca, a Roman author and philosopher in the first century A.D., discovered that a globe of water in proper position would make handwriting appear larger and clearer. Claudius Ptolemaeus (A.D. 100-170) wrote a treatise on optics dealing with indices of refraction. Lenses that appeared to have been ground and polished were found in the ruins of Nineveh, Pompeii, and Herculan-

Figure 8.1
Janssens's compound
microscope (replica)
constructed about 1590.
(Courtesy of Bausch
and Lomb, Inc.)

eum. The Arabs are known to have used lenses during the Middle Ages. Meissner, writing in the thirteenth century, spoke of glass spectacles that were recommended for the aged. In the same period Roger Bacon wrote in his *Opus Majus* of the remarkable optical advances made possible by refracting light. He mentioned the magnifying qualities of lenses along with their other properties.

Conrad Gesner, a Swiss naturalist, wrote about the use of magnification aids in scientific investigation in about 1558. His biological work included a book on shells in which he illustrated snails and other organisms with hard shells. Objects as small as the protozoan, Foraminifera, were sketched. These are barely visible to the unaided eye and could not be described in any detail without the use of magnifying lenses.

The Dutch microscope maker Zacharias Janssens combined lenses in an effort to improve magnifying efficiency and resolving power, and in about 1590 he produced the first compound microscopes (Fig. 8.1). Janssens had already worked with telescopes, invented first by Nicolaus Copernicus. His first compound microscopes were built on the order of a telescope and were about six feet long. Each had a barrel approximately an inch in diameter with a lens on each end. Galileo heard of Janssens's work and improved on the arrangement of lenses in his own numerous telescopes, but he did not further the development of the microscope.

Seventeenth Century Microscopists

Publications from the Academy of the Lynx included illustrations that were based on observations made with magnifying lenses. Figures of a honey bee, for example, were first presented in 1625, by Francesco Stelluti. He later improved the drawings and combined them with a poetical treatise in 1630. Dorsal and lateral views of the bee were illustrated with remarkable detail, and the magnification was given as five diameters. Between the two drawings of the whole bee, the mouth parts were sketched separately with a stated magnification of ten diameters.

Giovanni Borelli (1608-79; Fig. 8.2), a mathematician who had been a student of Galileo, also used lenses in biological investigations. In 1655, he saw "whales" in the blood of a nematode worm. He also described fibers of textiles, spider eggs, and other natural objects requiring magnification for critical observation. The Jesuit father, Athanasius Kircher (1602-80), described observations with the micro-

Figure 8.2
Giovanni Borelli, Italian
who made use of lenses
in the early
seventeenth century.

scope that led him to believe that disease and putrefaction were
caused by invisible living bodies. Some people have since suggested
that he could have seen bacteria, but Kircher did not provide illustrations to show what he had observed. His magnifying equipment
was presumed to be capable of only about 20 diameters, however, so
it is doubtful that he actually saw bacteria.

From these beginnings, the five classical microscopists—Malpighi,
Leeuwenhoek, Hooke, Swammerdam, and Grew—laid the foundation
for the development and the use of the microscope. It is interesting
to note that these five men, who did more than any others to bring
the microscope to a state of practical usefulness, were born within
a period of 14 years (1628-41). The momentum provided by their
combined contributions did much to promote mechanical improvements as well as extend applications of the microscope.

Malpighi

The first born of the five classical microscopists was the Italian,
Marcello Malpighi (1628-94; Fig. 8.3). He was the son of an independent land owner with modest resources. His university training was
obtained at Bologna where, along with other activities, he became
interested in lenses. The professor of anatomy, Massari, recognized

Malpighi as a man of great ability and invited him to his home where he could use the private library. This was before the time of public libraries and, even on a university campus, the only collections of books in particular academic fields were owned privately by professors in those fields. Along with other students, Malpighi dissected animals at the professor's home. This experience and the advice of another teacher led Malpighi to enter the study of medicine. For his thesis, required for the medical degree, Malpighi chose to study Hippocrates. He found great interest and pleasure in reading about the Greek father of medicine and in tracing a period in the early history of medicine.

Following graduation from the University of Bologna, Malpighi accepted a position as Professor of Medicine at the University of Pisa. He also carried on a private practice in this community. While at Pisa, Malpighi became acquainted with Professor Borelli, who was primarily a mathematician but was also interested in biology and had already made significant observations with the microscope. The two men worked together on anatomical and physiological subjects, including an investigation of the heart muscle.

After three years at Pisa, Malpighi returned to the University of Bologna to a career of teaching and critical microscopic investigations. He wrote letters to his friend Borelli and to other investigators describing his observations. One letter to Borelli described the minute structure of the lung and gave the first complete description of the capillaries. In a later letter on fat tissues he described the flattened red corpuscles in the veins of a hedgehog. This observation of red corpuscles was probably the first to be recorded, although Swammerdam was likely ahead of Malpighi in the actual observation. Malpighi later wrote more formal monographs on the subjects of his investigations, some of which were published in *Philosophical Transactions of the Royal Society of London*.

Malpighi described the minute structure of the liver and showed that the bile released from the gall bladder was actually secreted by the cells of the liver. The corpuscles that he described in the kidney now bear his name. Through careful dissection he showed that the spleen was not connected with the stomach. This organ had previously been thought to be the source of "black bile," which was believed to go directly into the stomach. Malpighi described the inner layer of the skin (Malpighian layer), the papillae of the tongue, and the structure of the cerebral cortex. When he held up an excised eye of a screech owl and looked through it as if it were a hand lens, he observed that the image was inverted.

Malpighi wrote an elaborate, detailed, anatomical description of a

Figure 8.3
Marcello Malpighi,
Italian classical
microscopist of the
seventeenth century.

silkworm, which was published as a monograph by the Royal Society in 1669. It included a description of the silk-forming apparatus, the respiratory system made up of air tubes, the central nervous system including the nerves to the eye and to other sense organs of the head, and the digestive tract. Tubules connected with the intestine were described and later were named after him (Malpighian tubules).

In addition to the extensive studies in animal histology and anatomy, Malpighi made detailed studies in plant anatomy. One of his largest and best monographs, entitled *Plant Anatomy (Anatome Plantarum)*, amounted to about 150 pages in its original form and was illustrated with 93 plates and figures. The bark, stem, roots, and seeds were described systematically, and plant processes such as germination and gall formation were discussed. In his study of the structures of plants, Malpighi anticipated the cell theory. Plants were described as being composed of separate structural units which he called "utricles."

Malpighi was also interested in the developmental anatomy of plants and animals. To follow the steps in animal embryology he used the chick, which was readily available and easily observed by opening eggs at successive stages of incubation. William Harvey had initiated the study of embryology (Chapter 6), and Malpighi supplied the next important step with a series of illustrations showing actual developmental stages. Malpighi began his description with the 24-hour stage. In two memoirs, both published by the Royal Society in 1672, he presented 12 plates containing 86 figures with some 24 pages of descriptions of chick embryology. Considering the era when his study was made and the instruments available to him, this series is remarkable. The original drawings, which were done in pencil and red chalk for one of the two memoirs on the chick, are still in the possession of the Royal Society. Because Malpighi was not able to see the developmental stages that occurred before the end of the first day (when the rudiments of the embryo were already established), he believed that some structures were preformed in the egg.

The types of microscopes used by Malpighi are not known precisely. Much of his work was done with simple lenses in sunlight. He is also known to have used an instrument with two lenses. His contribution to biology was in demonstrating how to use the microscope for critical probings into structures that represented a newly realized order of magnitude in the internal anatomy of plants and animals. He was a critical observer and a profound analyzer of minute structures and their relations to one another. His accurate observations and excellent descriptions and diagrams anticipated the work of the nineteenth century anatomists. Malpighi was not a prolific writer as compared with some contemporaries such as Athanasius Kircher, whose writings filled 44 folio volumes. Most of his known observations have been gleaned from letters to friends and short reports to the Royal Society, but some of his major works such as the anatomy of the silkworm reached monograph size.

Leeuwenhoek

Antony van Leeuwenhoek (1632-1723; Fig. 8.4) was born in Delft, Holland, the son of a basket maker. At 16 years of age he went to Amsterdam, where he became an apprentice to a linen merchant under whose supervision he learned to make and sell cloth. When this apprenticeship was completed, he had qualified as a draper and had been promoted to the position of bookkeeper and cashier of the establishment. After six years in Amsterdam, he returned to his home town, Delft, married, and remained there the rest of his life. He bought a house and shop and set up a business of selling cloth, but he also became qualified as a surveyor. In 1660 at the age of 28, he was made Chamberlain to the Sheriff of Delft and was charged with the care and cleaning of the Delft City Hall. He was also the wine

Figure 8.4
Antony van Leeuwenhoek,
Dutch microscope maker.

gauger and the town revenuer. Nothing was heard from Leeuwenhoek during the next 13 years but it became evident from his later
correspondence that he had devoted much time to his hobby of
grinding lenses.

As a lens grinder he was a perfectionist, always trying to make better
lenses with which to see more minute objects. The lenses (Fig. 8.5)
were mounted in copper, silver, or gold plates which Leeuwenhoek
extracted and prepared himself. He was independent and secretive,
resenting any interference even by people of high position, and being
particularly distrustful of people who offered advice. As time passed
he became so intent upon his lens grinding that he neglected his
shop and family and was ridiculed and called insane by his neighbors.

One man who appreciated Leeuwenhoek's discoveries with his
lenses was the Dutch biologist, Regnier de Graaf. He wrote a
letter dated April 28, 1673, to the Royal Society of London telling the
Society of the remarkable things Leeuwenhoek was doing. The
secretary of the Society, Henry Oldenburg, then wrote to Leeuwenhoek requesting an account of the discoveries. In response, Leeuwen-

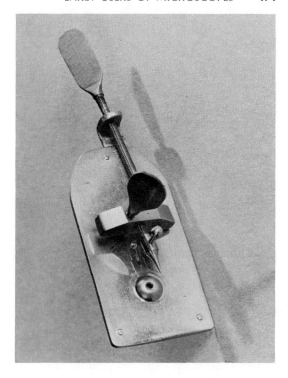

Figure 8.5
One of Leeuwenhoek's
microscopes. (Courtesy
of Bausch and Lomb, Inc.)

hoek sent a letter written in Dutch entitled (in English translation), "A specimen of some observations made by a microscope contrived by Mr. Leeuwenhoek concerning mould upon the skin, flesh, etc., the sting of a bee, etc." The letter was long, unorganized, and difficult to interpret, but Leeuwenhoek was thanked by the Society and invited to send more letters concerning his achievements. A second letter from Leeuwenhoek to the Royal Society describing more of his observations was dated August 15, 1673.

Leeuwenhoek's letters were published in part, in English translation, in the *Philosophical Transactions of the Royal Society of London* (1673 and 1674). They contained crude and scattered observations of mold, and of mouth parts and stinging parts of a bee and a louse. Copies were sent to members of the Society with the suggestions that those who so desired might send encouragement, suggestions, and problems to Leeuwenhoek. The communication from the Society to its members included praise for Leeuwenhoek with the comment that he had observed the structures he described more accurately than had any previous author. From the day of his first letter until the time of his death 50 years later, some 372 communications went from Delft to London, representing such widely separated fields as zoology, botany, chemistry, physics, physiology, medicine and other unclassified subjects.

All of Leeuwenhoek's known observations were described in letters originally written in his own Dutch language. He wrote with an informal conversational style. Many of his observations were left unpublished; some have been more recently translated and some were apparently lost. In 1680, during his 48th year, he was elected a fellow in the Royal Society. He also became a corresponding member of the French Academy of Science, but he never attended a meeting of either Society. He seldom left Delft except for seaside excursions on holidays.

As Leeuwenhoek's work became better known he was visited by many curious people including royalty and celebrities, but he was not gratified by their interest. He indicated in his letters that these intruders only annoyed him and interfered with his work. Peter the Great was impressed with Leeuwenhoek's microscopes and small objects made visible by them, and he requested a demonstration. As a special concession he was allowed to look through Leeuwenhoek's better microscopes. The Queen of England also paid Leeuwenhoek a visit and was given a demonstration. Leeuwenhoek refused to give or sell his microscopes to other people, and he jealously guarded even the more inferior instruments in his laboratory, especially when visitors were present. He was quoted by one interviewer as saying that he did not trust people, especially Germans.

Exact descriptions of Leeuwenhoek and his work are difficult to find. However, his biographer, Clifford Dobell, has listed two books carrying firsthand information: one was written by Thomas Molyneaux, an Irish doctor, and the other, written when Leeuwenhoek was an old man, was by Von Uffenback, a German biographer.

Molyneaux described one of Leeuwenhoek's lenses as being definitely ground (there had been some suggestions that the lenses were blown by a glass blower), and placed between two thin, flat plates of brass about an inch broad and an inch and a half long. Leeuwenhoek was reported to have told Molyneaux that he had other microscopes no one else had looked through that he saved for his own use. Molyneaux was impressed with the clearness of the image viewed through the instruments although he said the magnifications were no greater than those of microscopes he had seen in London. Leeuwenhoek was characterized as complacent and confident, but unschooled in languages. Molyneaux felt that Leeuwenhoek was seriously limited by not being able to consult the writings of others. Von Uffenbach wrote of Leeuwenhoek in satirical terms and scoffed at his ideas of blood circulation and his statement that flies have eyes made up of a thousand or more independent units.

In the course of his work Leeuwenhoek observed muscle fibers of a whale and dried scales from his own skin, an ox eye with uniform crystalline structure, and hair of sheep, beaver, and elk. He dissected a fly and marveled at the large size of its brain. His first major achievement to bring wide recognition was the discovery of microscopic organisms in water. Leeuwenhoek observed a sample of water from the tube in the earth which he used to measure rainfall and found to his surprise that this water was teeming with what he called "wretched beasties." When water from other sources, including the canals and swamps of Delft, was observed, more organisms of different shapes and sizes were seen. Leeuwenhoek never found out where they came from but he knew they did not come from the clouds in rain droplets, and he was against the idea of their spontaneous generation.

The sharp taste of pepper fascinated Leeuwenhoek. He thought pepper must have little barbs or points that jabbed his tongue. Dry pepper could not be observed with the type of microsope he used, so he soaked the pepper for several days to prepare it for observation. When he finally observed a drop of pepper infusion, he was so amazed with the numerous moving organisms that he forgot to look for the barbs on the pepper. This was a new way to culture microorganisms originating from dust particles in the air and on containers. Pepper infusions were employed by many students of microbiology in the years that followed Leeuwenhoek's discovery.

When Leeuwenhoek was in his fifties his teeth were well preserved, because, as he says, he rubbed them night and morning with salt, cleaned the spaces between with a quill, and wiped them with a cloth. Even after his teeth were cleaned in this manner, however, he observed that there was still a white material called tartar between the teeth. An examination of this material with the microscope revealed more "beasties." Some leaped around like pikes, he said, and others would move in one direction, then roll over and go the other way. Still others resembled corkscrews. Following the observation of specimens from his own mouth, he obtained samples from other people and made similar observations.

One morning after drinking hot coffee, he observed a sample of tartar from his teeth and to his surprise the "beasties" appeared to be dead. When he took a sample from between the molars, not touched by the hot drink, he found that the organisms were still alive. This led him to conduct an experiment which showed that the organisms could be killed in hot water. He drank hot coffee for another experiment in which he measured the quantity of perspiration that issued from his skin. Leeuwenhoek's curiosity led him to set up still another experiment in which he watched gunpowder explode under the microscope. The experiment left him blind for several days.

Leeuwenhoek made so many scattered observations that it would be difficult to tabulate all of them. Magnifications of the objects he observed ranged from 40 to 270 diameters. The higher magnifications were considerably greater than those obtained by the lower power of the present compound microscope which is remarkable, indeed, for handmade lenses. His more famous discoveries can be cited as follows:

1. Human sperm observed in 1677. A medical student named Hamm has been reported to have observed and sketched human sperm two years earlier (1675). Though Leeuwenhoek may not have been the earliest observer, he was probably the first to recognize the significance of the sperm. In addition to human sperm, Leeuwenhoek observed sperm from dogs, rabbits, birds, frogs, fish, and insects. His imagination, however, led him astray on one point. He reported to the Royal Society that he was able to identify male and female-producing sperm by inspection.

2. Observations and sketches were made of bacteria from the substance around his teeth and also from pepper and hay infusions. Among Leeuwenhoek's sketches the three major groups—bacilli, cocci, and spirilla—may be recognized.

3. Capillary circulation was observed in the tail of an eel, the web of a frog's foot, and the ear of a rabbit. The actual circulation of blood corpuscles was thus demonstrated. Corpuscles were described in fish, amphibia, birds, and mammals. In the course of his studies he proved that blood did not ferment.

4. Muscle fibers were observed in the iris of the eye, but their cellular nature was not recognized. Nerves in the brain were described as tubes.

5. Many independent organisms were observed. These included ciliated protozoa and rotifers from stagnant water and prepared infusions. Intestinal protozoa such as *Giardia* were observed in human feces, and marine protozoa including Foraminifera were found in the stomachs of shrimps.

6. Plant tissues and plant structures such as seeds were described.

7. The anatomy of insects, including the mouth parts of the flea, and the compound eye and stinger of a bee, was investigated. The general structure of the cochineal insect was also observed.

8. Hairs were observed and found to be cylindrical. Descriptions of dried epithelial cells were also included in the Leeuwenhoek communications.

In his letters to the Royal Society, Leeuwenhoek would frequently insert "I have observed" or "I imagine" or "I figure to myself," indicating that he was accurate and scientific in his approach. His letters also gave insight to his plain, almost childlike character by references to himself, his pets, and his habits. From these it may be concluded that he had a little, white, long-haired pet dog, and a horse that was a mare. He usually drank coffee for breakfast and tea in the afternoon, shaved twice a week, and got a rash on his hands when he sat in the sun. His choice cure for a "hangover" was a great many cups of extremely hot tea on rising. Although he was sincere to his beliefs and jealously defended what he believed to be true, he did on occasion change his opinions when more reasonable explanations came to his attention.

In 1716, when Leeuwenhoek was 84 years old, the University of Louvain in Belgium presented him with a medal in recognition for his work. The ceremony for awarding the medal was similar to that followed at the present-day universities when presenting distinguished service awards. As Leeuwenhoek grew older, his letters to the Royal Society were more concerned with his health and his diagnosis of the causes of his ill health. During the last six hours before he died on August 26, 1723, he left instructions for two letters to be translated from Dutch into Latin and sent with some microscopes to the Royal Society.

Dobell, the biographer of Leeuwenhoek, calls him the father of protozoology and bacteriology because he was the first man to see and correctly describe protozoa and bacteria. Some others have called him the father of microbiology, but Robert Hooke and Marcello Malpighi also have a claim to this title. While not the original inventor of the microscope, Leeuwenhoek developed his lenses independently without the benefit of other discoveries and treatises on the subject. He was the first describer of many tiny animals that he called "beasties" or "animalcules." Leeuwenhoek's contributions, either fully or in part, may be classified by today's standards as false, silly, fantastic, or true. Even if all his observations were false, however, the fact would still remain that this man possessed a remarkable curiosity and thirst for knowledge. Leeuwenhoek was comparatively uneducated, yet he appreciated the scientific approach and advanced his theories without help from others. Compared with his critical and well-trained contemporaries such as Malpighi and Hooke, Leeuwenhoek was an amateur and perhaps a "putterer." His observations were scattered and sometimes incoherent, but he had a gift for hard work and saw many things with his microscopes.

Hooke

Robert Hooke (1635-1703), who was described in Chapter 7 in connection with the early history of the Royal Society of London, contributed to the development of the microscope and also to its usefulness. His numerous demonstrations of successive developments in and new uses for the microscope before the Royal Society did much to popularize the instrument as a tool for biological research. His own microscope on which many new features were devised and tested is illustrated in Fig. 8.6. Also in his writings Hooke constantly urged more widespread and more critical use of the microscope.

Hooke's major written work, *Micrographia*, was published in 1665, when the author was 29 years old. This was primarily a review of a series of microscopial observations (57) in which Hooke described and discussed many demonstrations and inventions mostly associated with the development of the microscope. *Micrographia* also included discussions of three observations made with a telescope. The compound microscope was described in great detail and special attention was given to the matter of illumination. A method of focusing in sunlight by means of a globe of water or a burning glass on a background of oiled paper was suggested as a means of making the microscope more efficient. His scheme for using artificial lighting was a significant contribution to the mechanics of microscopy.

Hooke presented four important, fundamental concepts in *Micrographia*. First, a series of observations was made on the color of thin plates, and the phenomenon of diffraction was described. Second, he described combustion as depending on something in the air. He further associated combustion with respiration in the animal body, and a physical explanation for energy release was suggested. His third basic contribution was the correct explanation of the origin of fossils. The fourth was the biological observation of pores or boxes observed from sections of cork (Fig. 8.7), and the first use of the word "cell" to describe these open spaces.

In addition to the four fundamental contributions cited above,

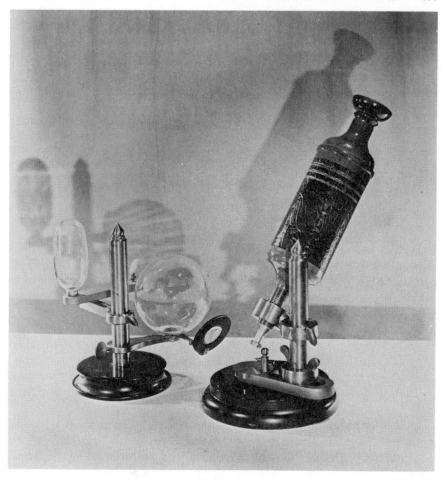

Figure 8.6.   Robert Hooke's microscope.
(Courtesy of Bausch and Lomb, Inc.)

Micrographia contained many more observations of great significance. Examples of included information having biological connota-
tions are as follows: a description of birds' feathers illustrated with
detailed drawings; a description and functional interpretation of
insect wings; descriptions of sponges, which Hooke observed to be
attached to their environment like plants but to have animal substance; a description of silk fibers; and a speculation that artificial
silk might eventually be made from some glutinous substance drawn
out mechanically in fine fibers. The many illustrations of microscopic

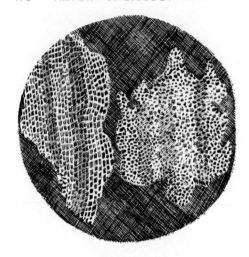

Figure 8.7
Pores or "cells" as seen
by Robert Hooke through
his microscope; *Left*,
thin slice of cork cut in
such a way that pores
appear rectangular;
Right, pores cut in
cross section. (From
Micrographia [1665]
by Robert Hooke.)

objects in *Micrographia* were drawn by Hooke himself. These include renditions of compound eyes of a fly, a louse, a silverfish, a flea, and a stinger from a bee.

Later writings of Robert Hooke, presented originally as lectures, included a series of papers devoted mostly to mechanical and physical subjects such as helioscopes, springs, pendulums, universal joints, gravitation, the earth's motion, and comets. The fifth of these treatises, published in 1678, dealt in part with a biological subject and was called *Microscopium*.

In this paper Hooke returned to microscopy, which he had laid aside for many years because of the weakness of his eyes. His return was prompted by the 1677-78 correspondence addressed to the Royal Society by Leeuwenhoek describing "little animals" in pepper water. Great excitement was aroused in the Society, and Hooke was asked to try to repeat the observations. He prepared cultures of organisms not only in pepper infusions but also in oats, wheat, and barley preparations and observed several sorts of tiny creatures. He also confirmed Leeuwenhoek's observations on blood, milk, and phlegm. Along with these observations, Hooke gave descriptions of the improvements made on the microscope.

Swammerdam

Jan Swammerdam (1637-80; Fig. 8.8) was born in Amsterdam, the son of an apothecary who had a home museum with zoological specimens. The home atmosphere helped to stimulate young Swammerdam's interest in nature. From his youth Jan was characterized as nervous, intense, and stubborn. When he was in his middle twenties he entered the University of Leyden and began his study of medicine. He continued his studies in Paris and was awarded the medical degree in about 1667. His interest in basic biology increased and he did not immediately enter the medical practice. Instead he devoted himself to research. He had remarkable skill in making dissections of minute animals and parts of animals. In Paris he became acquainted with the King's librarian, Melchisedec Thevenot (1620-92), one of the founders of the French Academy of Science (Chapter 7). Thevenot was impressed with Swammerdam's skill and promise as an investigator and loyally assisted him throughout his life.

In 1669, Swammerdam published a classical study of the metamorphosis of insects. Characteristics of metamorphosis were then used as a basis for classifying three types of insects: (1) no metamorphosis, insects which were complete when hatched from the egg and only needed to grow larger (e.g., silverfish); (2) incomplete metamorphosis, those which hatched without wings but developed wings later (e.g., grasshoppers, mayflies); and (3) complete metamorphosis, those with larvae, pupae, and adult stages (e.g., flies, bees, butterflies). Studies on metamorphosis led Swammerdam to visualize a preformation from instar to instar in the development of insects (see Chapter 11). He observed larval features in the adult stage and interpreted this as evidence for preformation. Unfortunately, this view of embryology was applied generally by other investigators to organisms which do not molt like insects and for which no similar observations were available. The theory was perpetuated for many years by less skillful microscopists who apparently saw things that were not there. Swammerdam wrote excellent descriptions of the development of gnats and dragonflies, and he also traced the development of the frog. In 1658, in connection with his study of the frog, he observed and described blood corpuscles, probably for the first time.

Swammerdam's remarkable observations were made possible by his skill in dissecting and interpreting. He developed new techniques to resolve specific problems. When studying the circulatory system, for example, he injected the blood vessels with wax to hold them firm for dissection. It was thus possible to observe the delicate valves in the lymphatic system. He dissected minute, fragile structures under water to gain added support and to prevent tearing or destroying the frail tissues. When the structures for which he was searching were obscured by fat, he dissolved the fat with turpentine. Micropipettes were used to permit inflation and injection under the micro-

Figure 8.8
Jan Swammerdam, Dutch
microscopist who made
critical observations
of minute anatomy of the
mayfly and other animals.

scope. With such techniques he studied respiration and muscular contraction in living specimens, thus contributing to the field of physiology.

Swammerdam not only had poor general health in his youth, but he contracted malaria at an early age. Before he had really started his life's work, he evidenced some early symptoms of mental instability; these became more serious as time went on. He did not work consistently in earning a living and this led to quarrels with his father. Though financially able to support him, his father took the position that he would refuse to nurture a mature son who was prepared but unwilling to practice his profession. In sheer desperation, Swammerdam was forced to leave his research periodically and attempt to practice medicine in order to have a source of income.

Following an agreement and limited financial help from his father, he went to the country, ostensibly to regain his health and develop a medical practice in a more substantial way. Instead, as soon as he was away from home, he spent long days and nights at his investigations. Accounts written at the time said he "worked like a madman," and his health suffered. He completed his classical anatomy on the mayfly in 1675. This was a remarkable investigation in which the minute internal structures of the mayfly were described with great accuracy and detail. It showed what could be done with the microscope in the hands of a skillful investigator and illustrator.

The father, not impressed with the scientific accomplishment and promise of Jan, withdrew all support. Swammerdam then tried to sell his books, specimens, and even his microscopes and dissecting tools to live. At this critical point the father died and a comfortable living was left to Jan. Now another circumstance interfered with his scientific career. He read the works of a religious fanatic, Antoinette Bourégnon, and became more interested in contemplation and religious devotion than in scientific accomplishment. He came to regard research in natural subjects as worldly and gave himself to adoration and worship. The remaining six years of his life were thus lost to science. He died in 1680, at the age of 43.

Swammerdam's scientific work, done in only about six years under unfavorable conditions, was remarkable indeed. Most of it was not published until after his death. Hermann Boerhaave collected and published much of the material in 1737 under the title *Biblia Naturae (Bible of Nature)*. This volume includes a remarkably fine section on the internal anatomy of the bee. Only when Swammerdam's scattered work was brought together, 57 years after his death, was its great value appreciated. Swammerdam had worked with structures smaller than those observed by Malpighi, and his drawings were superior to those of Malpighi. The *Bible of Nature* has been described as the finest collection of microscopical observations ever produced by a single worker.

Grew

Nehemiah Grew (1641-1712), son of an English clergyman, was an undergraduate at Cambridge when his father found himself on the losing side of an English civil war. It was expedient for the Grews to leave England, and Nehemiah transferred to the University of Leyden in Holland. There he studied medicine and graduated in 1671. He set up a medical practice in a small town but devoted more time and interest to plant anatomy than to professional medicine. As soon as possible after the war ended, he moved to London and continued his research on plants. In 1672, the first of his two great books was published under the title *Philosophical History of Plants*. Ten years later (1682) he presented the second major work to the Royal Society under the title *Anatomy of Plants*. The following comment appears opposite the title page: "At a meeting of the Council of the Royal

Society, February 22, 1682 Dr. Grew, having read several lectures on the anatomy of plants, some whereof have been already printed in diverse times and some are not printed, with several other lectures of their colors, odors, tastes and salts as also of the solution of salts in water and of mixtures, all of them to the satisfaction of the said society. It is therefore ordered that he be desired to cause them to be printed together in one volume. Christopher Wren PRS. (President, Royal Society)." The combined volume was published under the title, *The Anatomy of Plants with the Idea of a Philosophical History of Plants and Several Other Lectures Read before the Royal Society.* The first part is devoted to the anatomy of plants, followed by a general account of vegetation and several separate lectures on related subjects. The last part of the book contains a large number of diagrams of plants, mostly cross sections of parts of plants.

Grew's main contribution was a critical anatomical description of the parts of the flower. He observed that in some plants the pistil (seed-producing part) and stamens (pollen-producing parts), were in the same flower. This suggested that some plants might be hermaphroditic like snails. Twelve years after (1694) the publication of *The Anatomy of Plants*, which included descriptions of the flower parts, the German physician, Rudolph Camerarius, described sexual reproduction in plants. Grew had worked out the anatomy, and Camerarius supplied the functional aspects of the story of reproduction. The critical anatomical work of Grew was not fully appreciated until many years after his death.

Contributions of the Classical Microscopists

The combined contributions of the five classical microscopists and of the other lesser microscope makers and users brought the microscope into practical usefulness. They did all that could be done with the instruments available to them. No further major improvements occurred until the early 19th century, when progress was resumed.

More Recent Improvements

A microscope of the type used in the 1700s is illustrated in Fig. 8.9. The most serious difficulty encountered with early microscopes involved chromatic abberation. This occurred as lenses were made increasingly convex, and prisms were created that separated the different wavelengths in ordinary light and produced all the colors of the spectrum. Newton, who formulated a theory of light and color from studies on prisms, worked on the problem of chromatic aberration and concluded that it was insoluble. It is still insoluble as far as single lenses are concerned, but it was found that chromatic effects could be avoided by cementing lenses together so that one complemented the other in prismatic effects.

In 1766 the German mathematician Leonhard Euler described two methods of avoiding chromatic aberrations: (1) use monochromatic

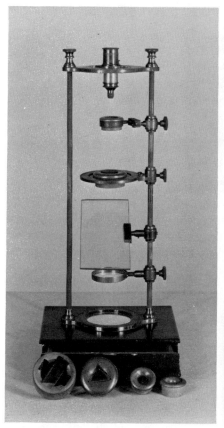

Figure 8.9.   Microscope of the type used in the 1700s. (Courtesy of Bausch and Lomb, Inc.)

Figure 8.10.   Original model of microscope constructed by Edward Bausch in 1872. (Courtesy of Bausch and Lomb, Inc.)

Figure 8.11
Ernst Abbe who made
many improvements on
the light microscope,
including the design of
the substage condenser.
(Courtesy of
Karl Zeiss Co.)

light, or (2) construct doublet or triplet lenses with different indices of refraction. Achromatic lenses with corrections for color aberrations were not used in microscopes until the nineteenth century however. G. B. Amici demonstrated.corrected lenses in 1827, and in 1830 microscopes with achromatic lenses appeared. The modern compound microscopes with ocular and objective spaced properly by a body tube were then developed. Apachromatic objectives with better correction for color aberration removed the color fringes around the outside of the optical field and represented another improvement.

Next the microscope was made suitable for the use of transmitted light. An optical axis and a movable mirror were built into the instrument (Fig. 8.10). When higher magnifications were realized with more convex lenses, it was found that a medium other than air between the lens and glass slide would improve the efficiency of the microscope. The immersion lens was first devised by John Dolland in 1844 with water as the medium to fill the spaces between the lens and the glass slide. This fluid improved the efficiency of the instrument considerably. Cedar oil, however, was found to have an index of refraction similar to glass, and thus it provided an even better medium. The oil immersion lens was developed for high-power magnification and good resolving power. With oil as the medium between the lens and the slide, the light rays follow from glass to oil and back to glass in the same plane. When the rays go from glass to air, they diverge in other directions and are not concentrated on the object.

Ernst Abbe (1840-1908; Fig. 8.11) designed the substage condenser in 1870. This glass lens, which is placed under the stage, brings the light rays together and directs them more precisely through the prepared object. It substantially augmented the efficiency and usefulness of the microscope for studying small microorganisms.

Further improvements centered around the manufacture of better lenses, which eventually made it possible to resolve objects at some 2,000 diameters. The light microscope now seems to have attained its maximum resolving power. The ultimate physical limits of refining the magnification and resolving power of these microscopes are associated inseparably with the light rays themselves. It is impossible, therefore, to make an instrument with which to observe an object smaller than the wavelength of light. Because light waves are too coarse to accommodate the study of extremely small objects, this limitation seemed insurmountable. Higher magnification and greater resolving power could be achieved only by using rays that are shorter than those of ordinary light.

Ultraviolet Microscope

The ultraviolet microscope operates with ultraviolet radiation which has a shorter wavelength than ordinary light. Since glass filters out ultraviolet, quartz transmitting or glass reflecting lenses must be used in the ultraviolet microscope. With ultraviolet, a magnification of about 6,000 diameters and double the resolving power of the light microscope can be obtained. Because man cannot see ultraviolet, photographs must be made on plates sensitive to ultraviolet. The ultraviolet microscope was useful for obtaining higher resolving power until the electron microscope was developed. In competition with the electron microscope, it is not practical for general use but has some special applications.

Television microscopes use ultraviolet. These microscopes are employed with television cameras and receivers through which living microorganisms are observed in color. Another specialized use is for studies of nucleic acids. If monochromatic ultraviolet of a wavelength absorbed by nucleic acids (260 mμ) is used, nucleic acid-rich structures may be photographed in unstained cells. In combination with a spectrophotometer, the ultraviolet microscope provides a method for the quantitative estimation of nucleic acids in cells.

Phase Contrast Microscope

Phase contrast microscopes represent a modification of the light compound microscope which makes it possible to see living objects that are not distinguishable with the ordinary light microscope. Objects that have the same density but different optical paths (distance times refractive index) were shown by Fritz Zernicke (1935) to be distinguishable in a particular optical system.

Light rays passing through an object of high refractive index will be retarded in comparison with those passing through the surrounding medium with a lower refractive index. Retardation or phase change for a given light ray is a function of the thickness and the index of refraction of the material through which it passes. In unstained living specimens, transparent regions of different refractive indices retard the light rays passing through them to differing degrees. Phase variations in the light focused on the image plane of the light microscope are not visible. The optical system of the phase contrast microscope converts phase variations into visible variations in light intensity or contrast. With the phase contrast microscope, activity in protozoa and living cells of other organisms can be observed. Such processes as cell division and the activities of parasites in and around cells can be studied with the phase contrast microscope.

Electron Microscope

The electron microscope incorporates the good qualities of the ultraviolet microscope and is capable of much higher magnification and resolving power. It utilizes electrons with waves very much shorter than those of ordinary light or ultraviolet. The electron microscope, first developed (in 1932) by M. Kroll and E. Ruska in Germany, makes it possible to resolve objects that are in a smaller order of magnitude than those observable with either ordinary light or ultraviolet. Electron waves, like ultraviolet, are not visible to the eye. It is therefore generally necessary to expose a photographic plate and observe the object through the medium of a photograph rather than directly on the stage of the microscope. A fluorescent screen has been developed, however, which provides an image on the microscope. This permits direct observation of the gross characteristics of the image and is used in focusing the microscope and making preliminary studies.

The electron microscope (Fig. 8.12) has proved especially valuable for the study of dead, fixed objects. Its potential usefulness for analysis

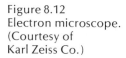

Figure 8.12
Electron microscope.
(Courtesy of
Karl Zeiss Co.)

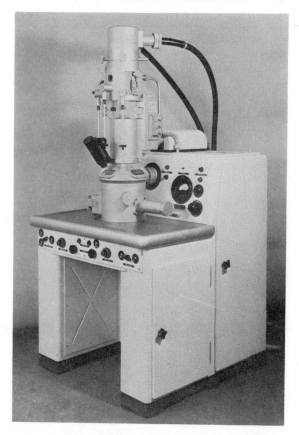

of living material is still in question. The tremendous voltage required to operate the electron microscopes kills living materials upon
penetration, and the objects that are seen may not be representative
of the living protoplasm. Furthermore, the techniques required to
prepare materials for investigations with the electron microscope are
exacting. Only small and discrete objects can be examined precisely.
Electron waves do not penetrate structures of any appreciable thickness. Therefore, if the objects to be studied are as large as protozoa
and many bacteria, they must be carefully preserved and cut into
extremely thin sections. Ordinary cutting instruments (microtomes)
cannot cut sufficiently thin sections. This adds further technical limitations to the usefulness of the electron microscope for living materials.
As new tools are devised and techniques are improved, however, it
will undoubtedly acquire greater value for biologists. Already viruses
are being studied with remarkable success.

Chronology of Events

1st Century	Seneca, globe of water used to magnify handwriting.
2nd Century	C. Ptolemaeus, treatise on optics.
13th Century	Meissner, spectacles for the aged.
1267-1268	R. Bacon, *Opus Magnus* including information about optical advances.
1558	C. Gesner used magnification in scientific investigations.
1590	Z. Janssens, first compound microscope.
1625-1630	F. Stelluti published the order of magnitude for his drawings of honey bee parts.
1632	Galileo Galilei, *Dialogue on the Two Chief Systems of the World.* Galileo made telescopes to observe distant objects and found that a different arrangement of lenses could be used to make tiny objects visible.
1643-1680	A. Kircher suggested from microscopic observations that disease and putrefaction were caused by invisible living bodies.
1655	G. Borelli observed structures in blood and other small objects with microscope.
1665	R. Hooke, *Micrographia* illustrating many microscopic objects.
1669-1672	M. Malpighi described microscopic observations of red corpuscles, silkworm anatomy, and numerous other objects in plants and animals.
1669-1675	J. Swammerdam published work on metamorphosis of insects (1669), anatomy of the mayfly (1675).
1672-1682	N. Grew, *Philosophical History of Plants* (1672), *Anatomy of Plants* (1682).
1673-1723	A. van Leeuwenhoek, numerous letters describing observations made with his microscopes.
1678	R. Hooke, lecture material dealing with biological observations that was published under the title *Microscopium*.
1694	R. Camerarius, sexual reproduction in plants.
1766	L. Euler, described methods of avoiding chromatic aberrations in compound lenses.
1830	G. B. Amici, achromatic lenses for microscopes.
1844	J. Dolland, water immersion which led to oil immersion lenses.
1870	E. Abbe designed substage condenser.
1932	M. Kroll and E. Ruska, first electron microscope.
1935	F. Zernicke, phase contrast microscope.

References and Readings

Allen, R. R. 1940. *The Microscope.* New York: D. Van Nostrand Co.

Beck, C. 1938. *The Microscope; Theory and Practice.* London: R. and J. Beck.

Belling, J. 1930. *The Use of the Microscope.* New York: McGraw-Hill Book Co.

Bradbury, S. and G. L'E. Turner, eds. 1967. *Historical Aspects of Microscopy.* Cambridge: Royal Microscopical Society.

Bradbury, S. 1968. *The Microscope Past and Present.* London: Pergamon Press.

Cohen, B. A. 1937. *The Leeuwenhoek Letter.* Baltimore: Soc. Amer. Bact.

Corrington, J. D. 1941. *Working with the Microscope.* New York: McGraw-Hill Book Co.

Crombie, A. C. 1961. *Augustine to Galileo,* vol. 2. Cambridge, Mass.: Harvard University Press.

Dobell, C. 1932. *Antony van Leeuwenhoek and his "Little Animals."* New York: Harcourt, Brace and Co.

Gage, S. H. 1964. "Microscopy in America." *Trans. Amer. Mic. Soc.* 83:9-125. (Supplement)

Grew, N. 1682. *The Anatomy of Plants with an Idea of a Philosophical History of Plants and Several Other Lectures Read before the Royal Society.* London: W. Rawlins (for the author and Royal Society of London).

Hooke, R. 1665. *Micrographia: Or Some Physiological Descriptions of Minute Bodies made by Magnifying Glasses with Observations and Inquiries Thereupon.* Originally printed by the Royal Society of London. Reprinted as vol. 13 of R. T. Gunther, *Early Science in Oxford.* Oxford: Clarendon Press, 1923. (Now available in Dover Publications, New York, 1961.)

Leeuwenhoek, A. van. 1680-95. *Arcana Naturae Delphis Batavarum,* Henricum Krooneveld. (Reprinted: Culture et Civilisation, Bruxelles, 1966.)

_____. *The Collected Letters of Antoni van Leeuwenhoek.* 1941. Amsterdam: Swets and Zeitlinger.

Munoz, F. J. 1943. *The Microscope and Its Use.* Brooklyn, N. Y.: Chemical Publishing Co.

Schierbeek, A. 1967. *Jan Swammerdam 1637-1680, His Life and Works.* Amsterdam: Swets and Zeitlinger.

Sjöstrand, F. S. and J. Rhodin. 1957. *Electron Microscopy.* New York: Academic Press.

Shippen, K. B. 1955. *Men, Microscopes and Living Things.* New York: Viking Press.

Swammerdam, J. 1680. *Ephemeri Vita: Of the Natural History and Anatomy of the Ephemeron.* Trans. from low-Dutch to English, with preface by Edward Tyson, and detailed drawings by Swammerdam. London: Henry Faithorne, 1681.

CHAPTER 9

SYSTEMATIZERS OF PLANTS AND ANIMALS

IN THE later part of the seventeenth and throughout the eighteenth century biology was dominated by systematists. New knowledge had been accumulated during the Renaissance and the next logical step was to classify it into useful systems. The movement to systematize and classify was prevalent in literature, theology, art, and other areas of thought, as well as in science. Newton's book *Principia,* published in 1687, was the key to the approach used during the period. It described the universe as fixed and static, with the earth and heavenly bodies precisely arranged in mathematical harmony. The same fixed and rigid pattern that was applied to the physical world was carried over to biology. Attempts were made to classify animals and plants according to a prearranged, immutable system. This method of biological classification persisted until the publication of Darwin's *Origin of Species*, in 1859.

For beginnings in the classification of living things, it is appropriate to go back to Aristotle, who provided descriptions of many animals and made keen distinctions among them. His groupings were based mostly on such characteristics as mode of reproduction and habitat. It was later shown that these characteristics are not valid in distinguishing natural groups. Aristotle, however, did provide a sound concept of the species as a fundamental unit of living things. The useful terms, genus and species, are translations of the Greek words *genos* and *eidos*, used by Aristotle. After the time of Aristotle, various attempts were made to classify animals and plants, but most of the systems proposed before the seventeenth century were confused and impractical.

German Fathers of Botany

Substantial progress in plant classification began in the sixteenth century under the leadership of the German "fathers of botany." An early representative of this group was Otto Brunfels (1489-1534). In middle life he forsook a life of church activity and entered upon the study of medicine. Medicine at that time was vitally concerned with plant products, and Brunfels soon became interested in more basic aspects of botany. An enterprising bookseller in Strasbourg commissioned him to prepare an herbal and secured the services of an artist, Hans Weiditz, to illustrate the work. The lasting value of Brunfels's herbal *Herbarum Vivae Eicones* or "Living Pictures of Herbs" derived largely from the original engravings of Weiditz. The text material was taken mostly from Dioscorides and other classical authors, and no original observations were made. Discrepancies were obvious when illustrations done by Weiditz, using German plants as models, did not match the original descriptions of plants from the Mediterranean region. Plants sketched in Germany but not corresponding with any description by Dioscorides were "nameless waifs."

Another member of the early German group of botanists, Jerome Boch (1498-1554), went to nature for his information. He traveled widely to observe plants in their native habitat, and he also studied in the garden of a patron of science in Zweibrücken, where he was employed as a school teacher. Boch was not satisfied with the alphabetical arrangement of plants employed by his predecessors and contemporaries. He worked out a more realistic system based on characteristics and relations of vegetative parts such as the shapes of stems and positions of leaves. Seasons of flowering and fruiting, soil conditions, and the characteristics of the localities in which the plants were found were also recorded. The first edition of Boch's herbal (1539) was not illustrated. Later editions, however, included drawings by David Kandel which were useful but inferior to the earlier drawings which Weiditz had prepared for Brunfels's herbal. The greatest value of Boch's work lay in the original descriptions and his beginnings in natural systems of classification.

Leonard Fuchs (1501-66), third member of the group, made notable contributions to classification. Despite the large amounts of time he devoted to his position as professor of medicine at Tübingen and as a practicing physician, he produced a botanical classic under the Latin title *De Historia Stirpium* (1542). A year later (1543) he published the German edition *Neu Kreuterbuch*. About 400 plants native to Germany and about 100 from other countries were described in this book. Descriptions were well considered and the plants were arranged under appropriate headings such as *Character, Form, Place*. The illustrated herbal of Fuchs was more elaborate, better documented, and more concise than any of the earlier botanical texts.

Another early German "Father of Botany," Valerius Cordus (1515-44), was a pharmacist as well as a botanist. His education was directed by his father, Euicius, a scholar and author of botanical works. Before the age of 20, Valerius had published a book, *Dispensatorium,* concerned mainly with plants of medicinal value. His great work, *Historia Plantarium,* completed in 1540, included descriptions of 446 species. Each plant was described in flower and in fruit. Internal as well as external morphology was included with the description. After observing and describing the plants in his own country, he went to Italy primarily to observe those described by Dioscorides, but he died in Rome at the early age of 29. His botanical work, completed in 1540, was published in 1561, several years after his death, by the Swiss naturalist and encyclopedist, Conrad Gesner.

Illustrations in the herbals did much to increase their popularity and practical usefulness. People who could not read the original Greek or Latin texts could learn much from the pictures. Later herbals, prepared with many pictures and minimal written material, could be used by uneducated women in identifying plants of medicinal value (simples) which they collected and sold to druggists. Virtually all plants had some value and the industry of collecting simples became economically important.

Although the Germans led the way in the development of botany, interest was soon evident in other countries in Europe. Surveys made in distant lands such as Mexico and India added new and strange plants to the botanical lists and collections. In an age of exploration and discovery it was quite natural for curious observers to undertake surveys of the flora of distant lands. Explorers of America and Asia returned to Europe with news about strange plants as well as actual specimens of leaves, fruits, and seeds.

A new kind of explorer thus developed, the scientific explorer, who was interested in new knowledge of natural history and was willing to undergo the hazards and hardships of an explorer for no other purpose than the accumulation of scientific facts. Established scientists such as physicians, professors, and curators of botanical gardens usually remained at home while younger, more adventurous scientists made the trips. Those at home had to make their descriptions more detailed and more accurate if they were to keep pace with new developments. The overall situation promoted a higher quality of botanical research.

Early Systematists in Other Countries

The leading English botanist of the period was William Turner (1510-68). He was a teacher, critical observer, and writer who demonstrated the true scientific spirit but because of political and religious controversy labored under great difficulty. When he was unable to work effectively in England, he carried on his botanical study at Bologna and later at Cologne. The first part of his *Herball* appeared in England in 1561. At Cologne, he arranged for publication of the second part, and the third part (1568), was added in England shortly before his death. Turner recognized that the species of plants that he was observing were not necessarily the same as those described by early Greek and Roman scholars and he corrected old descriptions on the basis of his own observations. Woodcuts, mostly from the work of Fuchs, were used to illustrate Turner's *Herball*.

In Switzerland, Jean (1541-1613) and Kaspar Bauhin (1560-1624) were the leading systematic botanists of the period. Jean undertook an illustrated history of plants that was published (1650) after his death under the title *Historia Plantarum Universalis*. Some 5,000 species were included and arranged in a more or less natural sequence. In determining the classification of each plant its structural parts, functional properties, and ecology were considered. This classification system was the first to be used extensively by others besides its author. Kaspar began describing local Italian plants while he was studying at Padua and he continued the study of plants in his home locality when he returned to Basel. He published (1623) a systematic synonomy of all previously described species under the title *Pinax Theatri Botanici.* Some 6,000 species were included. From his own observations he gradually added many new species to the list. Besides comparing living specimens, Kaspar, following the lead of Luca Ghini (1490-1556) and other Italian botanists, dried specimens, and preserved them in a herbarium for further critical study and comparison. More and more plants were becoming known and some method of systematic arrangement was essential. To fill this need, Kaspar Bauhin developed a system of making use of two names (genus and species) that led to the modern binomial system of nomenclature.

In Italy, Pietro Andrea Mattioli of Siena (1500-77) published (1544) a commentary on Dioscorides. It has a larger number of illustrations accompanying the descriptions of plants. The relationships between plants and other forms of life are also illustrated. For example, there are pictures of a woman milking a cow, a man churning butter, and so on. This work became immediately popular, many copies were sold, and several editions were subsequently reprinted. So great was the authority and influence of Mattioli that discoverers of new plants in distant parts of the world frequently sent information they had accumulated to him to be included in his next edition. Another Italian, Andrea Cesalpino (1519-1603), improved upon Bauhin's system of nomenclature. His principal descriptive work, *De Plantis* (1585), included flower and fruit characteristics which were used as a basis for classification.

Seventeenth and Eighteenth Century Systematists

Jung

A major contribution was made by Joachim Jung (1587-1657), who studied medicine at Padua and returned to his native Germany to teach botany and carry on botanical research. None of his work was published during his lifetime, but his manuscripts were preserved by his students and published as two small volumes in 1662 and 1679. Jung's material was written in a concise form and his published works demonstrated his remarkable facility for classification. Buds and flowers were used as characteristics for distinguishing plants, but Jung, like his contemporaries and some followers, did not understand the functional significance of these parts in sexual reproduction. The problem of reproduction in plants, which stimulated much interest in the seventeenth century, was resolved by Camerarius in 1694. His explanation, however, was not generally accepted until the middle of the eighteenth century.

Jung coined many terms and phrases that have persisted in plant taxonomy. Simple and compound leaves, alternate and opposite leaves, petiole, stamen, style, node, and internode, for example, were used by Jung. He distinguished disc florets from ray florets in composite flowers. He used a binomial system of nomenclature in identifying many of the plants he described. In this system, a noun was used for the major group, that is the genus, and an adjective for each smaller subdivision of the genus, that is, the species. Jung had come under the influence of the Cesalpino tradition while he attended the University of Padua in Italy. His application of the binomial system constituted an advance based on earlier beginnings and approached the modern system.

Ray

John Ray (1627-1705; Fig. 9.1) ranks with Linnaeus as one of the major founders of the biological science of systematics. He furthered the idea that species constitute the basic population unit. Ray arranged species and larger taxonomic groups according to what he declared was the "Creator's plan" of numerical series for grouping like with like. Some of his groupings, however, were in error. For example, he separated herbs and trees into two primary divisions of the plant kingdom without realizing that some herbs and trees are closely related. Although he claimed to be following the "Creator's plan," he erred in visualizing species as dynamic units subject to change in nature. Plants, he thought, must have a sexual method of reproduction, but he did not profess to understand the mechanism.

Ray was the son of a blacksmith and lived at Black Notley near Braintree in Essex, England. He received his early training at the Braintree School and entered Catherine Hall, Cambridge, at the age of 16. After three terms he transferred to Trinity College where he was tutored by James Duport, Professor of Greek. Ray was chosen a minor fellow at Trinity College in 1649, and a major fellow in 1651, when he obtained a university position as a cleric. In addition to his religious responsibilities, he tutored students in science and earned a reputation as an excellent tutor. As he grew older he developed a passion for natural history and spent more and more time and energy in biological projects. Among the several students attracted to him was Francis Willughby (1635-72), a wealthy young man with great ability and interest in natural history. During these years Ray and his students took excursions in England, Wales, and particularly the vicinity around Cambridge, observing plants and animals. In 1660 Ray wrote a flora of the Cambridge vicinity, the first work of its kind.

In 1661 Ray was obliged to give up his fellowship because of conscientious scruples which made him unable to subscribe to the Act of Uniformity. From then on he concentrated on pursuing his scientific interests. During the ensuing years his support came largely from his former student and close friend, Willughby, who became his constant companion. The two men traveled together and observed and collected plants and animals. They formed a partnership and agreed that Ray would specialize on plants and Willughby on animals. Ray set the pattern for his systematic studies when he based his classification of plants on the fruit, leaves, and flower parts. In 1663 Ray and Willughby began a tour of Europe to observe and classify all the plants and animals they could find. They returned with large collections of both plants and animals which they intended to use as the basis for a systematic description of the plant and animal kingdoms. Ray published (1667) a *Catalogus* of British plants, an extensive description of the flora of the British Isles. Following his election (1669) as a fellow in the Royal Society, he, with Willughby as coauthor, published a paper entitled "Experiments Concerning the Motion of Sap in Trees" in *Philosophical Transactions*.

Willughby had done much work on various groups of animals but had organized only the sections on ornithology and ichthyology when he died in 1672 at the age of 38. His vast collection of notes was left to Ray to edit and publish. The work on birds was published in 1678 under the title, *The Ornithology of Francis Willughby of Middleton in the County of Warwick, Fellow in the Royal Society*. Descriptions are illustrated by elegant figures resembling live birds. The work on fishes (1686) is illustrated with outline drawings of whole fish, rendered in sufficient detail of body surface and fins to identify the particular fish. Willughby also left Ray a substantial annuity with a charge to provide for the education of Willughby's two sons and to continue the biological projects which the two men had planned together. Ray fulfilled the charges faithfully.

Ray used his botanical collections as the basis for his publication (1682) entitled *Methods Plantarum Nova,* and for his greatest work, *General History of Plants (Historia Generalis Plantarum)*, printed in three volumes, one each in 1686, 1688, and 1704. In the book on methods, Ray described the true nature of buds and designated the

Figure 9.1
John Ray, early English
student of biological
systematics.

major divisions of flowering plants as the dicotyledons and monocotyledons. This difference is not a fundamental distinguishing characteristic among plants and proved a poor choice for a major criterion of classification.

In his momentous work, *General History of Plants,* Ray recorded information concerning structure, physiology, distribution, and habits of plants. He delineated some 18,600 plants, virtually all that were known at that time.

In 1684 Ray reactivated the work on mammals and reptiles that had been started by Willughby; the manuscript on fish was edited by Ray and published in 1686. The survey on the *Quadrupeds and Reptiles* (1693) is Ray's most important contribution to zoology. Although Ray was primarily a botanist, he was interested in animals, and his own observations enabled him to continue Willughby's investigations and to follow "nature's system" in classification. When he came to the mammals, Ray rejected Aristotle's classification, which had been based on three classes: animals with solid hoofs, those with cleft hoofs, and those with many toes. Instead he devised his own system with only two classes: those having toes covered with horny hoofs, and those in which the toes have only nails. After finishing the manuscript on the mammals and reptiles, Ray began working on the natural history and classification of insects. His best entomological work was that on the insect order Lepidoptera.

The studies of insects were made under difficult circumstances after Ray's health had partially failed. He had moved, with his family, back to his birthplace at Black Notley, where he planned to retire and live a leisurely life in his declining years, but he had too much ambition and interest in his work to retire. Because he was unable to move about as he desired, his children and friends did much of his insect collecting. Despite the obstacles, Ray maintained a wide correspondence and wrote a paper on the classification of insects which was prepared for publication in 1704.

Ray also made significant observations on fossils and other aspects of geology while on his numerous field trips. His theories and comments on these subjects were expressed in his paper *Physical, Theological Discourses* (1713), published after his death. In spite of prolonged ill health, Ray lived to the age of 78. The Ray Society, organized to promote the publication of Ray's works on natural history, was founded in his honor in 1844.

Ray's work as a botanist and zoologist was mainly concerned with descriptions and classification. He was content to devote himself to the business of observing, discriminating, defining, and arranging the flora and fauna of the areas he had visited in a systematic way. Ray insisted that distinctions among living things should be based on structure rather than less basic characteristics such as color, size, or habit. This was a most important change from the approach of most of his predecessors.

Tournefort

The French biologist, Joseph de Tournefort (1656-1708), also furthered the binomial system of nomenclature. Tournefort traveled extensively for the purpose of describing and collecting plants. He used the binomial system effectively, but his classification was artificial and inferior to that of Ray. The names of the genera were emphasized and he described various genera in great detail, but species and varieties were given less attention. Among his published works were *Éléments de botanique* (1694), *Institutiones rei herbariae* (1700), and *Histoire des plantes*. Tournefort was also inferior to Ray in his philosophical viewpoint and his understanding of basic biology. He denied sexuality of plants and made little effort to discover natural relations. His contribution, however, was a link in the chain that culminated in the work of Linnaeus.

Linnaeus

Carl von Linnaeus (Linné latinized, 1707-78; Fig. 9.2) was born in a rural district in Sweden near Uppsala and spent his early life in the modest home of his pastor father, who was a nature lover. Linnaeus gained an early appreciation of nature when, as a child and youth, he walked frequently in the woods with his father, observing plants and animals and discussing aspects of natural history. Most important for his later work, he learned early to observe and to go to nature herself as well as to books for facts on natural subjects.

When Linnaeus was 20 years old he entered the University of Lund in southern Sweden. A year later he transferred to the University of Uppsala near his home. Olaf Celsius, a Professor of Theology, but also a botanist, was impressed by Linnaeus's understanding and appreciation of nature and invited him to live at his home and use his library. While living with Celsius, Linnaeus carried out a study on the reproductive structures of plants and considered the possibility of distinguishing different plant groups on the basis of reproductive parts. The resulting paper was published (1730) under the title "Introduction to Floral Nuptials." The work was not original but represented a careful review of the anatomical researches of others. In this early study, Linnaeus became acquainted with such reproductive structures as stamen and pistil, which later became the basis for his classification of plants.

Professor Celsius introduced Linnaeus to Professor Rudbeck, botanist and chief scientist at Uppsala. In 1732, on the recommendation of Rudbeck, the Uppsala Academy of Science provided an opportunity for Linnaeus to make a field trip into Lapland, a broad, unexplored area that included territory now in northern Sweden, Finland, and the Russian peninsula. This was probably the first extensive field trip in the history of science that was made for the purpose of biological observations. Linnaeus traveled some 4,600 miles, partly alone and partly in the company of two Laplanders. His equipment consisted of a measuring stick, telescope, magnifying glass, knife, paper, facilities for dying plants, and the Swedish equivalent of about $100 to pay his expenses.

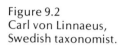

Figure 9.2
Carl von Linnaeus,
Swedish taxonomist.

After traveling for six months, he returned to Uppsala with specimens of plants and minerals, and vivid impressions of the vast and
little-known Lapland. In the course of his travels he had observed
many wild animals and discovered more than 100 new species of
plants. Three years later, Linnaeus was invited to make a botanical
survey of the province of Delecarlia (a region of Sweden). This time
he had assistants, better equipment, and more adequate expense
money. He later visited Germany, England, and France and continued
his project of describing and classifying plants.

Medical Training and Practice

When Linnaeus was 28 years old, he was courting Sara Lisa Moraeus,
whom he had met while on a lecture tour at Falun. He asked her
father, Doctor Johan Moraeus, town physician of Falun, for her hand
in marriage. Consent was given with the stipulation that Linnaeus
should go to Holland (an important medical center at the time),
study medicine, and become a physician before the marriage. The

father agreed to provide part of the money for the trip and school expenses. Linnaeus, who had already completed most of his medical training at Uppsala, accepted the challenge and went to Harderwijk, Holland, where he earned his medical degree in a short time. He remained in Holland for the next three years and devoted much time to observing plants and clarifying taxonomic problems which he had encountered in Sweden.

While in Holland Linnaeus became physician extraordinary to a wealthy banker and merchant of Amsterdam, Mr. George Clifford, who had an extensive estate that included museums, greenhouses, and a botanical garden with imported plants from all over the world. Many of these plants had come from Virginia and other parts of America. This was an ideal place for Linnaeus to try out his classification system. The garden is now maintained as a park called Linnaeushoff. One of Linnaeus's books, *Hortus Cliffortianus,* describing and illustrating the plants in Clifford's garden, was dedicated to Mr. Clifford. Hermann Boerhaave (1668-1738), the most famous physician in Europe at that time, also became interested in Linnaeus because of his publications and helped to promote his career.

After three years in Holland and seven publications (books) to his credit, Linnaeus was famous. He received offers of positions in various parts of the world. While Linnaeus was visiting at Oxford, J. J. Dillenius, a Professor of Botany, became so interested in the Linnean system of classification that he offered to share his own salary with Linnaeus if he would remain permanently and introduce his classification system at Oxford. Linnaeus also was offered a position with a collecting party going to the Cape of Good Hope. He accepted none of these offers but returned to his native Sweden and married Sara Lisa.

In Stockholm, he attempted to practice medicine but soon discovered that he did not care for the practical and professional aspects of medicine. Frequently, when he had no patients, he went to the slum districts and worked without pay among the poverty-stricken people suffering from tuberculosis and venereal disease. Although he worked hard and long and made a sincere attempt in the practice of medicine, he was not happy. An opportunity came to join the Swedish navy and eventually he was appointed head physician to the Swedish Admiralty, later becoming physician to the queen. The happiest time in his professional life came when he was offered and accepted the position of Professor of Physics and Medicine at Uppsala in 1741.

Professor of Botany

Later the chair of botany at Uppsala, formerly occupied by Professor Rudbeck, was available to Linnaeus, and he again had the opportunity to work in the libraries and gardens that he loved. His fame spread and many students from foreign lands came to Uppsala to work with the great systematist. Linnaeus sent his students out on expeditions patterned after his own Lapland adventure to find new plants and animals and to observe living things in their native habitats. Once established as Professor of Botany, he began a complete rearrangement of the University's botanical garden to fit his system of classification. The genus and species categories used by his predecessors had been vague, but Linnaeus clarified them and introduced, for the first time, the class and order categories.

Linnaeus died in his home, which was located in the corner of the botanical garden, in 1778. Nine years later a large garden was given to

Figure 9.3.   Linnaeus's Garden in Uppsala, Sweden, as restored by the Swedish Linnean Society.

the University by King Gustav III, and the home and garden of Linnaeus fell into decay. In 1917, the Swedish Linnean Society undertook the task of restoring the gardens used by Linnaeus. Plots were laid out according to Linnaeus's plan, which had been carefully developed on paper but had not been completed while Linnaeus was alive.

Today the gardens (Fig. 9.3) are arranged as Linnaeus planned them. Representatives of plant species are placed in plots according to taxonomic position. Classification is determined mostly by type of reproductive parts. To the right of the center path, as one walks from the front to the rear of the garden, are the annual plants, and to the left are the perennial plants. It is impressive to observe all of the plants in order, each identified by its Latin name on a white marker with the name of Linnaeus indicated after each species name. In the background the hothouses and head house are located as Linnaeus designed them. The home in the corner of the garden where Linnaeus lived for 35 years was restored by the Swedish government in 1937 and is now maintained as a museum.

Classification System

The system of classification used by Linnaeus was fixed and rigid, characteristic of the seventeenth and eighteenth centuries. Although natural relations were followed, Linnaeus himself was not impressed, at least in his early life, by the pattern of evolution that was illustrated in his garden. Philosophically he was a special creationist, and his objective was to classify plants and animals according to a workable and useful system without conscious reference to any evolutionary sequence suggested by the groupings. He was impressed, however, by the diversity which he noted first in plants and later in animals. Linnaeus had a rare genius for classification but he was not an original investigator. He was skillful in appropriating the best ideas of his predecessors (Table 9.1) and in synthesizing a useful system of classification.

Although all parts of the Linnean system had been used earlier, his major contribution was in promoting the binomial method and developing it to a practically applicable large-scale usefulness. The value of the binomial system of classification might be best appreciated by comparing it with the system it replaced. Although some beginnings had been made in systematics, the descriptive system of Aristotle was still generally in use. Twelve words were the minimum used as an

Table 9.1 Summary of contributions of Linnaeus's predecessors to binomial system of classification.

Date	Location	Contributor	Contribution
1530	Germany	Brunfels	Illustrated Dioscorides' plants.
1539	Germany	Boch	Original descriptions of plants.
1543	Germany	Fuchs	Described and illustrated plants.
1561	England	Turner	Corrected Greek descriptions.
1585	Italy	Cesalpino	Fruits and flowers for classification.
1623	Switzerland	K. Bauhin	Binomial system, genus and species.
1660-1713	England	Ray	Nature's system applied to plants and animals.
1662-1679	Germany	Jung	Terminology for plant characteristics.
1730-1778	Sweden	Linnaeus	Developed binomial system to practical usefulness.

identifying description of a plant. This was quite satisfactory when only a few species were known, but it became cumbersome and increasingly difficult as more species were identified. Obviously, it was not suited for use in keys.

Publications

The most famous of Linnaeus's written works was *Systema Naturae,* published originally in Holland. The first edition (1735) contained only about a dozen pages but the book was improved and enlarged through sixteen editions during Linnaeus's lifetime. Taxonomic groupings, class, genus, and species were used in the earlier editions and the order grouping was introduced in the 10th edition (1758). This edition presented the complete Linnean system and utilized binary nomenclature throughout. Phylum, family, and super- and sub-groups were added by later systematists. Present-day zoologists use the 10th edition as their basis for nomenclature.

Twenty-four classes of plants were recognized by Linnaeus. This is not sufficient at present but was adequate then. The lowest class, the Cryptogamia, was Linnaeus's "wastebasket" for all the lower plants that did not have recognizable reproductive structures. The life history of each plant had to be worked out before it could be properly identified and classified. *Species Plantarum* (1753) includes descriptions of all species of plants then known, and binomial nomenclature is used throughout. This book is the basis for modern plant taxonomy.

Linnaeus completed extensive descriptions of plants growing in his native Sweden, in the Netherlands, and also in various other parts of the world including America. He never visited America, but he observed imported American species in Clifford's garden and many American biologists sent him specimens to be named and described. Many plants and descriptions of plants from Virginia came from the collections of the American botanist, John Clayton. The plants studied by Linnaeus throughout his life were described in Linnaeus's later publications, *Genera Plantarum* and *Species Plantarum*.

In his later life, Linnaeus attempted to classify the fauna as well as the flora of the entire world. He was a better botanist than zoologist, but his binomial system was as well adapted to animals as to plants. In 1748, when he was 41 years old, he published a catalog of the fauna of Sweden in which he described some 300 species of beetles and a large representation of other animals. In the course of his work Linnaeus described about 4,000 species of animals compared with some 800,000 that have been described at the present time. Something over a million are now believed, by enthusiastic systematists, to be present on the earth.

Classification of Mankind

When Linnaeus was an old man, after he had been unusually successful in classifying plants and animals, he attempted to classify mankind. All living human beings were placed in the same genus and species, *Homo sapiens*. No other members of the genus *Homo* were found living upon the earth, although some fossil forms such as *H. neanderthalensis* have since been recognized. When Linnaeus classified the subgroups of man he did not employ the morphological type of characteristics which he had used so successfully on other animals but based his distinctions on attributes of mind and emotional status. He recognized four subgroups or varieties: *Homo sapiens americanus*, the American Indians, whom he characterized as contented; *H. sapiens europaeus*, the Europeans who, according to his observations were lively; *H. sapiens asiaticus*, the orientals, who were haughty; and *H. sapiens afers*, the African Negroes, who were slow. Obviously these criteria are subjective and are quite useless for valid classification. Shortly after this attempt by Linnaeus, Johann Friedrich Blumenbach (1752-1840), the German naturalist and father of physical anthropology, established the more objective and familiar classification in which five races were recognized. As distinguishing features, Blumen-bach used skulls and other parts of the skeleton. The more obvious characteristic, skin color, was soon recognized by others as fitting the same categories and Blumenbach's five groups became known as the white, red, yellow, brown, and black races. More recent attempts to categorize *Homo* have been based on gene frequencies, particularly those associated with blood antigens.

Indirect Influence

At the time Linnaeus was making his major contribution in taxonomy, he had no more progressive philosophical viewpoint than those of Cesalpino, Ray, and others of his predecessors. He was considerably less profound than Ray. Camerarius, a German scientist, had shown in 1694 that plants utilize sexual reproduction, but Linnaeus was not aware of all the fine points of Camerarius's contribution. Linnaeus did not conduct experiments, and he made few observations except those necessary to identify organisms. In his early life he followed and strengthened the dogma of constancy of species.

His viewpoint was much like that of many laymen today, that the worthiest task of the biologist is to attach a tag or scientific name to every plant and animal. Classification is necessary to enable biolo-gists to become acquainted with the fauna and flora, but when used alone it is a superficial approach to biology. The true biologist is interested in the natural relations among living things and not merely in "pigeonholing" each organism with its group. Classification should be a tool rather than an end.

There is evidence that Linnaeus developed a concept of evolution in his later life when he tried to hybridize different distinct species of plants. The results of that experiment and his comments on a natural origin of species were published in his *Systema Vegetabilum* (1774). Students who followed Linnaeus were too busy naming and classify-ing plants to appreciate or remember his later interests in evolution.

When Linnaeus died (1778), Joseph Banks, President of the Royal Society of London, offered to purchase Linnaeus's herbarium for 1,200 pounds. Linnaeus's son, who succeeded his father in the chair of botany at Uppsala and had control of the estate, refused the offer and retained the collection. The younger Linnaeus died in 1783, however, and his mother became heir of the collection. The new executor offered the herbarium, library, and manuscripts for sale to Banks at a price below the original offer for the herbarium alone.

At the time Banks received the letter from Uppsala, he was entertaining at his home a young medical student, James Edward Smith (1759-1828), son of a wealthy silk merchant. Banks showed his guest the letter and suggested that he ask his father to buy the collection. The young man complied, and soon owned a vast collection associated with the famous name of Linnaeus. The purchase elevated him socially, and he responded by touring Europe and lecturing about Linnaeus and the prize collection.

In 1785 Smith was elected a Fellow in the Royal Society. He disappointed some colleagues in the Royal Society when he promoted the organization of a new society honoring Linnaeus. In spite of some opposition, the new society was formally founded in 1788 and named the Linnean Society. The fear of the officers in the Royal Society was that the new society would become specialized in the narrow field of taxonomy and thus do injustice to the broad field of biology. These fears proved ungrounded, however; the Linnean Society became a broad-based and important scientific body devoted mainly to the development of taxonomy but also promoting natural history and evolution. The joint paper by Charles Darwin and Alfred Russel Wallace, which announced the theory of evolution by natural selection, was presented on July 1, 1858, at a meeting of the Linnean Society. The paper was published in the *Journal of the Proceedings of the Linnean Society* for August 20, 1858.

Other Plant Taxonomists

The influence of Linnaeus on classification was reflected in botanical gardens subsequently developed in Europe. Some of the most important were established by members of two families, de Jussieu and de Candolle. These are prominent names in the history of botany.

Jussieu

The original botanists in the de Jussieu family were three sons (Antoine, Bernard, and Joseph) of an apothecary who lived at Lyons, France. Antoine (1686-1758) was director of the Jardin des Plantes in Paris. He edited several treatises on plants including the third edition of J. P. de Tournefort's *Institutiones rei herbariae* (1719). Bernard de Jussieu (1699-1777) established the Royal Botanical Gardens, at the Trianon, Versailles, where he used a modification of the Linnean system of classification which he never published. He tried to work out a more natural system, and in some ways he improved on Linnaeus. Bernard revised (1725) Tournefort's *Histoire des plantes*. Joseph de

Jussieu traveled in South America and introduced many plants including the heliotrope into Europe.

Several other distinguished botanists in later generations carried the de Jussieu name. A nephew of Bernard, Antoine Laurent de Jussieu (1748-1836), continued the work on plant systematics and published his work, *Genera Plantarum* (1789), based on a modified Linnean system. He followed Ray's distinction between the monocotyledons and dicotyledons for the main subdivisions in the plant kingdom. As pointed out earlier, this is an artificial and superficial distinction for separating major divisions, but it did provide a single characteristic for distinguishing plant groups. To facilitate further segregation, Jussieu used the Linnean categories: class, order, genus, and species.

Candolle

The de Candolle family of botanists originated in Geneva, Switzerland, but most of the botanical work attributed to them was done in France. The most distinguished member of the family was Augustin Pyramus de Candolle (1778-1841), a profound botanist, who made careful investigations and wrote significant books and papers, some of which are still in use. Best known among these are *French Flora* and *Elementary Theory of Botany* (1813). His classification was more natural than the classifications of Linnaeus, because he placed more emphasis on fundamental morphological characteristics.

Later Developments in Classification

The influence of Linnaeus was perpetuated in classification systems devised for animals as well as in those for plants. Three French biologists, Buffon, Lamarck, and Cuvier, may be cited here for their contributions in natural history and particularly in the classification of animals. They also will be discussed in later chapters in connection with comparative studies and evolution.

Buffon

Georges Louis Leclerc, Comte de Buffon (1707-88; Fig. 9.4), was a prolific writer who attempted to classify all animals in the world. He did much to stimulate other people to make significant contributions to basic research, but his own works in this area were not especially valuable. His *Natural History* (1749-1804), which extended through 44 volumes, contained a theory for the formation of the earth, a history of the plant and animal inhabitants, and a description of many animals

Figure 9.4
Comte de Buffon,
French naturalist.

and plants. He directed the Jardin du Roi (later the Jardin des Plantes) in Paris, the largest botanical and zoological garden in Europe, and made it a center for research. Buffon had access to more biological material than had any previous collector and along with his numerous associates and employees classified many plants and animals.

Lamarck

Jean Baptiste Lamarck (1744-1829; Fig. 13.1) was more intimately connected with other developments in biology such as evolution (Chapter 13), but he made significant contributions in the classification of both plants and animals. He was a student of Buffon and succeeded him as director of the Jardin des Plantes in Paris. Lamarck's method of classification and concept of evolution were more modern than those of any of his predecessors. He visualized a progressive series in the animal kingdom similar to that described by Aristotle. Unlike Linnaeus, he based his distinctions on function rather than structure. In his ladderlike classification, he placed organisms which responded only by reflex at the bottom. Those with a sensory system were in the middle and those with intelligence were on the top of the ladder.

Figure 9.5
Georges Cuvier, French
taxonomist, comparative
anatomist, paleontologist,
and student of evolution.

This classification was associated with his view on the inheritance of acquired characteristics, which will be considered later in connection with evolution (Chapter 13).

Cuvier

Georges Cuvier (1769-1832; Fig. 9.5) was primarily a comparative anatomist (Chapter 10), but he also contributed to the classification of animals. Like Lamarck and other contemporaries, he was influenced in the direction of systematics by the worldwide explorations and collections that were in progress. Cuvier became the leading scientist of the Napoleonic era in France. Much attention was being given to orderly arrangement following the tradition of Linnaeus, and Cuvier found a place for which he was well prepared. One of his major contributions to classification was the establishment of a higher taxonomic category than the class of Linnaeus, comparable with the phylum in the present taxonomic system. Cuvier's animal kingdom was divided into four major groups which he called Embranchements: I Vertebrata, II Mollusca, III Articulata, and IV Radiata.

Embranchement I, the Vertebrata, included the classes: Mammalia,

Aves, Reptilia, and Pisces. Mammalia and Aves were identical with the classes of mammals and birds presently in use. Reptilia included the group now identified as Amphibia along with the animals now called reptiles. Pisces included the fish and some primitive chordates, such as the lamprey, which were known at the time. Embranchement II included the Cephalopoda, Gastropoda, Pteropoda (free-swimming marine animals), and Acephala (bivalve marine forms), all of which are now grouped with the Mollusca. Also in Embranchement II were the Cirripedia (barnacles) and Brachipoda (a primitive group that superficially resemble the mollusca but are now placed in a separate phylum). Embranchement III included Crustacea, Arachnida, and insects (which are now in the phylum Arthropoda), and the Annelida (now in a separate phylum). Embranchement IV, the Radiata, included Echinoderms and Polyps, which have radial symmetry, and a miscellaneous collection of aquatic forms called Infusoria. Internal parasites known at that time and other forms which did not fit anywhere else, were placed in this group.

Chronology of Events

B.C.
384-322   Aristotle, used Greek words for genus and species.
A.D.
1530   O. Brunfels, *Living Pictures of Herbs,* illustrated Dioscorides' descriptions of plants.
1539   J. Boch, herbal with original descriptions and a realistic system of classification.
1540 (1561)   V. Cordus, *Historia Plantarium.*
1543   L. Fuchs, German edition of herbal entitled *Neu Kreuterbuch.*
1544   P. A. Mattioli, commentary on Dioscorides.
1561   W. Turner, *Herball.*
1585   A. Cesalpino, *De Plantis,* classification based on fruit and flower characteristics.
1623   K. Bauhin, *Pinax Theatri Botanici,* binomial system.
1650   J. Bauhin, *Historia Plantarum Universalis.*
1660-1713   J. Ray, several significant books on plants and jointly with Willughby on animals.
1662-1679   J. Jung, works on plants published posthumously by his students.
1687   I. Newton, *Principia,* rigid system in nature represented key to late seventeenth and eighteenth century era.
1694   R. Camerarius, sexual reproduction in plants.
1694   J. P. de Tournefort, *Elements of Botany* (1694), *Institutiones rei herbariae* (1700); used binomial system of nomenclature.

1719	J. de Jussieu, director, Jardin des Plantes, Paris, revised Tournefort's early work.
1725	B. de Jussieu, established Royal botanical garden at Versailles, revised Tournefort's *Histoire des plantes.*
1730-1778	C. von Linnaeus, many publications in which binomial system was used for classifying plants and animals.
1749-1804	G. L. Leclerc, Comte de Buffon, *Natural History.*
1789	A. L. de Jussieu, published a modified Linnean classification system in *Genera Plantarum.*
1809	J. B. Lamarck, *Philosophical Zoology,* classification by function rather than structure.
1813	A. P. de Candolle, *Elementary Theory of Botany.*
1817	G. Cuvier, animal kingdom divided into four groups or Embranchements.

References and Readings

Arbor, A. 1938. *Herbals, Their Origin and Evolution (1470-1670).* London: Cambridge University Press.

Benson, L. D. 1957. *Plant Classification.* Boston: D. C. Heath and Co.

Berkeley, E. and D. S. Berkeley. 1963. *John Clayton, Pioneer of American Botany.* Chapel Hill: University of North Carolina Press.

Clausen, J. C. 1951. *Stages in the Evolution of Plant Species.* Ithaca, N. Y.: Cornell University Press.

Core, E. L. 1955. *Plant Taxonomy.* Englewood Cliffs, N. J.: Prentice-Hall.

Drachman, J. M. 1930. *Studies in the Literature of Natural Science.* New York: Macmillan Co.

Gage, S. H. 1964. "Microscopy in America (1830-1945)." Ed, O. W. Richards. *Trans. Amer. Mic. Soc.* 83:9-125.

Gourlie, N. 1953. *"The Prince of Botanists,"* Carl Linnaeus. *"The Prince of Botanists,"* London: H. F. and G. Witherby.

Hagberg, K. H. 1953. *Carl Linnaeus,* trans. from Swedish by A. Blair. New York: E. P. Dutton and Co.

Hawks, E. 1928. *Pioneers in Plant Study.* London: Sheldon Press.

James, W. O. 1957. "Linnaeus 1707-1778." *Endeavor* 16:107-112.

Keefe, A. M. 1966. "Our debt to the clerical botanists." *The Biologist* 48:45-61.

Klopsteg, P. E. 1960. "The indispensable tools of science." *Science* 132:1913-1922.

Lawrence, G. H. M. and K. F. Baker. 1963. *History of Botany.* Los Angeles: Clark Memorial Library. (Includes lecture by Lawrence entitled, "Herbals: Their History and Significance.")

Linnaeus, C. von. 1735. *Systema Naturae.* Lugduni Batavorum, Sweden. (Also Preface, *British Museum of Natural History,* 1956.)

_____. 1753. *Species Plantarum.* Stockholm, Sweden. (Facsimile ed., W. Junk, 1908.)

_____. 1754. *Genera Plantarum,* 5th ed. Lugduni Batavorum, Sweden. (Reprinted, London: Weinheim, 1960.)

Peattie, D. C. 1936. *Green Laurels, the Lives and Achievements of the Great Naturalists.* New York: Simon and Schuster.

Raven, C. E. 1942. *John Ray, Naturalist.* Cambridge: At the University Press.

Sachs, J. von. 1906. *History of Botany.* Oxford: Clarendon Press.

Simpson, G. G. 1961. *Principles of Animal Taxonomy.* New York: Columbia University Press.

Stearn, W. T. 1962. *Three Prefaces on Linnaeus and Robert Brown.* New York: Stechart-Hafner.

Svenson, H. K. 1945. "On the descriptive method of Linnaeus." *Rhodora* 47: 273-302, 363-388.

Tanner, V. M. 1959. "Carl Linnaeus' contributions and collections." *Great Basin Naturalist* 19:27-35.

CHAPTER 10

COLLECTORS AND COMPARERS

COMPARATIVE STUDIES of animals and plants were made by Greek, Alexandrian, Roman, and more recent investigators. Some well-known collectors and comparers of living things before the seventeenth century are listed in Table 10.1. Foundations for comparative anatomy were established by Vesalius and his followers, most of whom were located at Padua and Bologna during the sixteenth and early seventeenth century period.

Bartholomaeus Eustachius (1520-74) at Padua observed that, to understand the workmanship of nature, it is necessary to examine the anatomy of brutes as well as that of man. Another Italian anatomist, Carlo Ruini (1530-98), at Bologna made a careful study of the horse and wrote two books, one on diseases of horses and one on the anatomy of the horse (1598). The latter, published after the author's

Figure 10.1
Hieronymus Fabricius,
Italian anatomist; teacher
of William Harvey.

death, was a classic. Hieronymus Fabricius (1537-1619; Fig. 10.1) at Padua made many observations which compared structural parts of animals of different species. His broad interests led him to consider the functions of body organs and thus to compare physiological as well as anatomical characteristics. He was also a pioneer in the field of vertebrate embryology. Guilio Casserius (1561-1616) succeeded Fabricius in the chair of anatomy at Padua and became the most skillful of the early comparative anatomists. In a monograph on the larynx, Adrian Spigelius (1578-1625) compared this structure as it existed in about 20 different animals. Spigelius distinguished between human and animal anatomy. Through his influence, human anatomy became a specialized discipline for medical men quite distinct from the broad field of comparative anatomy.

The last quarter of the seventeenth century was especially productive for comparative studies. The most prominent leader in the field of anatomy in England was Edward Tyson (1650-1708; Fig. 10.2). In his

Figure 10.2
Edward Tyson, English
comparative anatomist.

Table 10.1   Some early comparers of animals and plants.

Date	Location	Contributor	Comparisons
A.D. 23-79	Rome	Pliny the Elder	Many living things.
120-200	Rome	Galen	Parts of animals.
1452-1519	Florence (Vinci)	Leonardo	Muscles and bones of man and horse.
1516-1565	Zurich	Gesner	Many animals and plants.
1517-1564	Le Mans, France	Belon	Bird and human skeletons.
1522-1605	Bologna	Aldrovandus	Many animals and plants.

early professional life he was a practicing physician and lecturer in London but he abandoned human anatomy in favor of comparative studies. His comparative work included critical studies of the female porpoise, male rattlesnake, tapeworm, roundworm (ascaris), peccary *(Dicotyles tajacu)*, and the male and female opossum *(Didelphys marsupialis)*. All through his work Tyson took every opportunity to draw parallels and compare structures among different animals.

Tyson's most significant single contribution was the anatomical study (1699) of an immature chimpanzee which he called "pygmy." This animal was not only dissected and described in great detail, but its anatomical parts were critically compared with those of man. Tyson made further comparisons with other primates and showed graduations among them. Similarities and differences among structural features of monkeys, apes, and men were pointed out and relations were suggested. Tyson was closely associated with London and Oxford scientists who were connected with the Royal Society, particularly Robert Boyle, Robert Hooke, and Christopher Wren. He contributed notes on Willughby's study on fishes, and John Ray acknowledged this assistance. Much of Tyson's work was published in *Philosophical Transactions,* but some major contributions were printed as separate monographs. Several of his unpublished manuscripts are now in the libraries of the British Museum and the Royal College of Physicians. Recognizing the value of Swammerdam's work, Tyson sponsored a partial English translation of the monograph on the mayfly.

Botanists of the Renaissance made extensive comparisons among plants, particularly those with medicinal value, and prepared a foundation for Ray, Linnaeus, and other taxonomists who based their distinctions on observable similarities and differences. Further contributions to the understanding of the comparative anatomy of plants were made by de Candolle and Lamarck in France. After the development of the microscope, the minute anatomy of both plants and animals was investigated and compared.

Comparative embryology and comparative physiology gained momentum in the nineteenth century. Special tools and techniques were required for embryological and physiological studies and therefore progress was slower in these areas than in the morphological sciences that depended, for the most part, on straightforward observations of accessible, tangible structures. Comparative morphological sequential studies of fossil plants and animals increased as more specimens were accumulated and methods of dating fossils were improved. Interest increased in comparing fossils with currently living forms of the same or related species to determine what evolutionary changes had occurred. Two prominent leaders of the eighteenth and nineteenth centuries who used comparative methods most effectively were the English surgeon, John Hunter (1728-93), and the French comparative anatomist and paleontologist, Georges Cuvier (1769-1832).

William and John Hunter

William (1718-83) and John Hunter (1728-93), sons of a Scottish farmer, were different in personality and philosophy and yet each made a notable contribution to the advancement of biological science. Both William and John were interested in human anatomy and problems of human reproduction, but William made more practical use of his knowledge by maintaining a private school where he taught anatomy and developing a practice as a physician and obstetrician. He was

Figure 10.3.   Hunter House, home of William and John Hunter, at Long Calderwood (now East Kilbride), Scotland, as it appears today. The rock part at left is a museum with books, pictures, furniture, and instruments associated with the Hunters.

characterized as one of the "most able anatomists in Europe." John became the best surgeon England had produced, but he also had broad interests and made great accomplishments in comparative studies. Both John and William were born in the "Hunter House" (Fig. 10.3), now maintained as a museum-memorial to the Hunter brothers. The home is located at Long Calderwood, about 2 miles northeast of the center of the new town of East Kilbride and about 7 miles from Glasgow.

William's Characteristics and Accomplishments

William Hunter was the seventh of 10 children born to Agnes and John Hunter. He was educated at the Parish School in Long Calderwood, now East Kilbride. At the age of 13 he enrolled at the University of Glasgow and for four years he studied humanities, Greek, logic, and philosophy. Then he was apprenticed to Dr. William Cullen in medical practice in Hamilton. At the age of 22 he went south to London where he distinguished himself and his original observations in anatomy, physiology, and pathology and exercised an important influence upon the standards of practice of medicine, obstetrics, and jurisprudence.

Through the early periods of civilization and into the time of the Hunters, midwives traditionally helped women during childbirth. At first midwives were women who had borne children themselves and were thus qualified by experience to assist their neighbors just as warriors or hunters who had been exposed to injury were considered qualified to render aid to their injured associates. As the organization of communities evolved, however, some women became professional midwives and performed the services regularly for gain. Thus the art of midwifery arose. Some midwives guarded their position jealously and effectively established a monopoly. Only in extreme cases when it was recognized that the efforts of the midwife were ineffective was a male physician invited to assist at childbirth. Mortality within the first few years of infancy was more than 50 percent. Thus it required courage and zeal for a man, even though well trained as a physician, to openly commit himself to a career in obstetrics during the eighteenth century. Nevertheless, William Hunter did just that in London. His research specialty was the anatomy of the human female genital tract and the course of pregnancy. Through his influence, infant mortality in his locality was reduced, and prejudice against scientific methods in childbirth was weakened. Despite the handicaps imposed by tradition, William Hunter moved in the best social circles and developed a lucrative practice in obstetrics, which he maintained simultaneously with his research and private instruction. He became Physician and Obstetrician to the Queen and was elected by his colleagues as President of the Society of Physicians.

In spite of professional distractions from his medical practice, William continued his research and published many research papers in *Philosophical Transactions* and other periodicals. He also continued his anatomy lectures, insisting that he could do infinitely more good by teaching his art than by practicing it. This judgment was well considered because no organized system of medical education was available in England or Scotland at this time, and, indeed, William's foremost contribution to medicine was made through his course of lectures which covered the entire field of medical knowledge. His course was attended by nearly all medical students in England and many from Scotland and as far away as the American colonies. William never married. During the last ten years of his life his health failed. In March, 1783, while giving a lecture contrary to the remonstrances of friends, he fainted and died a few days later.

John's Characteristics and Accomplishments

The youngest Hunter, John, was very different. In his childhood and youth he was allowed to run wild by his elderly father. Unsuccessful in school, and not willing to conform to social patterns in his home community, he was considered a worthless, drunken "roughneck." In his youth he discarded his wig, thus flouting good taste. He could not write; he could not read; he could not converse well and he came to despise all forms of scholarship to the embarrassment of his polished brother and other educated relatives. At the age of 17, and virtually illiterate, he went to stay with his sister in Glasgow and worked with his brother-in-law as a cabinetmaker. In this occupation he showed great skill with his hands and in the use of tools. He soon demonstrated excellence as a craftsman. He eventually did learn to read and write but never did become polished in these skills.

Three years later (1748) William invited John to be an assistant in his private anatomical laboratory. John was interested in the new opportunity and rode on horseback to London to begin the new assignment. After a brief apprenticeship, his special task was to perform dissections while William lectured about the demonstrations before his classes. John also discharged routine duties, including the cleanup and other menial tasks that had to be done in an anatomical laboratory. He was described as a pleasure-loving young man who went by the nickname of Jack and had the bad habit of swearing. John showed remarkable skill at dissection, and soon he was making many of the preparations that William otherwise would have made himself. By his second winter in London, John had acquired sufficient anatomical knowledge to be entrusted with all preparations and demonstrations in the classes. Through his knowledge and skill John won a reputation among practitioners and students as an excellent technician and in a short time he became supervisor of the laboratory.

In the summer of 1749 William obtained permission for John to study at Chelsea Military Hospital under William Cheselden, and John succeeded with distinction. When Cheselden retired in 1751, John became a student of Percival Pott at St. Bartholomew's Hospital. In 1754 he entered as a surgeon's pupil at St. George's Hospital, where he was a resident surgeon for one year. He then entered into a full partnership with William in a private anatomical school. Having been recognized as a competent surgeon, he proceeded to show more interest in basic biology than in the narrow practice of surgery.

William decided that perhaps he might yet promote a university education for John and convert him into a cultured gentleman. He provided an opportunity for John to go to Oxford. On June 5, 1755, John entered St. Mary's Hall at Oxford University. He stayed at Oxford less than one term and let it be known that he did not appreciate the type of classical training provided at that institution. Later he said to Sir Anthony Carlisle, "They wanted to make an old woman of me, trying to stuff me with Latin and Greek."

In London again, he enjoyed lively company, went often to the theatre, and moved in elite social circles. He was now well known among physicians and was recognized as an investigator in basic aspects of biology. At the anatomical school he gave lectures and became an accomplished teacher of anatomy. His originality of mind and inquiring spirit generated highly productive biological research. He outstripped his brother in contributing to basic biology and medical science. John traced the descent of the testes in the male foetus, followed the nasal and olfactory cranial nerves, studied the formation of pus, and observed the nature of placental circulation. His medical and anatomical investigations and basic biological research gained him a worldwide reputation.

In 1759, John became concerned about his own health. He had symptoms of a lung disease similar to that from which his elder brother, James, had died. His prescription for himself was a warmer climate and a change in pace in work load. He joined the military service as a surgeon. In 1760, when England was at war with Spain, the unit to which Hunter was attached was sent to protect Portugal.

No activity was experienced and John had time to observe and collect animal and plant specimens. He studied lizards, snakes, and other animals and plants, and conducted simple physiological experiments. Apparently his leisurely and interesting activities in the fresh air and sunshine of Portugal proved to be a favorable treatment for the lung condition.

In the next year John was a staff surgeon on a military expedition to Belle Isle, an island in the Atlantic at the entrance of the Strait of Belle Isle, between Labrador and Newfoundland. The armament sailed in 1761 under command of General Hodgson and Commodore Keppel. Belle Isle was sieged in a short and bloody battle. It was here that John initiated his important studies on gunshot wounds. Results of these investigations were finally published in 1794, a year after John's death, under the title, *A Treatise on Blood Inflammation and Gunshot Wounds.* John had the responsibility of practicing surgery to a considerable extent following the engagement at Belle Isle.

The state of surgery was primitive. There were no anesthetics, so surgical treatments were restricted to operations of necessity such as removing foreign objects and repairing wounds, amputating gangrenous limbs, opening abscesses, setting fractures, and cutting for bladder stones. William and John Hunter were among the first to introduce the scientific approach into medical practice. John brought about a scientific approach to surgical problems. His lectures, even more than his writings, revolutionized surgery. His main interest, however, remained in the more general aspects of biology, such as comparative anatomy, embryology, histology, and general biology.

When the Seven Years War was over in 1763, John (Fig. 10.4), now a leading surgeon, returned to London. His former place in William's school was filled and he set up private practice in an office at Golden Square. In addition to his surgical practice, he taught anatomy and surgery to his own private students. When the army awarded him half pay as a retirement benefit, and he could again indulge his private interests in collecting and comparing specimens, he returned to his studies of comparative anatomy, physiology, and embryology. Dead animals from the King's menagerie in the Tower of London were turned over to him for study. Animals were also purchased or traded from traveling wild beast shows. Specimens and demonstrations of surgical procedures were preserved in Hunter's home, which eventually grew into Hunter's museum.

John Hunter was elected a fellow in the Royal Society of London (1767) and later he became a member of the Academy of Medicine at Paris. His first contribution to *Philosophical Transactions* was an

Figure 10.4
John Hunter, English
comparative anatomist.

essay, published in 1767, on postmortem digestion of the stomach. The essay explained digestion as a result of the activity of gastric juice. In the same year John joined the Surgeons Corporation and on December 9, 1768, he was elected surgeon at St. George's Hospital. Much of his money, time, and energy went into his museum. Many specimens he provided himself, through his own collecting trips and dissections, but he also purchased numerous articles for the museum.

In spite of his fame and fortune, John never became polished in manners and behavior. He was impatient, blunt, and at times inconsiderate of others. In this respect he was different from his wife Ann Home, whom he married in 1771 when he was 43 years old. She was a sister of John's house pupil, Everard Home, and daughter of a surgeon. Ann was characterized as: beautiful, refined, and accomplished in music and art. Being a slow reader, John read comparatively little and usually could not recall and express the information already in the literature on a given subject. He had acquired most of his knowledge himself, and he took pride in his independence from books by attaching much importance to personal and independent observation and investigation. Always speaking the truth in his blunt way, he was pitiless in exposing errors, even his own. He was a true scientist and his work was some of the finest ever accomplished in the field of comparative studies.

Philosophically, John Hunter was a vitalist. In searching for the underlying principle of life, he concluded that life, even in its simplest forms, was independent of physical structure. He spoke of the "latent heat of life" as the life-giving property found in all living things.

John was a hard worker and was rigidly economical with his time. On a typical day he arose at 6:00 A.M. and dissected until breakfast at 9:00 A.M. He then visited patients and performed surgery between 9:30 and 12:00 noon. In the early afternoon he held office hours and visited more of his hospital patients. At 4:00 P.M. he had dinner, after which he rested for an hour or so. In the evening he conducted experiments and wrote notes on the day's activities. His colleagues mentioned midnight visits to his home when he was still at work.

During his lifetime he dissected and studied representatives of some 500 different species of animals, some of which he worked over repeatedly. In the course of his anatomical studies, he conducted pioneering work on the lymphatic system. From his experimentation he concluded that digestion did not occur in snakes and lizards during hibernation, and he demonstrated that enforced vigorous movement during hibernation was fatal.to the reptiles studied. From experiments on dogs, he showed that tendons will reunite after being divided, and thus he prepared the way for the modern practice of cutting tendons for treatment of distorted and contracted joints. He also made critical studies of blood coagulation.

In his search for facts he was persistent, patient, and cautious. A careful experimenter, he not only gave due attention to details, but always had a clear objective in performing an experiment. One of his favorite axioms was: "Experiments should not be often repeated which tend merely to establish a principle already known and admitted. The next step should be the application of the principle to useful purposes."

Some 50 volumes of notes were said to have accumulated during his lifetime. They were not always orderly but represented scattered ideas and comments which he put down as they occurred to him. Two of these volumes were published after his death by his brother-in-law, Everard Home. The brother-in-law, however, claimed them as his own, and when he became fearful that the deception would be discovered, he destroyed the remaining 48 volumes. The destroyed manuscripts included Hunter's 86 surgical lectures and other highly valuable materials, a serious loss to surgical and biological literature.

Hunter's preserved works are illustrated with beautiful drawings. The best series of drawings available before the nineteenth century on the embryology of the chick was prepared by John Hunter. Because he worked on the chick in the days before microscopes had reached a high level of development, he thought how much easier it would be to study embryology if the specimens were larger. This led him to work with duck eggs, and he waited 30 years for an ostrich to lay an egg for him, but found to his disappointment that the developing duck and ostrich were no larger than the developing chick.

The following titles from among John Hunter's many written contributions to medicine and biology will indicate his breadth of interest: *A Treatise on the Natural History of Human Teeth, Observation on Certain Parts of Animals, Directions for Preserving Animals and Parts of Animals for Anatomical Investigation, The Works of John Hunter, Observations and Reflections on Geology,* and *Memoranda on Vegetation.* Essays and accounts of observations on natural history, psychology, and physiology were also included in his bibliography.

In January 1780, John read a paper before the Royal Society on the structure of the human placenta. He claimed in this paper the discovery of certain features of the utero-placental circulation which his brother, William, had previously reported. This led to a dispute between the two brothers which continued until William died. John had made many dissections in William's laboratory for which William had received credit. In this particular case it was difficult to determine which of the brothers should have credit for the contribution. The Royal Society, however, printed the paper of John.

John Hunter's death in 1793 was due indirectly to an experiment which he performed on himself beginning in 1767, when he was involved in a controversy concerning the nature of the venereal diseases, syphilis and gonorrhea. He thought they were different symptoms of the same disease, and in order to demonstrate the point and observe the course of the disease, he inoculated himself with the pus of a syphilitic patient. His theory was that a "virus" was responsible for the combined symptoms of both diseases; that the virus when placed on a secreting surface, i.e., a mucous membrane, would produce the symptoms of gonorrhea, but when the same virus was placed on a nonsecreting surface, such as the exterior of the skin, the symptoms of syphilis would occur. The causative organism in either case was believed to be the same, and the manifestation of the disease was supposed to depend on where the organism became located in the body.

The pus with which Hunter inoculated himself evidently was from a patient who had both syphilis and gonorrhea. The symptoms that John studied in himself were characteristic of both diseases. Therefore, he concluded that the two diseases were different manifestations of the same thing, as he had previously postulated. His report, entitled *Treatise on the Venereal Diseases*, was printed in his own publishing house in 1786. Without the aid of a bookseller it sold 1,000 copies in a year. His erroneous conclusion misdirected the study of syphilis for more than half a century after his death. Although he had made a careful study of syphilis, he did not know some of its most serious effects; for example, that it was a disease of the blood vessels as well as the epithelial tissues. He died October 22, 1793, following a dispute with a colleague at a board meeting of the Royal Society. The immediate cause of death was apoplexy; but, more correctly stated, the cause of death was untreated syphilis.

Hunterian Museums

Two major Hunterian museums now house collections of William and John Hunter. In the main building at the University of Glasgow, a museum is maintained in honor of the Hunter brothers. This museum includes mostly items belonging to William Hunter. Another more elaborate museum now located in the main building, second floor, of the Royal College of Physicians and Surgeons in London includes the extensive collections of John Hunter.

Hunterian Museum, Glasgow

During his lifetime, William Hunter built a library of books and manuscripts. He also collected and preserved an adequate supply of subjects for dissection and research by his students. These formed the nucleus of his museum. To his purely medical collections he added zoological, botanical, and geological specimens, ethnographical and archaeological material, coins, and paintings. The ethnographical material included carved wooden idols and clubs acquired by Hunter from Captain James Cook.

By his will, William provided that the collections be eventually turned over to the College of Glasgow. Appropriate facilities were required to house the specimens. In 1807 the building of the first Hunterian Museum was completed, and most of the collections were shipped by sea from London to Glasgow. The cabinet of ancient coins, one of the most valuable coin collections in the world, was transported overland escorted by "six trusty men, accustomed to the use of arms." This Hunterian Museum, now housed in a prominent part of the main building, University of Glasgow, contains William Hunter's wax model of the human body, surgical instruments and laboratory equipment, many art treasures, and furniture from the original Hunter home. A portrait of William Hunter by Allan Ramsey is in the room with Hunter relics. Many collections of various objects have been brought together from other museums and additional objects have been presented by other donors. Vaccination lancets used by Edward Jenner, student of both William and John Hunter, and the seal of Edward Jenner are a part of the collection at Glasgow.

Hunterian Museum, London

John had purchased (1764) ten acres of land and had built a country home at Earl's Court, Kensington. He provided space for caged live experimental animals and laboratory facilities in his home. Bees had become especially interesting to him, and he kept several hives. His house soon proved too small, and he built an addition where he kept his increasing numbers of specimens and pets. He subsequently invited his best students to live in his enlarged home; Edward Jenner (1749-1823; Fig. 10.5) was one of the students to whom Hunter became strongly attached and he came to live with John in 1770. The home museum grew, and John, recognizing its value as a teaching medium, preserved and displayed many dissected forms from which comparisons could be made. In 1783 John purchased a home and more land on the east side of Leicester Fields in London. Here he built a new home for his museum of comparative and pathological anatomy with a lecture theater for classes and meetings. He also bought a house on Castle Street for his work on human and comparative anatomy. The country home property at Earl's Court was used for live animals and for other projects in connection with his research and growing museum.

The museum was designed to illustrate the entire phenomena of life including all organisms in health and disease. John had intended to provide a catalogue of his demonstrations in each department of the museum, but this was not completed in his lifetime. He did prepare notes that were of great value to other anatomists on matters pertaining to dissection, preservation, and embalming. At the time of John Hunter's death, more than 10,000 preparations were in the museum, representing all areas of science that had been of interest to him. The museum was especially noted for the logical and natural arrangement of specimens. Groups of animals were arranged in proper taxonomic order, and anatomical demonstrations were in

Figure 10.5
Edward Jenner, English
physician who pioneered
in the discovery of
smallpox vaccination.

logical sequence. Thus the museum served as a learning aid for serious students and also satisfied the curiosity of casual observers.

John had a menagerie of animals. He never missed an opportunity to secure the body of a rare beast for preservation and/or dissection. He was interested in fossils and all sorts of other objects of interest from nature. One of his most prized and most expensive acquisitions was the skeleton of Charles Byrne, who was 7 feet 7 inches tall. The giant had at one time been exhibited in London under the title, "O'Brian, the Irish Giant." British newspapers had given the exhibit wide publicity and it became a well-known attraction for Londoners. Hunter became interested in procuring the skeleton for a museum specimen. Some advanced plans were made through agents while Byrne was alive but appeared to Hunter to be in declining health. Byrne learned of the exchanges, and as he had a horror of being dissected made other plans to be sure his body would not fall into the hands of the doctors. Several stories have been recorded about what happened at the time of Byrne's death in 1783 although the details of these stories do not agree. It seems clear that through fees to some and bribes to others along with eloquent persuasion, the skeleton became a museum specimen.

According to the directions of John's will, the museums and specimens were offered for sale to the British Government. The offer came at an unfortunate time during the French Revolution when the French had declared war on England. When William Pitt, the Prime Minister, was informed of the matter, he said, "What! Buy preparations? Why, I have not enough money to purchase gun powder." The vast collection remained locked up for six years under the supervision of a caretaker, William Clift (1775-1849). Finally in 1799, six years after John's death, influential people who realized the value of the collection persuaded the government to purchase the museum. By Act of Parliament, the purchase was made and the vast collection was entrusted to the Master Governors and Commonalty of the Arts and Sciences of Surgeons of London, which later became the Royal College of Surgeons of England. The purchase price was $75,000, which was about one-fifth of the total sum which John had actually spent in the collection of materials and the preparation of the museum.

The museum is now maintained by the British Government at the Royal College of Surgeons in London. According to the terms of John's will, the collection must be open at least four hours in the afternoon two days each week for inspection by Fellows of the College of Physicians, members of the Corporation of Surgeons, and persons introduced by members of these groups. A catalogue of the exhibits must be available, and an official must be in attendance to explain the displays. All of these conditions have been fully met except in times of extreme emergency. The greatest disaster to the museum occurred in May 1941 when the College was severely damaged by bombing. The museum suffered great damage and much of the collection was destroyed. Rebuilding began in 1952 and the present Hunterian Museum, completed in 1962, was again available to serve its original purpose. Specimens with black labels are original and those with red labels are those now included to restore the logical sequence of Hunter's ideas.

John Hunter greatly influenced the development and general aspects of biology, and, in particular, comparative anatomy, through his own contributions, those of his students, and his museum. Eminent comparative anatomists such as Georges Cuvier and Richard Owen made use of the preparations in the Hunterian Museum. For his personal achievements, Hunter won the Copley medal and other distinguished awards. His publications brought him lasting recognition.

In his approach to comparative anatomy, man was the central object and other types were compared to man. This, however, is not true comparative anatomy. Evolution had to be considered as a factor. In this respect some of his predecessors such as Edward Tyson were ahead of him.

Comparative Studies in France

Comparative anatomy developed to a high level in France during the revolutionary era of 1790 to 1815, with Georges Cuvier (1769-1832; Fig. 9.5) as its leader. From an early age, Cuvier showed a great interest in natural history. He read books and studied illustrations of Buffon, Lamarck, and others, and collected, dissected and made drawings of seashore animals. A renowned agricultural investigator, Alexander Henri Tessier, saw Cuvier's drawings of starfish, molluscs, and marine worms and sent them to the French zoologist, Etienne Geoffroy Saint-Hilaire. Impressed with the interest and brilliance of Cuvier, Saint-Hilaire persuaded Cuvier to come to Paris where he was employed successively at the central school, College of France, and as Professor of Comparative Anatomy at the Jardin des Plantes. This latter appointment was remarkable for a man who had never dissected a human body. Within a short time, however, Cuvier was recognized as the greatest anatomist in France.

Explorations into distant and previously unknown parts of the world had made vast collections of animals available in Paris for study and comparison. Cuvier took advantage of the wealth of specimens and worked out a systematic organization of the animal kingdom (Chapter 9). His comparisons of living animals led to an interest in fossils, and he worked out a theory of catastrophism (Chapter 13) which accommodated his religious views with his observations on the succession of different but related animal forms on the earth over a period of time. He became a distinguished leader in the field of vertebrate paleontology as well as in comparative anatomy and natural history.

One of Cuvier's first contributions to comparative anatomy was on fossil elephants and their relation with living elephants. Cuvier had made a systematic study of all living animal groups and also an exacting study of elephants. During this period many fossils were being discovered and some fossilized bones of an elephant were given to Cuvier. His knowledge of elephant bones was so great that he was able to reconstruct an entire elephant from a few fragments of bone.

To appreciate the contribution of Cuvier in comparative anatomy, it is necessary to recall that the study of human anatomy, and that of comparative anatomy of animals in general had been completely separated by Spigelius, who followed Vesalius at the University of Padua. Spigelius did not recognize the practical importance of studying the two together. Between the time of Spigelius and Cuvier, John Hunter in England was the only one to realize the values to be derived from studying human anatomy as a part of comparative anatomy. With his knowledge of comparative anatomy of all animal groups, Cuvier decided he should systematize and classify the entire animal kingdom, as some of his predecessors, particularly Linnaeus, had done. But whereas Linnaeus had classified animals according to their external features, Cuvier developed his classification according to the structure and relationships of their internal parts. He thus became the first to use comparative anatomy in a system of classification. In this project Cuvier seems to have been guided by an idea of Aristotle that a function or need existed for every part of each animal. From his findings he published *Natural History of Animals* (1798), which included his plan for dividing the animal kingdom into four great divisions (Chapter 9).

In dividing the animal kingdom in this way Cuvier was again guided by an idea of Aristotle, that the animal kingdom was a graded series based on the level of "vegetative" and "animal soul." Cuvier recognized two main sets of animal functions: heart and circulation, which were thought to represent the center for "vegetative functions" (growth, reproduction); and the brain and spinal cord, which were thought to direct the "animal functions" (active movement, muscular system).

A quotation from Cuvier's book shows how Cuvier used comparative anatomy in his work on classification.

> As no living organism is entirely simple in its character, it is insufficient to try to distinguish it merely by one single feature of its make-up. We must therefore, nearly always, rely on a combination of several distinct features if we wish to differentiate any particular individual from its fellows, which later may in part, if not entirely, through their very similarity show traits pertaining to the first specimen. In a like manner, such original sample may present characters common to itself as well as to all the others and will then be distinguishable only through those features not possessed by the remainder. Some beings are very closely related to one another in the biological sense and possess many characters in common, whilst others show far less a degree of similarity in detail. The larger the quantity of beings requiring identification, the greater the number of characters will it be necessary to take into

consideration, even perhaps down to the point of making a complete and exhaustive specification of every single detail of the living form. (Cuvier 1817)

Cuvier's classification system was used for many years after this time and much of existing modern classifications can be traced to Cuvier and through him to Aristotle.

Another Cuvier theory that can be traced to Aristotelian principles is the theory of the correlation of parts. Cuvier was not content to study the various parts of the animal body independently; he wanted to know what relations existed between them. He could see a close relation between the internal and external structures of an animal and a correlation between structures and the animal's habits of life. Organs, he observed, could not function alone and were useless when isolated; they were only parts of a whole being. This indicated that even though whole organisms functioned in a complicated way, it was also a recognizable and even a predetermined way. If the correlations were understood, any fragment of any part of any species could be recognized by careful observation. Carnivorous animals, for example, would have good vision, power for rapid motion, claws for catching and holding prey, and sharp teeth placed in certain positions in the jaw for best tearing action, whereas herbivorous animals would not have aggressive power, sharp teeth or claws, but would have hoofs and appropriate teeth characteristics for feeding on plants. Cuvier was also able to show that the internal organs were constructed differently for carnivorous and herbivorous animals because of the different types of food utilized by each.

Cuvier was so thoroughly convinced of the validity of his theory of correlation of parts that he thought a single representative bone was sufficient to indicate what a particular animal was like. He concluded that certain kinds of teeth were always correlated (on the same animal) with hoofs. One whole group of extinct animals (chalicotheres) is now known to have teeth of the type which *must* (according to Cuvier) be associated with hoofs, but the animals actually had large claws rather than hoofs. Adaptation and natural selection would now explain the observations for which Cuvier invoked a prearranged pattern.

Cuvier's interest in prehistoric reptiles was stimulated when a skull with a crocodilian appearance was sent to the Jardin des Plantes by French soldiers during the French Revolution. It had been discovered some years before in a quarry near Maastricht in the Netherlands under land belonging to a Dr. Goddin. Cuvier identified it as the skull of a giant lizard which was later found to be a relative of the modern genus, *Veranus*. Another strange fossil, described by a Florentine named Collini in the early 1800s, had reptilian characteristics except for its wings. Cuvier tried to obtain the fossil for study but was unsuccessful. Collini's published drawings enabled Cuvier to identify it as a flying reptile which he named pterodactyl (Gr. *pteron*, wing; *daktylos*, finger).

Besides his accomplishments in science, Cuvier became an eminent politician and statesman. Since he was a good organizer and prodigious worker, he was able to act simultaneously as a scientist and political servant. To save time and facilitate his work, he fitted out a separate desk with writing material, manuscripts, and reference books for each project upon which he was working at a given time. In moving from one project to another in the course of a day's work, he could proceed immediately without wasting time in shuffling papers and gathering equipment. This precision and organization characterized all of his works.

One of the controversies generated by Cuvier's theory of correlation of parts that attracted international interest was carried on with the American statesman and scholar, Thomas Jefferson. Jefferson, like Benjamin Franklin, had become closely associated with French diplomats and scientists of the period. Jefferson associated a huge fossil claw (Megalonyx) with a prehistoric lion. Cuvier worked out a systematic correlation of parts with functions and identified it with an animal related to the modern sloth, but much larger and living on the ground instead of hanging upside down from limbs of trees.

Cuvier brought his observations and theories concerning fossils together in a book, entitled in English translation *Research on Fossil Bones* (1825), which formed the foundation of paleontology. This work had developed from earlier beginnings such as his work on elephants which showed the relations between fossil and living forms. Cuvier did not realize the great lengths of the geological time periods, and his view of catastrophism was not compatible with evolution. He was, however, the leader of the period in the fields of comparative anatomy and vertebrate paleontology.

In philosophical viewpoint, Cuvier was ahead of John Hunter and Richard Owen because he visualized a more natural plan of creation and a natural explanation for fossils. His premature hypothesis of catastrophism, proposed and established prior to wide acceptance of the evolution concept, was better than no hypothesis at all. Unfortunately, his great influence and brilliance as an orator, writer, and illustrator caused his mistakes to be perpetuated for generations after he was dead.

Comparative Anatomy in Nineteenth Century England

Richard Owen (1804-92), an assistant and later a conservator in the Hunterian Museum, carried on the Cuvier tradition in comparative anatomy and became known as the "Cuvier of England." After leaving grammar school at the age of 16, Owen became an apprentice in surgery, studied medicine for a year at the University of Edinburgh, and went on to study with John Abernethy at St. Bartholomew's Hospital, London. When Abernethy was elected President of the Royal College of Surgeons (1826), Owen was employed to assist William Clift and to take over preparation of the catalogue of the physiological series in the Hunterian Museum, which became a five-volume monumental work (1833-38). Soon after this appointment, Owen gained a lectureship in comparative anatomy at St. Bartholomew's Hospital and carried on a medical and surgical practice. He married Caroline Clift, daughter of William Clift, first conservator of the Hunterian Museum.

Owen used the collections in the Hunterian Museum and in the Natural History Section of the British Museum, where he became director (1856), made investigations and dissections, and became distinguished as a comparative anatomist. In 1836 he gave the series of lectures on comparative anatomy required by the will of John Hunter. This was followed by many other series of lectures during the eighteen years he served as Hunterian Professor. His reputation grew as his publications continued to appear. Honors were bestowed on him, including the Royal and Copley medals of the Royal Society, honorary degrees from Edinburgh and Oxford, and positions of recognition and honor in scientific societies. During the time Owen served as Conservator of the Hunterian Museum, more than 300 publications were credited to him. These cover a wide range of subjects from the minute structure and development of teeth to memoirs on prehistoric animals. His great thrust was on the anatomy and physiology of vertebrates. He published (1858) a book entitled *On the Classification and Geographical Distribution of the Mammalia*. Owen was also interested in paleontology and published the book entitled *Paleontology or a Systematic Summary of Extinct Animals and Their Geological Relations* (1860).

Owen was interested in evolution and was at first favorably disposed to Darwin's theory (Chapter 13). He even claimed that he had to some extent anticipated Darwin's theory in his own early writings. Owen could not, however, accept Darwin's premise that man is involved in biological evolution. In one of his early papers prepared as a lecture, Owen asserted that man was clearly marked off from all other animals by the anatomical structure of his brain. In his later writings Owen took a firm position against Darwinian evolution. His place as the leading comparative anatomist of the period, however, is well established.

Summary

Comparative studies began in the early history of biology and were used effectively in classical investigations of Vesalius and Harvey. The comparative method, however, did not receive real impetus until the latter part of the seventeenth century when Tyson extended his work in human anatomy to comparative anatomy and made a critical anatomical study of an immature chimpanzee for comparison with structures of man. John Hunter dissected, studied, and compared representatives of some 500 species of animals, using man as the central focus in all comparisons. He arranged specimens in his museum to facilitate serious study of comparative anatomy, surgery, and comparative physiology. The French comparative anatomist and paleontologist, Cuvier, worked out a systematic organization of the animal kingdom and developed a theory on catastrophism to account for fossils. He visualized a more natural plan of creation than did his contemporaries. Cuvier approached comparative anatomy using his theory of correlation of parts. Richard Owen, one time assistant in the Hunter museum, carried out significant work on comparative anatomy and physiology of vertebrates.

References and Readings

Atkinson, D. T. 1956. *Magic, Myth, and Medicine.* Cleveland: World Publishing Co. (Chapter 19 on John Hunter)

Cole, F. J. 1944. *A History of Comparative Anatomy, From Aristotle to the Eighteenth Century.* London: Macmillan and Co.

Coleman, W. 1964. *Georges Cuvier, Zoologist.* Cambridge, Mass.: Harvard University Press.

Cuvier, G. 1817. *The Animal Kingdom Arranged According to its Organization.* Paris: Deterville.

Dobson, J., ed. 1969. *John Hunter*. Edinburgh: E. and S. Livingstone.

———. 1970. *A Descriptive Catalog of the Physiological Series in the Hunterian Museum of the Royal College of Surgeons of England*. Edinburgh: E. and S. Livingstone.

Fuchs, L. 1542. *De Historia Stirpium*. Basel: Isingrin. German ed. 1543, *Neu Kreuterbuch*.

Gloyne, S. R. 1950. *John Hunter*. Edinburgh: E. and S. Livingstone.

Gross, S. K. 1881. *John Hunter and his Pupils*. Philadelphia: Presley Blakiston.

Illingworth, C. 1969. *The Story of William Hunter*. Edinburgh: E. and S. Livingstone.

Kobler, J. 1960. *The Reluctant Surgeon: The Life of John Hunter*. London: Heinemann.

LeFan, U. 1960. *A Catalog of the Portraits and Other Paintings*. Edinburgh: E. and S. Livingstone.

Mattioli, P. A. 1568. *Dioscurides Codex Vendobonensis Medicus Graecus*. Akademische Druck u. Verlag Sanstalt, Graz, Austria (1970).

Oppenheimer, J. M. 1946. *New Aspects of John and William Hunter*. New York: Henry Schuman.

Ottley, D. 1837. *Life of John Hunter*. (With Palmer's edition of *Hunter's Complete Works*, 4 vols.)

Owen, R. 1882. *Experimental Physiology*. London: Longmans, Green and Co.

Paget, S. 1847. *John Hunter: Man of Science and Surgeon, 1728-1793*. London: T. Fisher Unwin.

Rogers, G. 1956. *Lancet*. New York: G. P. Putnam's Sons. (Historical novel about John and William Hunter.)

———. 1957. *Brothers Surgeons*. London: Putnam.

Taylor, T. 1874. *Leicester Square; Its Associations and its Worthies*. London: Bickers and Sons.

Teacher, J. H. 1900. *Catalogue of the Anatomical and Pathological Preparations of Dr. William Hunter in the Hunterian Museum*. Glasgow: J. MacLehose and Sons.

Tyson, E. 1680. *Anatomy of a Porpess Dissected at Greshem College*. London: Benj. Cooke.

———. 1699. *Anatomy of a Pygmie Compared with that of a Monkey, an Ape, and a Man*. London: Thomas Bennet.

———. 1751. *The Anatomy of a Pygmie Compared with that of a Monkey, an Ape and a Man with an Essay Concerning the Pygmies, etc. of the Ancients*. London: T. Osborne.

CHAPTER 11

OBSERVATIONS ON REPRODUCTION AND DEVELOPMENT

Early descriptions of reproduction and development were based on supernatural interpretations. Natural explanations finally came with the observations and reasoning of Greek philosopher-scientists. Three books (Regimen) dealing with obstetrics and gynecology originated from the Hippocratic school. In these books the four humors were discussed in relation to the formation of human embryos and suggestions were made concerning methods to be employed by the physician at childbirth. Aristotle (Chapter 3) used the Greek word "embryon" (embryo) in his general treatise on embryology (De Generatione Animalium) which was the first and remains one of the great classics on the subject. Some problems described by Aristotle remained unsolved for some 2,000 years. He questioned, for example, whether the embryo was preformed and therefore only enlarged during development or whether it actually differentiated from a formless beginning to a complex, organized individual. Aristotle himself believed the egg to be composed of undifferentiated material which after fertilization became organized as it developed to the adult stage.

Two schools of thought (Table 11.1), each centered around one of Aristotle's alternative proposals regarding mode of development, were in conflict as attempts were made to trace the tangible changes that occur in the embryo. Individuals ascribing to preformation believed that each egg contained a miniature individual that required only a suitable environment for growth to the adult stage. The opposing, epigenesis school, held that the egg was undifferentiated, and a step-by-step process resulted in succeeding stages of development.

Table 11.1   Chief proponents of preformation and epigenesis

Preformation			Epigenesis		
Contributor	Date	View or Data Source	Contributor	Date	View or Data Source
Joseph	1626	Miniature chick in egg.	Aristotle	B.C. 384-322	Marine organisms.
Power	1664	Complete heart in chick egg.	Harvey	A.D. 1651	Development from undifferentiated egg.
Swammerdam	1666	Frog development.	Wolff	1786	Chick.
Malebranche	1674	Box within box theory.			
Leeuwenhoek	1677	Observed sperm, animalculist.			
Hartsoeker	1694	Homunculus idea.			
Bonnet	1745	Aphid, ovist.			
von Haller	1758	Chick, ovist.			

Preformation

Preformation, as a philosophical concept, had its roots in antiquity. It is essential to the usual interpretation of the biblical scheme of creation. Empedocles, Plato, and the Church Fathers all regarded it as a part of their own systems. Aristotle considered the male semen to be nothing but fluid and therefore was against the idea of the animal existing ready-made in the semen, but he lacked proof for his belief. Had he known of spermatozoa in the semen he might have been a preformationist.

Joseph of Aromatari was probably the first seventeenth century writer to claim that the rudiment of the chick embryo was actually visible in the egg before incubation. In a letter dated October 31, 1626, Joseph said briefly that the chick is fashioned in the egg before the egg is incubated by the hen. Henry Power, in 1664, supported this view and stated that as soon as the pulsating particle appears in the chick, the microscope most distinctly shows it to be a complete heart with both auricles and ventricles. He believed that the complete circulatory system was fully developed on the second day of incubation and held that it was not discernible merely because the circulating liquid was colorless and not yet converted by the heart into red blood. Jan Swammerdam made observations and prepared drawings of early frog development from which he stated (in 1666) a theory of preformation.

The basic assumption of the preformation theory was that the early embryo had a complete set of all organs, but they could not be seen because they were very small and/or liquid. The developmental process was assumed to consist of a gradual solidification of each part accompanied by an increase in size. Furthermore, the seventeenth-century naturalists suggested that the total preformed germs of all plants and animals were formed by special creation within the original parents of each species. No new organisms were created but preformed individuals were merely unfolded generation after generation.

Writing in 1669, Swammerdam expressed a belief in preformation, and used this theory to explain certain passages in the Bible. He had just discovered that the larva of a butterfly was present in the "egg" (chrysalis) and he promoted this observation as an indication that all animals are preformed. In the development of insects generally, Swammerdam observed an unfolding process from stage to stage, or instar to instar, and he visualized the process of embryology as a simple enlargement from minute but preformed organisms to adults. The French philosopher, Nicolas de Malebranche, in 1674, developed Swammerdam's thesis into a conception of an endless series of embryos, each encased in the others like a nest of boxes. Swammerdam had suggested further that the whole human race had been comprehended in the loins of Adam and Eve and consequently the race would be faced with extinction when the original supply of "germs" was exhausted.

Marcello Malpighi (1682; Fig. 8.3), less philosophical and more scientific than his contemporaries, made accurate microscopic observations of chick embryos at various developmental stages and set them forth with terse descriptions and clear sketches. Malpighi described development in broad outline as occurring gradually, but he was concerned about some details of the process. He did not believe all parts were formed gradually; the heart, for example, he supposed to exist fully formed from the beginning, but he observed that it did not begin to beat until the 38-40 hour stage. His theoretical views in some instances seemed to be opposed to what he might have readily observed in his own investigations. This may be explained partly by his inability to observe the early stages of chick embryology with his optical equipment. He saw nothing during the first 24 hours of incubation. On one occasion he observed an unincubated egg (that had been in the sun) and found an embryo already developing. This and his lack of observations of early stages led him to believe that parts of the embryo were preformed although philosophically he was an epigenecist. Malpighi's two papers on chick embryology were entitled, in English translation, *On the Formation of the Chick in the Egg* (1673) and *Observations on the Incubated Egg* (1675).

Preformation was generally accepted from about 1675 until the end of the eighteenth century. Only one parent could be custodian of preformed germs. Attention was at first drawn to the female contribution whereas that of the male remained obscure and most naturalists of the period assumed that the female contributed the germ and the male semen initiated the germ's growth. When sperm were observed in semen some naturalists took the position that the male contributed the preformed individual and the mother only provided nutrients and an appropriate environment for development. During this period the main question was whether the miniature organism was in the egg or in the sperm and the two schools of thought, ovists and animalculists, developed.

Ovists versus Animalculists

Albrecht von Haller (1708-77; Fig. 11.1) and Charles Bonnet (1720-93), foremost among the ovists, contended that miniature but complete organisms were contained in eggs. They presumed that the eggs were all created together and had existed since the beginning of the world. Generation, in their interpretation, consisted of two distinct processes: (1) the production by the female under the influence of the male, of an "ethereal ferment," that reacted on the dormant and miniature foetus and prepared it for expansion; and (2) the generation

Figure 11.1
Albrecht von Haller,
Swiss physiologist
and embryologist.

process proper, or the nutrition and development of the animated foetus.

In 1758, von Haller completed his investigation on the development of the chick. He concluded that it was then almost demonstrable that the embryo could be found in the egg, and that the mother contained in her ovary all the essentials of the foetus. According to von Haller, one of the most powerful arguments in favor of the ovist, as opposed to the animalculist theory, is the fact, first demonstrated for Aphididae by Leeuwenhoek (1695), and afterwards confirmed in detail by Bonnet (1745), that an unfertilized egg can develop into a perfect individual. It is now known that aphids are atypical in their ability to reproduce by parthenogenesis, that is, during certain seasons of the year female aphids reproduce without fertilization.

Bonnet used his own observations on parthenogenesis of aphids as evidence for ovism and later endeavored to explain the phenomena of regeneration on the basis of preformation. Germs of higher animals, he said, are confined to definite organs of the body. In worms and polyps, however, the germs are scattered throughout the tissue generally, and hence they are able to reproduce the organism at any point, and to replace lost organs. When the question was asked, "Why should the animal reproduce a head at the front end only, and a tail at the back end only?" he fit the observations to his theory by multiplying or reducing the germs in various regions of the organism. The theory became complex, indeed. Bonnet's explanation of the mechanism was stated concisely: "It is well known that the eggs of virgin hens grow, and it is now demonstrable that the germs preexist in them. Hence the germ grows also, but it encloses others which grow with it and through it." According to Bonnet, the germs had a preexistence and were indestructible. Resurrection of the body was explained by assuming it to be composed of an essential, imperishable basis or framework that was unaffected by development or by death.

Leeuwenhoek, an animalculist, was the first to direct attention to the importance of spermatozoa in animal development. His views on preformation, however, were not clear and often appear to have been inconsistent. He once pointed out that insect larvae are not insects, although insects proceed from them; and likewise, spermatozoa are not children, although children proceed from them. He would not admit that a frog's egg contained an immature frog, but insisted that the parts of the embryo appeared step by step in the fertilized egg. He noted that the frog embryo does not resemble a frog. After fertilization the frog must be locked up in a fertilized egg, but it was not present in the unfertilized egg nor in the form of a miniature frog in the sperm. Leeuwenhoek did not describe homunculi in spermatozoa, but philosophically he was a preformationist and an animalculist; at least he professed belief in an intangible preformation in the sperm. A list of animalculists could be compiled, most of whom based their views on philosophical speculation and linked their theories with theologi-

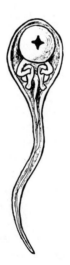

Figure 11.2
Sketch of homunculus,
miniature individual
imagined to be preformed
in the sperm cell
(after Hartsoeker).

cal dogma. Curious sketches and elaborate descriptions of the homunculus (Fig. 11.2) were recorded during the seventeenth and eighteenth centuries by Niklaas Hartsoeker (1656-1725), Dutch histologist, and others.

Experimental Evidence

The preformation theory could be neither confirmed nor refuted by direct observation because the "germ" could not be seen in either sperm or egg. At this stage of the miniature organism's life, the parts were considered to be too small or too liquid to be observed. The fact that fixation techniques such as hardening the embryo with alcohol made it possible to see microscopically the parts of the embryo at a slightly earlier stage than in the living form seemed to add indirect supporting evidence in favor of preformation. The theory of preformation was tied in with spontaneous generation. If the germs for all subsequent generations were encapsulated within the sex organs of the first individuals of each species, no place was left for spontaneous generation. Redi's experimental proof against spontaneous generation of flies (Chapter 7) and the independent reports of Malpighi and Swammerdam that insects found in plant galls developed from eggs deposited in the plants by adult flies led to further denial by naturalists of all forms of spontaneous generation.

In the next century, Etienne Saint-Hilaire incubated chicken eggs for three days under normal conditions and then subjected them to abnormal treatment such as shaking, perforation, and coating the shell with varnish or wax. Monstrosities of various kinds were produced. He regarded this as experimental evidence against preformation, but others argued that the organism could be preformed and still become abnormal because of injury.

Karl von Baer's celebrated work, *Developmental History of Animals* (1828), added the final stroke against the old preformation doctrine. He observed eggs and actual developmental stages and showed that differentiation occurred progressively. The wealth of careful, sound observations contained in his work reduced to negligible proportions the rhetorical and argumentative methods of the preformationists.

A more modern interpretation of preformation has been developed recently by geneticists in connection with the gene theory. They suggest that development is the expansion of an organism's potential that is preformed and embodied in the gene complement of the individual. It is evident, however, that organisms are more than bundles of genes. When interactions among gene products are incorporated into the theory, a more tenable interpretation of preformation is derived. The germ cell is not a simple, unorganized unit but a highly complex microcosm, the architecture of which can, in a measure, be deduced from its reactions during development.

Epigenesis

The theory of epigenesis was based on the premise that each organism arose gradually from undifferentiated material in the egg and was entirely new. Aristotle originated the epigenesis theory from his observations on the chick, but his main interest was in the philosophical implications. If all parts were made by the fluid semen, he considered it evident that no part could exist in it from the first. Aristotle assumed that a vital external force directed the organization of living material.

Epigenesis as a theory is also associated with the work of Harvey, but many of Harvey's ideas and opinions were taken directly from Aristotle. Harvey spoke of the egg in terms of its proceeding from the male and female, being endowed with qualities of both, and hence producing a foetus that resembled both parents. He did not support preformation, which he considered inconsistent with true generation, but he suggested that the fertilized egg had the potential to produce all parts of the embryo. Harvey is believed to have coined the word "epigenesis" to refer to the process that derives the embryo from an apparently undifferentiated and homogeneous egg by a gradual process of differentiation and growth. His observations were made on the chick and other animals without the aid of the microscope (Chapter 6). Harvey attempted to explain generation by assuming the existence of a First Cause or Generative Principle, with the power of initiating growth. His most positive suggestion was that all living things started from eggs.

Notwithstanding the powerful combination of Aristotle and Harvey, epigenesis found no seventeenth century supporters. Casper Friedrich Wolff (1738-94), a century after Harvey, became the most celebrated proponent of epigenesis. Interestingly, as the basis of his work on epigenesis, he used the same sort of chick material from which Malpighi had found some evidence for preformation. Wolff found Malpighi's account of the embryology of the chick to be well adapted for illustrating development of epigenesis. He was able to observe microscopically the building up of the chick embryo and he saw no evidence of an encapsuled chick in the egg. Hence he found nothing to substantiate the idea of enlargement of a preexisting miniature. Rather, he postulated a continuous growth process and gradual development toward a more complex form. Wolff had already made excellent observations of the development of plants, particularly the flower parts. In his study of the chick he demonstrated the precise development of the intestine and other internal structures. The primitive kidney or Wolffian body bears his name. The great service that Wolff performed for embryologists earned him no immediate followers. His thesis did not persuade his fellow scientists because he was ahead of his time and was utilizing modern methods of biological research. His contemporaries attached more importance to abstract reasoning than to observation.

The eighteenth century history of embryology, and particularly the significance of the epigenesis theory, might have been different if John Hunter's records had been available, but much of Hunter's work had been lost entirely to succeeding generations. Fortunately, his detailed sketches of chick embryology (prepared between 1773 and 1780) were preserved, but they were not available until 1840. Besides the studies on chick embryology, Hunter's critical studies of the development of insects, which might have modified the preformation-epigenesis problem, had been partly lost and partly overlooked. Hunter had proposed three possible explanations for development: preformation, metamorphosis, and a modified form of epigenesis, in which the structural parts were present at the beginning but were altered in form and function as development proceeded. Embryology did not come into its own as a science until the nineteenth century.

Early Theories of Generation

Several theories of procreation were coexistent with the epigenesis vs. preformation controversy during the seventeenth century. Some of them influenced the biological thinking of the period, but all are obsolete now and have only historical significance. Four of the more important theories, pangenesis, precipitation, seminism, and pan-spermy, will be identified.

Pangenesis, the oldest theory of generation, postulates minute granules that develop in particular areas of the body and represent the parts in which they originate. At the time of sexual maturity they come together in the reproductive organs and carry information that is transmitted in inheritance. The whole organism would thus take part in the generative act. Nathaniel Highmore (1613-85), English physician and anatomist, and one of the first authors to produce a reasoned theory of pangenesis, published a work on generation (1651) in which he speculated that the genital organs collect atoms corresponding to every part of the body. These undergo a concentration process in the gonad and develop into germs. Pangenesis also had an evolutionary aspect because it suggested a means by which animals become different from others of their kind and eventually give rise to new types. The theory was resuscitated by Charles Darwin in *The Variation of Animals and Plants under Domestication,* Volume 2 (1868), to account for the origin of variation. He theorized that body cells secrete minute corpuscles or "gemmules" that record growth patterns for the area they represent. These gemmules were believed to be carried by body fluids and the bloodstream to the reproductive organs and there packed into the eggs or sperm. In the new individual they were considered to determine the visible characteristics and the growth pattern.

The theory of precipitation was based on the notion that the embryo is formed suddenly at the moment of fecundation, by precipitation of materials already present in the ovum. Seminism was based on the idea that the generative principle resides in the male and female "semen." The adherents of seminism, however, did not attempt to explain the origin of these primary substances. An early view of seminism was pronounced by Aristotle, who believed that the male semen represented the impulse or efficient cause, and the female "semen" was the substance upon which it operated in producing an embryo.

According to the theory of panspermy, generation depended on a primordial, indestructible, and unorganized substance or principle comparable with air, water, and earth but endowed with life. A revision of this theory centered around the idea that a widespread distribution of germs accounted for apparent cases of spontaneous generation.

Development and Reproduction in Plants

Theophastus (Chapter 3) had made careful studies of seed germination and development in plants. Nehemiah Grew (Chapter 8) had described the anatomy of plants including the reproductive parts, and John Ray (Chapter 9) was interested in reproduction in plants. Sexual reproduction, however, was not described for plants until the end of the seventeenth century, when problems encountered by the earlier investigators could be subjected to direct observation and experiment.

Rudolph Camerarius (1665-1721), Professor of Medicine at Tübingen, designed and carried out experiments on plant reproduction between 1691 and 1694. He removed the anthers from the castor oil plant *Ricinus* and observed that the seeds produced by the emasculated plants were empty and did not develop. He then removed the stigmas and found that no seeds were produced. These plant parts, he reasoned, must be sex organs. He cited Swammerdam's case of hermaphroditism in snails and singled out this mode of reproduction as an exception in animals but a common method in plants. In one of his experiments with maize, he thought he had prevented all pollen from reaching the silks (female parts) but 11 kernels were produced. He explained these kernels as resulting from uncontrolled pollen and designed more critical experiments to check this explanation.

Camerarius's report on these experiments was included in a letter addressed to Gabriel Valentin, Professor of Botany at Giessen, dated August 25, 1694. A part of the letter entitled "De Sexu Plantarum" (The Sex of Plants) contained the most profound observations on sexual reproduction in plants made before the time of Kölreuter, nearly a hundred years later.

Hybridization in Plants

The discovery of sexual reproduction in plants made possible an experimental approach to plant hybridization. Camerarius himself is credited with the first artificially produced plant hybrid on record, from a cross between hemp and hop plants. Thomas Fairchild in 1717 was reported by his contemporaries to have cross-pollinated a carnation with a related flower called a "pink" (Dianthus). The hybrid, showing characteristics of both parents, was called Fairchild's Sweet William and by some, "Fairchild's mule." No record of this example of hybridization was left by Fairchild himself, but good evidence from his contemporaries indicated that the reported results were actually obtained. Furthermore, the experiment apparently was designed, and the results were not accidental. Following this beginning, many artificial pollinations were performed between different related plants.

One of the most important biological researchers in the eighteenth century was the German botanist, Joseph Kölreuter (1733-1806; Fig. 11.3), who found that hybrids between plant varieties might resemble one or the other parent or appear intermediate between them. One of his most valuable observations came from making reciprocal crosses in plants that demonstrated the equality of the contributions from the two parents. That is, the same results were obtained by mating a male from variety A with a female from variety B as by mating a female from A with a male from B.

Kölreuter had broad interests and sound judgment. His experiments were well designed and carefully conducted. From one series of crosses between tall and dwarf varieties of tobacco, he obtained results that reflected the modern principle of quantitative inheritance. Progeny in the first generation were intermediate in height between the two parents. In the second generation, variation was continuous and a fairly normal distribution was observed. Some mature plants were as large as the tall parent and some were as small as the dwarf parent. The size of most of the hybrids, however, fell between the extremes of the two parents. Kölreuter could not explain these results; in fact, it was not until the early part of the present century that an adequate explanation was obtained through the multiple gene hypothesis.

Fertilization and Developmental Mechanics

Hieronymus Fabricius (1537-1619; Fig. 10.1) was the first author to propound a reasoned scheme for the mechanics of generation. He wrote two treatises on embryology: *On the Form of the Foetus* (1600) and *On the Formation of the Egg and the Chick* (1621, published after his death). Fabricius did not believe that the spermatic fluid reached the genital organs or took any part in generating the animal. Harvey pondered the subject, and, being unable to visualize a completely natural explanation for development, he followed the theme of his predecessors and applied a mystical interpretation to some aspects. His *The Development of Animals* (1651) was published six years before his death. It was long and cumbersome and did not lead immediately to any significant advance in embryology. Harvey simply did not have access to the techniques demanded for a comprehensive study of development.

The best remembered part of Harvey's book was the dictum, "All creatures come from an egg" *(Ex ovo omnia)*. Harvey was far ahead of his time in this suggestion, but his concept of an egg was quite different from the modern definition. He called any embryonic mass an egg. Harvey had not seen a mammalian egg and he did not know with certainty what an egg was, but nevertheless the concept he introduced has proven to be true for all higher animals. Harvey considered all animals to arise from "eggs." He also indicated that semen had a vitalizing role in development.

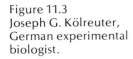

Figure 11.3
Joseph G. Kölreuter,
German experimental
biologist.

Nicolaus Steno (1638-1686), Danish anatomist who later became a church authority, first (1667) introduced the term ovary for the female reproductive gland. In 1672, Regnier de Graaf (1641-73) reported that he had succeeded in tracing the mammalian egg (which he maintained existed already in the ovary) down the fallopian tubes to the uterus. He had studied sections of mammalian ovaries and had observed the fluid-filled spaces, now called Graafian follicles. These he erroneously considered to be eggs. Although he saw only the follicles and not the eggs themselves, the spaces do correspond in number with the embryos in the uterus and de Graaf was able to describe in general terms the process of ovulation. Thus, a century and a half before the mammalian egg was observed, the generative process was associated with changes occurring in the ovary.

Leeuwenhoek (in 1677) observed human sperm with his single-mounted lenses, and in the years that followed he observed the sperm of several animals (Chapter 8). He also observed the association of sperm with eggs in frogs and fish and considered the sperm to furnish essential life-giving properties, whereas the egg was assumed to merely provide the proper environment for nutrition and development of the embryo. Leeuwenhoek, stating his views on the nature of fecundation in 1683, maintained that the eggs were impregnated by seminal animalculae. He demonstrated that living spermatozoa could actually be seen with the microscope in samples taken from the uterus of a dog after coition.

Hartsoeker, an animalculist, stated (in 1694) that each spermatic worm of a bird enclosed a male or female bird of the same species, and in copulation a single worm entered an egg where it was nourished and grew. Andry (1700) and others also believed that the function of the egg was to receive, enclose, and nourish the spermatic worm. Confusion on the subject prevailed during the eighteenth century, mainly because the Graafian follicles were thought to be eggs and efforts to trace them led only to frustration and imagination. There was a difference of opinion as to when and where spermatozoa could be found in the female reproductive organs, and how many spermatozoa were necessary to produce fecundation. This period was characterized by speculation that tended to obscure simple truths with elaborate theories.

Segmentation or cleavage of the ovum was first observed by Swammerdam in the seventeenth century, but the results of these investigations were not published until 1738. Swammerdam saw the first furrow of the frog's egg in cleavage. Lazaro Spallanzani (Fig. 15.2) described and sketched (in 1780) the furrows formed by the first two cleavage divisions in the egg of a toad. Spallanzani was also first to demonstrate artificial fecundation. He showed that frog eggs fertilized artificially would develop like those naturally fertilized, whereas those left unfertilized would not develop but would undergo decomposition. Artificial fecundation was successfully demonstrated on several kinds of animals including various amphibia, insects, and dogs. When Spallanzani came to explain the results, however, he maintained that semen deprived of animalculae still possessed fecundating properties.

Nineteenth Century Embryology

The discovery of the true ova of mammals by von Baer in 1828 had an important bearing on the history of embryology. It established the egg as the morphological unit at the root of development of all animals. Von Baer had discovered the answer to Aristotle's challenge. Under his microscope he could see that the mammalian egg was an undifferentiated mass containing no preformed creature. He was the first to trace the egg to the embryo and earned the respected title "Father of Modern Embryology."

Von Baer

Karl Ernst von Baer (1792-1876; Fig. 11.4) was born on an estate in Esthonia, Russia (Esthonia was a part of Russia at the time of von Baer), to a poor family of German nobility. He enrolled in 1810 at the University of Dorpat, as a reluctant premedical student. In his *Autobiography,* he stated later that he would rather have pursued other courses of study, but that medicine seemed the only door to an academic career in science. His thesis on endemic diseases among the Esthonians (1814) won for him a Doctor of Medicine degree, but he had no laboratory experience whatsoever. He felt that he must learn to work with his hands but found no opportunity immediately available to him.

Following graduation he traveled in Austria and Germany and became located for a time in Würzburg, where he bought a leech in a drugstore to be used for his first dissection. At Würzburg he studied physiology and comparative anatomy independently until he had pawned all of his possessions and exhausted his funds. From there he walked to Berlin, where he eked out an existence and continued his studies of medicine during the winter of 1816-17.

Von Baer finally obtained a position as Professor Extraordinary in Anatomy at Königsberg, and during the next several years he worked at the zoological museum. Here he established himself as a brilliant and skillful embryologist. In the years that followed, he was elected dean of the medical faculty, published several anatomical and physiological addresses, was promoted to professor ordinarius, contributed to a physiology text, and was elected rector of the university.

In 1828, von Baer published three treatises: The *Epistle,* which set forth his discovery of the ovum; the *Commentar,* which extended the *Epistle;* and the famous *Über Entwickelungsgeschichte der Tiere* (*Developmental History of Animals*), which became a standard text of embryology. Four major advances in embryology were included in that book: the description of the mammalian ovum; the proposition known as the "germ-layer theory"; the general correspondence between stages in the development of the embryo and stages in the (phylogenetic) history of the race; and the discovery of the notochord. Von Baer did not recognize the egg as a cell. The cell theory was not formulated until some ten years later. He did not theorize extensively but he described what he saw and interpreted his observations.

With his germ-layer theory, von Baer simplified a complicated aspect of development. Following Wolff's earlier suggestion, he recognized four germ layers, an outer, an inner, and two middle layers. Later

Figure 11.4
Karl Ernst von Baer,
German embryologist.

(1845), however, Remak observed that the middle part of an embryo was a single rather than double layer, making only three primary germ layers: ectoderm, endoderm, and mesoderm. Von Baer illustrated the development of the embryo from leaflike germ layers that may form tubes. From the ectoderm, he described the formation of the neural tube and outer skin. From the endoderm came the digestive tube and associated glands, and from the mesoderm connective tissue, muscle, and other visceral parts.

In comparing corresponding stages in the development of different kinds of embryos, von Baer advanced four propositions which were later incorporated into Haeckel's biogenetic law: (1) in development, general characters appear before special characters; (2) from the most general characters are developed the less general and finally the special; (3) in the course of development, an animal of one species diverges continuously from one of another species; and (4) a higher animal during development passes through stages that resemble stages in the development of lower animals. Von Baer observed corresponding stages in the development of different animals, but he did not relate these observations to evolution.

In addition to these specific contributions, von Baer also influenced embryology in a more general way. With his book, he established a higher standard for investigations in embryology, and he made embryology a comparative science. His book included an explanation of the cause of amnion formation, a description of the development of the urogenital system, and a description of the development of the lungs. He also discussed the various stages in the formation of the digestive canal and the nervous system.

In his later years von Baer returned to Russia as a member of the Academy of St. Petersburg. He also was elected a member of the Royal Society of London, from which he received the Copley Medal, and he was chosen a corresponding member of the Paris Academy of Sciences. Von Baer founded the Society of Geography and Ethnology of St. Petersburg and became a cofounder of the German Anthropological Society.

Although Karl von Baer's fame is due almost wholly to his embryological work, he was a naturalist with wide interests. At one time he headed a scientific expedition to Lapland and Nova Zembla; later he joined another expedition to the North Cape and the Sea of Azov. He inspected the fisheries of the Russian Empire, especially of the Caspian Sea. For three years he toured the museums of Europe. With his sight and hearing becoming defective, he retired to Dorpat, and there he continued his studies until his death in 1876 when he was 84 years old.

Haeckel's Biogenetic Law

Ernst Haeckel (1834-1919) made the connection between development of the individual and evolution of the race (the biogenetic law) and popularized the phrase "ontogeny recapitulates phylogeny." On the surface, the evidence for recapitulation seems convincing. Distinct pharyngeal clefts in early stages of vertebrate embryology bear a resemblance to the respiratory structures of fish. The heart and central nervous system of higher vertebrates resemble, at least superficially, the structures of lower forms at comparable stages of development. A cartilaginous endoskeleton and notochord that occur in the early development of higher vertebrates have counterparts in lower forms. Haeckel placed particular emphasis on morphological characteristics of contemporary types and gave little attention to paleontologi-

Figure 11.5
Wilhelm Roux, pioneer
in the field of experimental
embryology.

cal data. When the history of animals (phylogeny) is considered in a time sequence, many new variations seem to be superimposed on the old patterns.

More recent contributions in the field of developmental biology provide well-founded reasons for treating Haeckel's recapitulation theory with caution. Child, for example, showed (1915) that axial gradients can be followed in development. An organizer substance, described by Spemann in 1918, confirmed interdependencies of parts of living organisms in space and time. In amphibian material, for example, the dorsal lip of the blastopore was found to induce differentiation in surrounding tissue. The modern gene theory has provided the beginning of an explanation for a sequence of events as well as for end products of ontogeny. In perspective, modern concepts of embryology are better served by the interpretation of von Baer, which explained resemblances in terms of processes, than by that of Haeckel, which was based on precedents and recapitulation.

Francis Balfour's (1851-82) two-volume *Comparative Embryology* covers the development of vertebrates and invertebrates and makes critical comparisons. The great mass of information that had been accumulated by the Haeckel school was sifted, digested, and molded by Balfour into an organized whole.

Roux

Wilhelm Roux (1850-1924; Fig. 11.5), a student of Haeckel, formulated a mechanistic conception of embryonic development and found evidence that seemed to support the old theory of preformation. He theorized (in 1881) a sort of selection mechanism among structures of the body which he called "functional adaptation." Every organ of the body and even every cell was believed to have a given structure but to be capable of changing to conform with function. To test his theory experimentally, Roux studied developmental mechanics in embryos. Some embryonic structures Roux found to be predetermined. Certain parts of the embryo were observed to develop in a regular pattern before they became functional. Roux visualized a complex machine already operating in the fertilized cell, which suggested preformation.

To be transmitted from cell to cell, Roux reasoned, the machine must in some way break down into its component parts during the process of cell division. This must occur in such a way that the parts could be transmitted to another cell and there reassembled in some way to again become functional. Before cell division (mitosis) and reduction division (meiosis) were fully understood, Roux developed models (Chapter 17) to show how the units (now known to be chromosomes) must line up in the center of a cell and go through a duplication process in cell division and a segregation process in reduction division.

Roux's famous experiment (1888), which seemed to demonstrate preformation, was carried out by using a hot needle to kill one of the two cells (blastomeres) that arose from the first cleavage of a fertilized frog egg. Only half an embryo developed and Roux concluded that determination had already occurred in the two-cell stage. One cell, he believed, had already received the determiners necessary to form half of the embryo. When one cell was destroyed, only the other half could develop. Later (1891) Hertwig performed essentially the same experiment on sea urchins, but he succeeded in separating two blastomeres. Both lived and each formed a whole embryo, indicating that the two cells were not predetermined. A similar separation was later performed on the same material Roux had used; that is, frog blastomeres and the presumed evidence for preformation was negated.

Roux was a pioneer in the field of experimental embryology, but unfortunately he gave up experimental work and became more and more of a theoretician. He continued to make valuable contributions, however, because many of his theoretical models led to further experimentation and his influence was thus extended to modern developments in embryology.

Cellular Fertilization

The assertion that fecundation is accomplished when one animalcule gets into one egg was originally made by Leeuwenhoek in 1683, but verification did not come for a century and a half. The delay can be ascribed partly to the lack of microscopic techniques. Pringsheim in 1855 may have been the first to see nuclear fusion as evidenced in the plant Vaucheria. He reported that numerous male elements and one female gamete were present. It is probable that he saw only insemination and not nuclear division. O. Hertwig (1875) suggested that such fusion would occur and would be followed by nuclear division. The phenomenon of penetration by a single sperm, the fate of the sperm within the egg, and the equal participation of the egg and sperm nuclei in fertilization were completely demonstrated in animals by the researches of O. Hertwig, A. Weismann, and H. Fol between the years 1875 and 1879.

Chronology of Events

B.C.
460-370	Hippocrates, Regimen, obstetrics and gynecology.
384-322	Aristotle, treatise on embryology, view of epigenesis.
380-287	Theophrastus, seed germination and development in plants.

A.D.
1600	H. Fabricius, mechanics of development in chick.
1626	Joseph of Aromatari, miniature chick in egg.
1651	W. Harvey, epigenesis, "all creatures come from eggs."
1651	N. Highmore, pangenesis.
1664	H. Power, "complete heart" in chick egg.
1666	J. Swammerdam, preformation theory from frog development.
1667	N. Steno, introduced term "ovary" for female reproductive gland.
1672	R. de Graaf, Graafian follicles thought to be eggs.
1674	N. de Malebranche, "box within box" theory.
1677-83	A. von Leeuwenhoek, observed sperm and commented on fecundation.
1682	M. Malpighi, microscopic observations of chick embryology.
1691-94	R. Camerarius, sexual reproduction in plants.

1694	N. Hartsoeker, sperm of a bird carries miniature member of species (homunculus idea).
1717	T. Fairchild, hybridization in plants.
1738	J. Swammerdam, cleavage in frog (Date of publication of Swammerdam's *Biblia Naturae* by Boerhaave).
1745	C. Bonnet, ovist; aphid development.
1758	A. von Haller, ovist; chick embryology.
1760	J. Kölreuter, equal contribution of the two sexes in plants, shown by comparable results from reciprocal crosses.
1780	L. Spallanzani, cleavage in toad.
1786	C. F. Wolff, epigenesis; chick embryology.
1828	K. von Baer, *Developmental History of Animals;* observed mammalian egg.
1855	N. Pringsheim, cellular fusion in Vaucheria.
1868	C. Darwin, cited pangenesis in *The Variation of Animals and Plants under Domestication.*
1874	E. Haeckel, corresponding stages of development related to evolution ("ontogeny recapitulates phylogeny").
1875-79	A. Weismann, H. Fol and O. Hertwig observed cellular fertilization.
1880	F. Balfour, *Comparative Embryology.*
1881-88	W. Roux, "functional adaptation" in embryology; experiments with frog blastomeres; chromosome mechanics.
1875-91	O. Hertwig, separated blastomeres in sea urchin two-cell stage; fertilization and cleavage.
1915	C. M. Child, axial gradients in development, *Individuality in Organisms.*

References and Readings

Adelmann, H. B. 1942. *The Embryological Treatises of Hieronymus Fabricius of Aquapendente.* Ithaca, N. Y.: Cornell University Press.

———. 1966. *Marcello Malpighi and the Evolution of Embryology.* 5 vols. Ithaca, N. Y.: Cornell University Press.

Aristotle. *De Generatione Animalium.* Translated by A. L. Peck, 1943. London: W. Heinemann.

Aromataria, Joseph of. 1625. *Epistola de Generatione Plantarum ex Seminibus.* Venice: Translated in *Philosophical Transactions,* Royal Society of London, Vol. XVIII, 1694.

Baer, K. E. von. 1828. *Ueber Entwicklungsgeschichte der Tiere Beobachtung und Reflexion.* Königsberg: Gerbruder Bornträger.

Camerarius, R. 1694. "De Sexu Plantarum." Tübingen: *Academicae Caesareo Leopold,* Vol. VIII.

Cole, F. J. 1930. *Early Theories of Sexual Generation.* Oxford: Clarendon Press.

Gasking, E. 1967. *Investigations into Generation 1651-1828.* Baltimore: Johns Hopkins Press.

Harvey, W. 1651. *Exercitationes de Generatione Animalium.* London: Octaviani Pulleyn.

———. 1847. *The Works of William Harvey.* Translated by R. Willis. London: Sydenham Society.

Huxley, J. S. and G. R. de Beer. 1934. *The Elements of Experimental Embryology.* Cambridge: At the University Press.

Malpighi, M. 1673. *Formatione Pulli in Ova; 1675. De Ovo Incubato.* London: Joannem Martyn, Printer for Royal Society. (Followed by seven letters between Malpighi and Oldenburg with illustrations.)

Meyer, A. W. 1936. *An Analysis of the De Generation Animalium of William Harvey.* Stanford: Stanford University Press.

_____. 1939. *The Rise of Embryology.* Stanford: Stanford University Press.

_____. 1956. *Human Generations, Conclusions of Burdach, Döllinger, and von Baer.* Stanford: Stanford University Press.

Needham, J. 1934. *A History of Embryology.* Cambridge: At the University Press.

Radl, E. 1930. *The History of Biological Theories.* Translated from German by E. J. Hatfield. London: Oxford University Press.

Willier, B. H. and J. M. Oppenheimer, eds. 1964. *Foundations of Experimental Embryology.* Englewood Cliffs, N. J.: Prentice-Hall.

CHAPTER 12

EXPLORERS SEEKING PLANTS, ANIMALS, AND FOSSILS

THE EIGHTEENTH century world had been enlarged through the discovery and exploration of America, the opening of the Orient to travel, and the discovery of Australia and the islands in the South Seas. Many opportunities were available for exploration and adventure. Early biological explorers such as Ray, Tournefort, and Linnaeus (Chapter 9) had demonstrated the professional rewards that could come from traveling. Governments, especially those of sea powers, were aware of the values to be gained, both monetary and scientific, from exploring new worlds. Sufficient background had been established in classifying and comparing living things to make exploration meaningful. Expeditions were undertaken in the latter part of the eighteenth century for the purpose of observing living things in their native habitats and collecting animals and plants. Some exploring projects were supported by academies and other institutions and some were undertaken by individuals, at their own expense, for scientific purposes and adventure. Stories of exciting experiences, as well as valuable collections of animals and plants and other evidence of concomitant scientific accomplishments continually stimulated additional excursions.

During the latter part of the eighteenth and into the nineteenth century, several well-organized expeditions were dispatched to various parts of the world by governments, particularly the British government, to draw maps and make observations. Although based on a small island, Great Britain was a strong sea power and had interests around the world. France, Italy, and other countries of continental Europe were involved with internal problems, and their biologists of the period were more inclined to make their observations and contributions at home. Most notable for their biological accomplishments were the voyages of the *Endeavour,* the *Investigator,* the *Beagle,* and the *Challenger* (Table 12.1). Naturalists on these and other less famous explorations did much to describe the fauna and flora in different parts of the world, including that of the ocean at various depths and at the bottom. One important measure of the value of the expeditions was the maturing influence they had on the naturalists who participated. Scientists such as Joseph Banks, Robert Brown, and Charles Darwin received, through their travels as young men, the training and enthusiasm for natural history that influenced their later accomplishments.

Table 12.1   Major biological explorations supported by British Government

Vessel	Dates	Main Areas Explored	Captain	Chief Biologist
Endeavour	1768-71	South Seas, Australia	James Cook	Joseph Banks
Investigator	1801-05	Australia, Tasmania	Matthew Flinders	Robert Brown
Beagle	1831-35	South America, around world	Robert Fitz-Roy	Charles Darwin
Challenger	1872-76	Atlantic and Pacific, bottoms, continental and island shore-lines	George S. Nares	Charles W. Thomson

The *Endeavour* to the South Seas

In November, 1767, the Royal Society of London sent a memorandum to the king calling attention to a significant astronomical phenomenon that was to occur on June 3, 1769. On that day the planet Venus would pass over the disc of the sun. It was pointed out that information of value to navigation could be obtained by observing this event from an appropriate place near Tahiti. An expedition to the South Seas at the appropriate time to witness this event was suggested. Because of the difficulties encountered on previous expeditions to the South Seas, the British Government was reluctant to finance more exploring parties, and yet the navy depended on maps, charts, and weather data. After the king's interest in the project was aroused, 4,000 pounds were placed at the disposal of the Royal Society for a trip to the appropriate place with the request that the Admiralty and the Royal Society work together in preparing for the expedition. A 308-ton vessel three years and nine months old was purchased and renamed The Endeavour Bark. James Cook, a naval warrant officer, was commissioned a lieutenant and appointed commander of the expedition. Cook and an astronomer were then officially appointed by the Royal Society to make the observation of the transit of Venus.

Banks, Chief Biologist on *Endeavour*

The Royal Society granted the request of one of its wealthy young fellows, Joseph Banks (1743-1820; Fig. 12.1), to go along at his own expense on the expedition and to take seven additional naturalists and technicians to assist him in making biological studies. Later, two others were added to the biology section, making a total of ten. Equipment for biological studies, including a library, was provided personally by Banks at a cost of some 10,000 pounds. Thanks to Joseph Banks, the *Endeavour* was the best equipped ship sent out up to that time for the purpose of natural history studies.

Banks had been trained in botany and had previously acquired some experience as an explorer. On his own initiative, while attending school at Eton, he had hired women who collected plants of medicinal value (simples), for sale to druggists, to teach him the names of the plants and show him where and when they could be found. During one vacation period from Eton he discovered at his home a battered copy of Gerard's *Herball*, which he studied thoroughly and carried back with him to school. Soon he was able to teach the women much more about plants than they could teach him.

Figure 12.1
Sir Joseph Banks, English
botanist who sailed on
the *Endeavour*.

When 17 years old, Banks entered Christ Church College at Oxford but found, to his disappointment, that the Professor of Botany did no teaching. The professor was said to have given only one lecture in 35 years. Through Banks's influence and insistence, a young botanist was found at Cambridge and employed to teach botany at Oxford. In April, 1766, three years after completing his schooling at Oxford, Banks was engaged as a naturalist on the Fishery Protection vessel, *Niger*, dispatched to the shores of Labrador and Newfoundland on a 7-month voyage. While there, he made excursions inland observing and collecting plants. He wrote on animals, plants, and people and included several pen sketches of people. On the return trip to England, the ship stopped at Lisbon where Banks added more plants to his collection. He later made a collecting tour of the western part of England.

Cook, Captain of *Endeavour*

The *Endeavour* sailed from Plymouth, England in August, 1768, under command of James Cook. Explorations were made on several of the

South Sea Islands, mainly Tahiti, New Zealand, New Guinea, and Australia. On April 19, 1770, the east coast of Australia was discovered. In his journal entry of April 26, from the deck of the *Endeavour*, Banks compared the country to, "the back of a lean cow, covered in general with hairs, but nevertheless, where her shaggy hip bones have stuck out further than they ought, accidental rubs and knocks have entirely bared them of their share of covering."

Many landmarks in these areas now bear names that date back to this exploring party (Fig. 12.2). Cook's Reef, a part of the Great Barrier Reef some distance off the eastern shore of Australia, was named after Captain Cook. Endeavour Strait, the discovery of which proved the separateness of Australia and New Guinea, was named after the ship. Botany Bay was so named because it yielded many specimens to the naturalists. Banks described what later became known as Botany Bay from the deck of the *Endeavour* on April 28, as "an opening like a harbour." The name "New South Wales" was first used by Cook to identify southeastern Australia.

Near the mouth of the Endeavour River, the *Endeavour* struck a reef

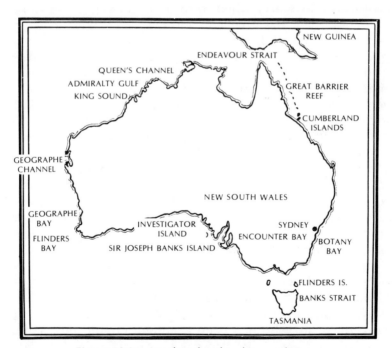

Figure 12.2.   Landmarks of early expeditions
around Australia.

and a month was required for repairs. This was a "veritable botanical holiday." Banks took notes about the fauna and flora of what he called "New Holland." Eventually the ship was repaired and followed its course to scheduled landings where Cook, Banks, and other members of the party became well acquainted with Australia. The land was generally low, thickly covered with long grass, and seemed to promise great fertility. A limiting factor, especially in some seasons, was lack of water. In some areas the soil was barren. Banks noted a lack of edible vegetables, few fruits, and unusual hardiness of timber. Only two of the plants were considered to be useful. A scarcity of quadrupeds, but several varieties of bats, were noted. Many pages in the journal were devoted to descriptions of Australian birds and insects. In reply to a letter from John Hunter (Chapter 10) about the prospects for a colony in Australia, Banks said (later in life) that the climate, and soil, are superior to those of the areas already settled by Europeans.

One of the reasons for the success of the expedition was the high quality of leadership provided by Captain Cook. Recognizing the importance of good health on the part of crew members, Cook insisted on sanitation and proper nutrition. Sauerkraut was provided in the rations to prevent scurvy when fresh fruit and vegetables were not available. Discipline was excellent, and there was little illness or dissent in spite of cramped quarters and poor living conditions. Other attempts had been made previously to explore the uncharted South Seas, but in most cases the men had been overcome by scurvy and homesickness before they arrived at the place designated for study.

One great advantage enjoyed by Cook that had not been available to his predecessors was a satisfactory method of determining longitude. It had been possible, for a long time, to determine the north-south position (latitude) of a ship at sea, but there had been no method for determining the east-west position (longitude). A table worked out by an astronomer to determine longitude from the position of the stars was used for the first time on the voyage of the *Endeavour*. With this device, it was possible for Cook and his crew to determine their position accurately at all times. This enabled them to prepare an account of their route with detailed locations of the coastlines visited. Development of new devices, good leadership, and a series of favorable circumstances not only made the voyage of the *Endeavour* possible, but contributed greatly to its success.

Extensive notes were compiled by Cook, Banks, Daniel Solander, a naturalist on board who had been a student of Linnaeus, and other members of the party. A detailed day-by-day account of the *Endeavour's* voyage (J. K. Hawkesworth (1773), *Captain Cook's Voyage Round the World*) begins with the entry of Friday, May 27, 1768 when Lt. Cook (later promoted to captain) went on board and took charge of the ship *Endeavour* and ends with the entry of June 12, 1771 when the ship returned to Dover. Most of the notes, along with the collections of plants, were eventually deposited in the British Museum, where they became part of the vast holdings at that institution. Interesting descriptions of strange animals such as the kangaroo are included in the notes.

Cook and Banks are considered the two greatest men in Australian history; Cook was the navigator and discoverer, and Banks was the "godfather" and "patron saint" of New South Wales. Banks demonstrated skill and leadership during his three years in the South Seas. When he returned to England in 1771, with volumes of notes and vast collections of plants and animals, he was immediately recognized as a leading biologist. A year after the return of the *Endeavour* (1772), Banks and Solander made a voyage to Iceland where they gathered more plants, and wrote, in their own handwriting, more volumes of notes now in the British Museum.

Captain Cook made two more exemplary voyages in the South Pacific following the voyage on the *Endeavour*. Although he was not primarily a biologist, he contributed indirectly, and in some cases directly, to the development of biology and other areas of natural science. He commanded (1772-75) the expedition to the South Pacific of two ships, the *Resolution* and the *Adventure*. On this expedition Cook disproved the rumor of a great southern continent, explored the Antarctic Ocean and the New Hebrides Islands, and discovered New Caledonia.

Cook sailed again on the command ship *Resolution* in 1776 and, after a year in the South Pacific, rediscovered the Hawaiian Islands which he named Sandwich Islands in honor of John Montague, fourth Earl of Sandwich, a lord in the Admiralty and one of Cook's patrons. He sighted the islands on January 18, 1778, landed at Waimea on the Island of Kauai, and went on to the Northern Pacific in search of a water passage from the Atlantic to the Pacific Ocean.

Later he returned to the Hawaiian Islands and on January 17, 1779 anchored off the Island of Hawaii. His ships were welcomed by the chiefs of native tribes. Captain William Bligh went ashore to explore the island, and, on his return to the ship, reported that volcanic material was spread over the island but he could find no volcanos. He expected classical cones and did not recognize Mauna Loa, Mauna Kea, and Kilauea as parts of a volcanic rift thousands of miles long, mostly under the sea. Cook's ships departed on February 4, 1779 under

Figure 12.3.   Cook's monument, located near the place where
Captain James Cook was killed on the Island of Hawaii.

friendly relations with the natives. When, however, bad weather
forced the ships to return on February 11, relations with the natives
soon deteriorated. In a skirmish on February 14, Cook went ashore to
confer with the aging chief, and Cook and four marines were killed.
Captain Cook was buried on the Island of Hawaii. A monument (Fig.
12.3) is a memorial to his accomplishments.

Banks and the Botanical Gardens at Kew

In 1773, Banks was appointed by King George III as Royal Scientific
Adviser. One of his new responsibilities involved directing the Royal
Botanical Gardens at Kew, which were developed into a station where
plants from the entire British Empire were studied and tested. Rare and
little-known plants were gathered from all over the world and subjected to observation and experimentation at Kew. When plants were

found that might have practical values in particular locations or under special growing conditions, they were sent to subsidiary botanical gardens in other parts of the vast British Empire for further testing in particular environments.

A major source of materials to be included in the initial layout at the Kew gardens and a framework for the plan of the gardens was the book by J. Banks and D. Solander, *Illustrations of Australian Plants Collected in 1770 During Captain Cook's Voyage Round the World in HMS Endeavour.* This is a large (about 24 x 14 inches and 5 inches thick) volume with plants listed in taxonomic groups followed by drawings of plants giving stems, leaves, flowers, and fruits. Some detailed drawings are also included to show internal structures, particularly reproductive parts. Plants brought from Australia and other parts of the world by later voyages were planted in the Kew gardens.

After his visit to the South Seas, Banks requested every vessel leaving Great Britain's ports to bring or send him specimens from places they visited. They brought plants, animals, fossils, examples of workmanship of native peoples, and many objects of scientific value and interest. Notes and correspondence show, for example, that William Bligh (remembered from Mutiny on the *Bounty,* 1789) sent "an animal, the colch (Koala), in spirits" (unpublished correspondence); Paterson sent, "an animal and its young in spirits"; George Bass sent, "two skins and a skull of the wombat in spirits, . . . a new animal found on the island where *Sidney Cove* was wrecked, . . . a bird of the species of the Bird of Paradise"; "the Golden Grove transport brought fourteen kangaroo skins." (Journal of Banks)

Banks's extensive correspondence included continued reminders about the needs for agricultural crops in different parts of the British Commonwealth. He requested live specimens of hop-plants, cotton, indigo and "seeds of all sorts" to be grown in the Kew gardens. In one of Banks's memoranda dated 1805 he suggested "a project for supplying His Majesty's Colony in New South Wales with a circulating medium." Banks proposed the decimal system of coinage and a Government Bank of Exchange. (England finally adopted the decimal system for money exchange on February 1, 1971). Banks served as president of the Royal Society of London for some 40 years and was recognized as one of England's greatest benefactors for science.

As a part of Banks's ambitious program as Director of the Gardens, expeditions were dispatched to different areas of the world to discover new plants and to locate environments where certain types of plants might be raised successfully and have practical value. The most significant accomplishment of these expeditions in this respect was that made by the voyage of the *Investigator,* commanded by Matthew Flinders and carrying Banks's young friend, Robert Brown (1773-1858), as chief naturalist.

The *Investigator,* to the South Seas

Matthew Flinders (1744-1814) had already made successful explorations in the South Seas, and he had been acclaimed for sailing around Tasmania and proving that it was an island not connected with Australia. As his next project he proposed to sail around Australia and map the coastlines. Banks, who was approached by Flinders with this proposal, was immediately interested and enthusiastic. He began to arrange for such a trip by persuading the First Lord of the Admiralty to fit out a suitable ship for the voyage. A 334-ton vessel was chosen, renamed the *Investigator,* and placed under the command of Flinders.

Visualizing an expedition of biological discovery similar to his own voyage on the *Endeavour,* Banks also arranged, at his own expense, for the young Scottish botanist, Robert Brown, and a draughtsman, Ferdinand Bauer, to go with Flinders. Assistants were also provided, and equipment and supplies for botanical work were prepared, although it was understood that biological observations and collections represented only a secondary objective on this expedition. The first objective and justification for the voyage, as far as the Admiralty was concerned, was the exploration of the Australian coast. Brown was given no specific instructions but was left to make his own decisions as to how he would proceed with his natural history investigations.

The *Investigator* sailed from England in 1801. When only a short distance out at sea, the crew received forebodings of trouble ahead. The ship developed leaks that were a source of continuing trouble and anxiety to members of the party. Only through the dogged persistence of the crew did the ship reach its destination. New territory was charted along the south coast of Australia and the continuity of the Australian coastline was demonstrated. Many places were named after familiar English locations. The name of Flinders was added to those of Cook and Banks for names of geographical features, such as Flinders Bay and Flinders Island. Main streets in several cities in Australia are now named after Flinders, who was first to use the name "Australia" for the great island continent. The Investigator Islands, south of Australia, were named after the leaky ship which caused the party so much trouble.

Flinders's Return to England

Because of the ship's dilapidated condition and the lack of facilities to make repairs, the ship was abandoned after two years. Flinders packed the maps, notes, and some biological specimens, and took passage for England in 1803 on another boat. When this ship was a week out from Australia it crashed on a reef and was wrecked. At the risk of his life, Flinders made his way back to Australia in a small boat and arranged for the rescue of the other survivors, who by then had been marooned on a sandbar for about five weeks. Most of the expedition's maps and notes were saved, but the plant and animal specimens were lost.

In a smaller ship, the *Cumberland*, which had assisted in making the rescue, Flinders continued his journey to Europe. At Mauritius, the *Cumberland* stopped for repairs and Flinders became a prisoner of the French, who were at war with England. Flinders, carrying maps and messages from Australia, was considered a spy and held as a prisoner for seven years, until 1810. His health was broken, and he died four years later. The published account of his explorations, entitled *Voyage to Terra Australis,* appeared when he was ill and unable to read the finished work.

Biological Accomplishments of Brown

Brown and Bauer remained in Australia until 1805 in order to complete their biological projects. They then returned to England with great stores of herbarium specimens (some 4,000 species), and seeds to be planted in the garden at Kew. Funds were not available from the Admiralty in these troubled war years to publish the complete botanical report of Brown. The large volume of material was eventually condensed and presented as an appendix to Flinders's *Voyage to Terra Australis*.

Robert Brown grew in professional stature while on the expedition, and he was recognized as the leading botanist of his time when he returned to England. He became the librarian of the Linnean Society, a position that he held until 1822, when he was elected a fellow in the Linnean Society. He also served as librarian to Banks and managed Banks's vast collection of specimens, books, and manuscripts. When Banks died, the collection was bequeathed to Brown with the stipulation that it should go to the British Museum when Brown was through with it. Seven years after Banks's death, in 1827, the collection was turned over to the British Museum. Brown went with it as the first caretaker. Brown was elected (1811) a fellow in the Royal Society and from 1849 to 1853 he served as president of that society.

Brown did not publish books, as some of his predecessors and contemporaries had done, but he published a long series of scientific papers of great value. His work on the flora of Australia and New Zealand is a classic. No new system of classification was developed. Instead, Brown followed the pattern of others, mostly that of Jussieu. Several orders and families of plants were analyzed and revised. Brown made a special study of the adaptations of members of some families to different climates and other environmental conditions. His associates recognized him as a keen observer and sagacious scientist. Among the special contributions to his credit are the discovery (Chapter 14) of the cell nucleus and an analysis of the sexual process in higher plants. He saw the pollen tubes in the pistils and the gross mechanism of fertilization in the ovules. Brown also studied fossil plants under the microscope.

Charles Darwin and the *Beagle* Voyage

Charles Robert Darwin (1809-82) was the son of a physician, Robert Darwin, who lived at Shrewsbury in England. His mother, Susannah Wedgwood, eldest daughter of the famous potter Josiah Wedgwood, died when Charles was eight years old. His grandfather, Erasmus Darwin, was a well-known eighteenth century doctor and writer. Charles's early education was obtained at a boarding school located a mile or so from his home.

At 16, Charles complied with his father's desire and went to Edinburgh to study medicine. He did not take to this study, and later characterized medicine as a "beastly profession." Lectures on geology and zoology were termed by him as "incredibly dull," but they were, nevertheless, more pleasant than those in some other subjects. At Edinburgh Darwin acquired much valuable knowledge, if not from the lectures, at least from personal observations. He became friendly with fishermen at the seashore and frequently accompanied them as they trawled for oysters. In hours stolen from his studies he observed marine organisms and made a small discovery concerning a seaworm which he had the pleasure of describing before a learned society. He loved hunting and learned how to preserve bird skins for further study and comparison. Vacation periods were spent with his Wedgwood cousins in North Wales, hiking and shooting.

Dr. Darwin, observing that his son would not succeed in medical school and was rapidly becoming "an idle sporting man," proposed that Charles leave the study of medicine and become a clergyman. Charles was agreeable and in his 20th year he began the study for the ministry at Christ's College, Cambridge. Here his school work was again neglected while he collected butterflies, rode horses, shot birds, and led a gay social life. The Reverend John Henslow became acquainted with Darwin, and with sympathy and understanding influenced him greatly. As they became more intimate and Darwin's admiration for Henslow increased, the Reverend suggested books for Darwin to read. One with which Darwin was particularly impressed was the work of Alexander von Humboldt (1769-1859), *Personal Narrative of Travels to the Equinoxial Regions of America During the Years 1799-1804.* This book gave a glowing account of a five-year scientific exploration in South America, with detailed information about the distribution of plants in relation to soil, climate and latitude. The distribution of volcanos and their relation to cracks in the earth's crust were also discussed.

After passing his B.A. examinations, Darwin stayed on at Cambridge for two more terms, planning to complete the work for the ministry. Through Henslow, Darwin became acquainted with the geologist, Adam Sedgwick, with whom he later studied geology.

Invitation to Travel on *Beagle*

While on a geological field trip in Wales on August 29, 1831, Darwin received a letter from Professor Henslow informing him that a certain Captain Robert Fitz-Roy was looking for a biologist to serve without pay as a naturalist on the *H.M.S. Beagle,* an Admiralty vessel that was scheduled to undertake a five-year survey around South America and other more distant parts of the world. Henslow had been invited to make the trip but had declined and had recommended Darwin for the position.

The *Beagle* was dispatched (1831-36) by Great Britain to sail around South America and to continue around the world. The mandate given her crew was to draw maps, measure distances, and circumnavigate the globe (Fig. 12.4). The expedition was said to be organized entirely for scientific purposes, particularly those of practical concern to the British Navy. The geographical area to be given major attention was Patagonia, or the southern part of South America within the present

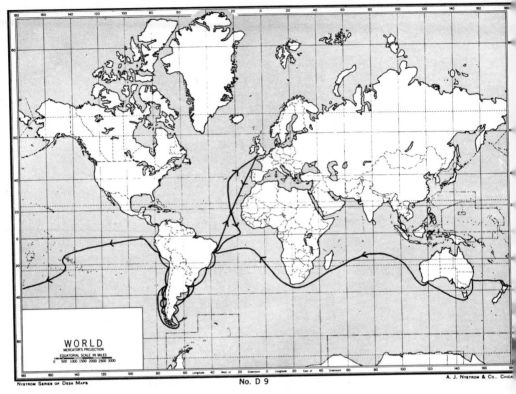

Figure 12.4.   Route followed by the *Beagle* in circumnavigating
the world. The voyage began at Plymouth, England, in
December 1831, and ended at Plymouth in October 1836.

countries of Chili and Argentina. Surveys were also to be made
along the shores of Chili and Peru and among the islands in the
Pacific and Atlantic.

Charles had mixed feelings about such an adventure, but went
home to discuss the matter with his family. The entire family opposed
the notion at first. His father and sisters mentioned the poor living
conditions, wild places, and wild company, and they were concerned
about the effects of such experiences on Darwin's position with the
clergy. Charles himself had doubts, but the anticipation of high
adventure strengthened his desire to accept the proposition. Because
his father was not enthusiastic, however, Charles had nearly decided
not to accept when his Uncle Joe (Wedgwood) persuaded him to
reconsider the matter. They went together to Charles's father, who
readily consented because of his esteem for Joseph Wedgwood.
Uncle Joe reasoned that this might provide motivation for young
Charles who was, at this time, indifferent about a career and was not
applying himself to his studies. The voyage, he explained, by neces-

sitating personal application and responsibility, would almost certainly prove more beneficial than harmful.

The decision made, Charles had little time to prepare for the voyage. He purchased shirts, slippers, Spanish language books, two pistols, a rifle, and a microscope, placed these items in a carpet bag, and set out for Plymouth where the ship was being fitted. With this meager equipment, plus a few other books, particularly Lyell's *Principles of Geology*, volume I (1830), Milton's poems, and a Greek testament, Charles believed himself ready for a long journey. While waiting for the final preparations of the ship to be completed, he wandered around Plymouth indulging in dreams and speculations about the world of which he had read in von Humboldt's exciting narrative of travels in South America.

The *Beagle*, Darwin observed, was a solidly built vessel of 242 tons, rigged as a barque, and she carried two whaling boats. She was of a class that had been nicknamed "coffins" because of their tendency to sink. A total of 76 people was signed on board including crew, navy personnel, specialists, and artists. Captain Fitz-Roy, only four years older than Darwin, was well prepared as navigator and commander. He had made a previous voyage around South America on the *Adventurer*. At the lower tip of South America he had taken on board three natives whom he intended to educate in English language and customs and hopefully to use them as interpreters on succeeding expeditions. They were included on the *Beagle* roster.

Departure from Plymouth

After a few false starts, the *Beagle* put out to sea in December, 1831. Charles had heard of seasickness, but he had never dreamed that it would happen to him. Certainly he had no forewarning that the condition would plague him for five years. He felt well during the first day on board while the ship was in the harbor, but with the first effects of the open sea, he retired to his hammock. He was intermittently ill for the rest of the voyage, causing Fitz-Roy at times to consider returning him to England for the sake of his health. The enforced idleness during his periods of illness gave him time for reflection and enabled him to recall facts he had observed in their proper perspective.

Darwin had planned to use his idle time to compensate for his inattention to studies at the university, by working on languages, mathematics, and classics. Prescribed courses, however, were as distasteful as ever, and he spent increasing amounts of time on other things, particularly geological studies. His devotion to geology, incited partly by his enthusiasm for Lyell's *Principles of Geology,* caused him at first to neglect botanical and zoological observations. Though he collected industriously, his ignorance of anatomy and classification left him with woefully inadequate and incomplete collections. Some of his collections were so worthless that upon his return to England no institution would accept them. He was far from systematic in making collections and was forced at times to borrow specimens from Fitz-Roy's private collections to make his own appear respectable.

Tropical Islands and South America

On St. Jago Island, his first encounter with a tropical land, he observed and collected enthusiastically but indiscriminately, considering everything to be important. He emphasized geology and could not overlook the abundant marine zoological specimens; but he neglected other less obvious fields. In his *Geological Observations,* published later, he described the rock formations and speculated on the origin of this volcanic island.

Throughout the remainder of the voyage to South America, Charles worked industriously whenever his health permitted, occupying his time with notes and specimens. He dragged a net behind the ship, and although he obtained some unique specimens of marine life, he also incurred the displeasure of the officers and men whose job it was to keep the decks clean. During this period, he discovered that the "dust at sea," a phenomenon discussed by many previous voyagers, was in reality the bodies of small "Infusoria," wafted about by the winds.

On February 17, 1832, the *Beagle* crossed the equator and on the last day of February made port at Bahia in Central Brazil. Here, free and dependent upon his own resources, Charles observed and collected along the shore. When he moved inland, he encountered a squabble over his passport but was finally allowed to explore into the interior. Here he observed parasol ants and large spiders, and was fortunate in being able to observe a vampire bat in the act of feeding upon a goat. He also observed numerous fossils and strange living animals of South America, and he indulged in speculation as to whether a tree large enough to hold a prehistoric gigantic sloth had ever existed.

The *Beagle* continued its voyage to Patagonia, where Darwin joined General Rosas in an expedition into the interior which was organized to put down a native revolt by exterminating the natives. Darwin was shocked and distressed by the poverty of the natives and the cruelty of the soldiery, and he was impressed by the proficiency of the natives, particularly in the art of tracking. In Patagonia he observed some of the better known venomous reptiles and had leisure enough to attend several bullfights and operas. He also became acquainted with the native herdsmen (Gauchos) of the pampas country. When they gave him puma meat to eat, he became ill because he thought it was unborn calf, which he had heard was considered a delicacy by the Gauchos.

The ship paused at the southern tip of South America to let off the three young Tierra del Fuegians that Fitz-Roy had taken aboard on the previous trip of the *Adventurer* for the stated purpose of training them as interpreters. The children had gone to England where they attended school and learned to speak English, to wear clothes, and to behave according to the conventions of British people. They were now being returned to their homes. Here Darwin commented on the degenerate condition of the natives, and he wondered if they had remained in this state since the creation of the world.

The trip around the horn left Darwin so ill that he did not care to see anyone or anything, but he was revived by an earthquake at Concepcion. Seeing "geology in action" and obtaining confirmation of his theory concerning the importance of earthquakes in the rise and fall of land levels had a soothing effect on his stomach.

Discoveries on Galapagos Islands

H.M.S. Beagle reached the Galapagos Archipelago, often called "Isles of Enchantment," on September 15, 1835, and on the 17th dropped anchor in a harbor on Chatham Island. Darwin went ashore expecting to find palm trees, gaudy-colored birds, jungles, and parrots. Instead he found a volcanic desert with unexpected types of wildlife. As he visited the various islands, Darwin was amazed at the strikingly similar geologic formations and peculiar faunas on the different islands. His observations of some dull-looking finches, giant tortoises, and garden thrushes, started a train of thoughts which culminated in the *Origin of Species* (1859), which was published 24 years later.

The Galapagos Archipelago consists of ten principal islands (Fig. 12.5), which are located some 600 miles west of Ecuador, South America. Albemarle, the largest of the Galapagos Islands, is approximately 80 miles long and has an elevation of 4,000 feet above sea level. Other islands, named Chatham, Indefatigable, James, Narborough, and Charles are approximately 10-20 miles long and rise to some 3,000 feet above sea level. The entire Archipelago has a water area of some 23,000 square miles and a total land surface of about 2,800 square miles. Albemarle Island comprises more than half of this total land area.

The islands are volcanic in origin and are generally jagged with irregular contours. Crater lakes are present on Chatham and Albemarle Islands. Darwin stated (1845) that the entire Archipelago probably contains 2,000 craters. Most of the islands are geologically recent and smoke still issues from the mouths of some active volcanos. Some islands (e.g., Chatham) are older than the others and their ground is smooth with gentle contours. The Galapagos Islands have a cool climate despite their equatorial location. Their relatively low temperatures are caused by the Humboldt Current which bathes the west coast of South America. Rainfall is intermittent but the general rainy season occurs from December to March.

From the kelp-draped rocks of the littoral zone to the high open country of clubmosses and grasses, Darwin found a constantly changing floral environment. The islands support three general types of floral zones: arid lowlands, a transitional zone, and fertile humid

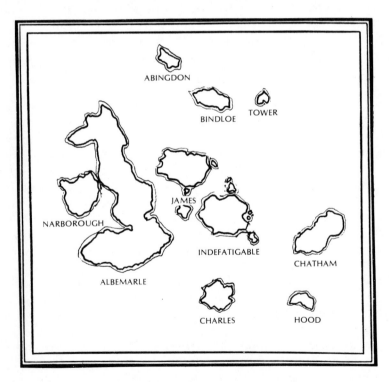

Figure 12.5. Galapagos Islands.

forests at the higher elevations. The larger islands, James, Charles, Indefatigable, and Albemarle, also include a barren, rocky seashore at their lowest elevations and moss-fern areas at their highest elevations. The low elevations of the smaller islands preclude the damp forest and moss-fern open areas. The Galapagos shores consist of low cliffs and black lava boulders, with scattered white sand beaches and an occasional dense mangrove swamp or a few tall cliffs. Recent lava flows have left certain areas bare, jagged, and lifeless. The most noticeable plants of the arid lowlands are tall tree-cacti, dildo trees (West Indian cactus), and torch thistles. Cacti are tall, with the fleshy, spine-covered pads of the prickly pear rising to about 30 feet above the ground. Other vegetation consists of bushes and trees, some being thorny and others smooth.

The transitional zone is extensive on the larger islands and much reduced on the smaller islands. Shrubs and cacti become infrequent and the forest trees with epiphytic lichens become more common as one moves inland from the coast. The forest zone is characterized by rich black soil, high humidity, and tall trees covered with ferns, orchids, lichens, and mosses. This rich floral growth is possible on wind-swept volcanic terrain because of the increased amount of rain received in the area. The floral zone occurs only in the larger islands. High altitudes are characterized by open country with grass, ferns, clubmosses, and occasional bushy thickets.

Darwin found only two species of terrestrial mammals, a mouse and a rat, on the Galapagos Islands. He felt quite certain that the mouse, *Mus galapagoensis*, was indigenous to the islands but later admitted the possibility of its importation from North America by ship. Darwin observed that the rat found on James Island was probably transported from Europe to the island in a vessel, and consequently he did not attach much importance to its presence.

Birds are easily the most abundant of the animals on the islands. Darwin was able to observe 26 kinds of land birds, all peculiar to this archipelago except the bobolink, *Dolichonyx oryzivorus,* which breeds on the North American continent as far north as the 54th parallel. The remaining 25 birds include: a buzzard-like hawk, two species of owls (one short-eared owl and one barn owl), a wren, three tyrant flycatchers, a dove, a swallow, a frigate bird, several boobies, and three species of mocking thrushes. All these species were analogous to, but distinct from, South American species. The remaining land birds were a group of 13 species of finches, related in form of body, plumage, short tails, and structure of their beaks. Darwin observed 11 kinds of water birds and found only three (one was a rail confined to the damp marsh areas of the forest zone) to be new species. The gulls inhabiting the islands were peculiar to these islands but allied to a species in South America.

Darwin commented that the reptiles gave the most striking character to the zoology of the islands. He found a small lizard belonging to a South American genus and two species of the genus *Amblyrhynchos*, which were confined to the Galapagos Islands. He further noted a species of snake that was identical to one *Psammophis temminkii* from Chile. Three species of giant tortoises that Darwin observed on the islands were to give him his first idea that new species could be formed by a process of natural selection.

Sixteen kinds (and two distinct varieties) of land snails were procured and all, except one Helix which was found also at Tahiti, were found to be peculiar to the Galapagos Islands. Darwin took great pains to collect insects on the Galapagos Islands but was discouraged by their scarcity even in the damp forest areas. He did, however, manage to obtain some Diptera, Coleoptera, and Hymenoptera. Fifteen kinds of seafish were obtained and all were found to be new species. Four similar species were known to live on the eastern coast of America.

At the time of his visit to the Galapagos Islands in 1835, Darwin did not attach much importance to the animal life on these islands, but the work and observations he completed there provided the foundation for his later theories on evolution. When introduced to Mr. Lawson, Vice-governor of Charles Island, Darwin conversed with him concerning the possibilities of finding a live volcano on that island. Lawson stated that no live volcanos were available but the animal life presented interesting problems. Lawson claimed that if animals from different islands were lined up before him, he could tell at a glance from which island in the group each animal had come. Striking examples were the dull-colored finches and the giant tortoises. The importance of this did not impress Darwin at first. He went on, collecting blindly and putting specimens from various islands into the same sack. In his notes, he recorded the graduation in beak size and diversity of structure among the finches. He observed that the shape and color of tortoise shells were different on the different islands.

Darwin finally awakened to the potentialities of the situation because of the specific differences in the garden thrush. Separate species were observed on different islands. Why, he wondered, do islands that are formed of the same rocks, and of equal height, and nearly identical climate, have such different tenants? This observation was later used to support the theory of origin of species by natural selection.

On to the South Seas

Four years now had been spent in traveling around South America, and the mission of mapping the coastline had been accomplished. The *Beagle* thus could commence the second part of its assignment, a cruise to the South Seas and around the world.

In Australia, Darwin noted the effects of food supply on human and animal populations. He observed the wide differences in appearance between animals, such as the platypus, in Australia and the animals that inhabited other parts of the world. A study of coral reef structure in the vicinity of Australia gave him material for a later monograph on the structure of coral reefs and atolls.

During the sea voyage and on land expeditions, Darwin collected more specimens than he could carry conveniently on the ship. From time to time he sent back to England, in care of Professor Henslow, cases of specimens that included rocks, snails, fossil bones, spiders, and barnacles. He also sent long letters describing his scientific observations which Professor Henslow read before various scientific societies.

Darwin noted that Lyell's book, *Principles of Geology,* advanced the theory that great geological changes had occurred previously on the earth and that existing agencies for change were responsible for the past changes. This book did much to help Darwin comprehend the past periods of time involved in the history of the earth.

Return to England

After almost five years at sea, the *Beagle* returned to England, arriving on October 2, 1836. Charles Darwin (Fig. 12.6) was immediately acclaimed for the fruits of his collecting, and, to his surprise, he already was well known in scientific circles. The next few months were busy ones. Darwin took lodgings at Cambridge for a short time while he worked on his collection. Now 27 years old and a recognized naturalist, he settled down to his life's work. He was considerably changed from the indifferent student of five years before.

He distributed his collections to the various museums and supervised the publication of the official account, *Voyage of H.M.S. Beagle.* Although busy with this work, he took time in 1839 to court and marry his cousin, Emma Wedgwood. With the coming of a settled home life, he again turned to his scientific labors, but now his health began to fail. He wrote to a friend that it was a bitter mortification for him to digest the fact that "the race is for the strong." Seeking the solitude of the country, Charles Darwin took his wife and family to the little village of Downe in Kent.

Darwin acquired a name as an author when he published *A Naturalist's Voyage Round the World,* compiled from his *Beagle* notes. He also published some observations in the book *Coral Reefs,* a classic example of the use of the scientific method. He had asked himself why coral reefs sprang so steeply from the ocean floor, and why they rose only a few feet above the waves. Their conical shape led him to surmise that they were built on submerged volcanic peaks, but upon investigation he found that coral (animals) could not live below a depth of 150 feet. How was the gap bridged? Darwin showed that the foundation upon which the coral was built must have been near the surface at one time and that it sank gradually while the creatures always worked near the surface of the water. Thus eventually the coral layer was several hundred feet thick and extended into the depths of the ocean. The work demonstrated Darwin's curiosity and relentless pursuit of facts. Actually, the theory does not apply to all coral formation; but it has permanently affected the science of geology. Darwin's geological observations and interests also resulted in two publications: *Volcanic Islands* and *Geological Observations upon South America.*

The *Challenger*

By the latter part of the nineteenth century, governments, more or less routinely, dispatched highly organized scientific expeditions to different parts of the world. Teams of scientists and technicians, rather than individuals, made observations and collections while other teams analyzed the results, wrote reports, and classified museum specimens. Efficiency was enhanced by this cooperative effort and previously inaccessible places were explored with improved facilities. But the adventure associated with exploring and the recognition of individual naturalists, were, to a large extent, lost.

The high point of the nineteenth century's organized scientific exploration and discovery was reached in the expedition of the *Challenger* (1872-76). This voyage was organized by the British Admiralty to study oceanography, meteorology, and natural history. The ship was a sailing vessel equipped with steam. It carried special apparatus for sounding, dredging, and studying the ocean at different depths, particularly at the bottom. Chemistry and biology laboratories were established on the ship. The purpose of the expedition was to study the physical and biological aspects of the Atlantic and Pacific Oceans, the islands, and the shorelines of the continents. The six naturalists on board worked under the supervision of Charles W. Thompson (1830-82). George S. Nares was the ship's captain.

Figure 12.6
Charles Darwin, British
naturalist and explorer.

During the four years of the expedition, vast collections of plants and animals were made. These were studied and classified by a whole army of investigators under the leadership of John Murray (1841-1914). The results, published by the British Government in 50 folio volumes, demonstrated the importance of the physical world in relation to living things, perhaps the first real ecological insight. Among the organisms described in these volumes were several important groups of marine protozoans, particularly Foraminifera and Globigerina. Remains of these protozoans from past geologic ages were found on the ocean bottom, where they created thick layers of chalk. Living forms were also observed, and the deposits were considered to represent comparable forms that had lived in past periods of biological history.

Interesting studies and collections were made from the plankton, or living community, at the surface of the ocean. Many hitherto unknown organisms were found. In the Sargasso Sea, for example, where seaweeds had become established in a vast area of the open water, a wide variety of organisms abounded. Representatives of most of the animal phyla were found in this habitat. The adaptations that the different animals had made to fit the environment became the basis for many studies. Shrimps and crabs, for example, were well concealed in the yellow seaweed because they had become yellow like their environment.

Explorations by Other Governments

Great Britain was the foremost sea power, with colonies around the world during the latter part of the eighteenth and the nineteenth century. In the latter part of the nineteenth century other countries supported expeditions with implications for biology. The United States Government Steamer *Tuscarora* investigated the floor of the Pacific. Other American and Norwegian expeditions studied biological and physiographical features of oceans, islands, and coastlines around the world.

Exploration by Individuals

In addition to the government-organized expeditions, which made massive contributions to biology, many individuals explored, collected specimens, and wrote of their adventures. Alfred Russel Wallace (1823-1913) will be particularly cited here because of his achievements as an explorer and his great contributions in the areas of evolution and geographical distribution of animals.

A. R. Wallace

At the age of 14, Alfred joined his brother, William, in London and set out to learn surveying. The following year he started an apprenticeship as a watchmaker, but before this was completed he returned to surveying and formed a partnership with his brother. While traveling on surveying projects, Wallace became interested in botany and studied plant taxonomy. He was also interested in astronomy and agriculture. Surveying was not in enough demand (in 1843) to occupy the full time of the two Wallaces, so Alfred began writing. In 1844, he became a schoolmaster in Leicester. During this period he read the treatise, "On Population," written by Thomas Robert Malthus (1766-1834), and he found it interesting and suggestive of basic biological relations. At Leicester he met Henry Walter Bates (1825-92), the naturalist, who stimulated in him an interest in entomology.

Wallace and Bates decided to go on a collecting trip to the Amazon region and planned to sell the specimens they collected to defray the costs of the trip. They sailed from England in April, 1848. The following year Wallace's younger brother, Herbert, joined them in South America, but he died a year later with yellow fever. Bates remained in South America for seven years, but Wallace returned to England in 1852. On his way home misfortune struck again. The ship burned, and most of the specimens and notes that he had collected for his own use and to sell were lost. His impressions of the Amazon region, however, remained with him—the majesty and variety of the equatorial forest, the beauty and strangeness of the butterflies and birds, and particularly the concepts he evolved following his contacts with primitively savage human beings—all were so vivid that he began to write about them. During the next few years, Wallace settled in London, worked on the notes and collections he had been able to save, and wrote two books entitled *Travels on the Amazon* and *Palm Trees of the Amazon*.

The trip to the Amazon had stimulated Wallace's interest in further travel and explorations, and in 1854, he decided to visit the Malay Archipelago. This region, he believed, offered the richest possible field to the biological collector. He remained in the Archipelago for eight years, studying the animals on every important island in the group. He noted differences in animal life between the eastern and western parts. The dividing line was a narrow but deep strait between Borneo and Celebes, and between Bali and Lomboc, now known as Wallace's Line. East of Wallace's Line are the Australian-type animals and to the west are the Oriental-type animals.

In 1855, Wallace wrote an essay entitled *On the Law Which Has Regulated the Introduction of New Species*. The theme of this essay was: every species has come into existence coincident both in time and space with a preexisting, closely allied species. During the next three years, Wallace pondered the problem of how such changes could have been brought about, and in 1858 he made a great contribution to the theory of evolution which is discussed in Chapter 13.

Early in 1862, Wallace left the Archipelago and returned to England. He brought with him many specimens, some of which were alive. Again he settled in London, worked on his collection, and wrote about his travels and observations.

Wallace did not have a position with any institution, but made his living and paid the expenses of his travels by selling specimens, giving lectures, and writing books and popular articles. Although he sold many of the specimens which he brought back to London, he kept a large collection for his own investigations. Much of his early work was on taxonomy, but as he grew older his interest turned more to problems of evolution and geographical distribution. His book (1859) on

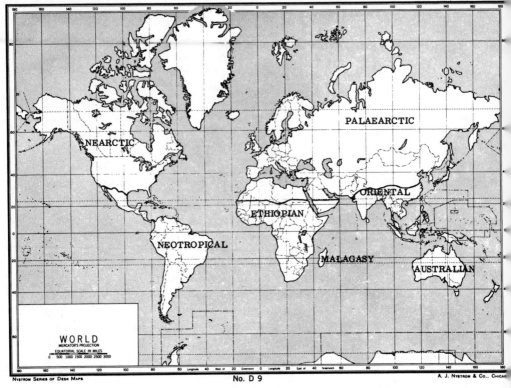

Figure 12.7. Zoogeographical regions of the world.

the Malay Archipelago included many illustrations of natural history as well as vivid accounts of his travels.

At the age of 46, Wallace married and settled in London to a life of study and writing. About this time, he was awarded the Royal Medal by the Royal Society. Wallace wrote his mature views on evolution in a book entitled *Contributions to the Theory of Natural Selection* (1870).

The book entitled *Geographical Distribution of Animals* (1876) established Wallace as the leader in this field of biology. He divided the land masses of the world into six zoogeographical regions (Fig. 12.7) on the basis of geography and mammal inhabitants: (1) Palearctic, all of Europe and most of Asia; (2) Nearctic, North America, in which such animals as the elk, fox, and bear are found (these two realms are quite similar and are usually listed as subdivisions of the Holarctic); (3) Ethiopian, South Africa, the home of the gorilla, giraffe, lion, and hippopotamus; (4) Oriental, South Asia, inhabited by the orangutan, Indian elephant, and flying fox; (5) Australian, the continent of Australia and surrounding islands, where marsupials are found; and (6) Neotropical, South America, the home of the tapirs, sloths, monkeys with tails, and vampire bats. Other plants and animals in addition to the mammals are distinctive in these major geographical areas.

Wallace published *Island Life* (1878), and *Land Nationalization* (1881). During his later years, he made many short collecting journeys on the continent of Europe, especially to Switzerland. He spent 1886 lecturing in the United States, and on his return to England he wrote *Darwinism* (1889), *Studies, Scientific and Social* (1900), *Man's Place in the Universe* (1903), and a two-volume autobiography that was published under the title *My Life* (1905).

Summary

Through exploration of America, Australia, and the South Sea Islands, previously unknown animals and plants were collected and new biological relations were observed. Widespread interest in newly discovered areas led to expeditions sponsored by learned societies and governments. The British government with its great navy and far-flung empire took the lead in mapping shorelines and investigating new lands. Extensive collections of specimens and notes were returned to England, and, more important, distinguished biologists were trained. From the voyage of the *Endeavour,* Joseph Banks emerged as a leading biologist; from the *Investigator,* Robert Brown; and from the *Beagle,* Charles Darwin. The accomplishments of the *Challenger* voyage depended on the cooperation of many investigators working as teams rather than on the single individual. Other countries, particularly Denmark and the United States, participated in later biological explorations. A. R. Wallace, a distinguished free-lance naturalist-explorer, made contributions in the areas of evolution and geographical distribution of animals.

References and Readings

Austin, K. A. 1964. *The Voyage of the Investigator 1801-1803; Commander Matthew Flinders.* London: Angus and Robertson.

Bailey, E. B. 1963. *Charles Lyell.* Garden City, N. Y.: Doubleday and Co.

Baker, S. J. 1962. *My Own Destroyer: A Biography of Matthew Flinders, Explorer and Navigator.* Sydney: Currawong Publishing Co.

Banks, J. Catalog of books brought from Iceland and given to the British Museum by Joseph Banks, Esq. Handwritten list of 117 titles under the heading, "Books Printed in Iceland."

Banks, J. and D. Solander. 1905. *Illustrations of Australian Plants Collected in 1770 During Captain Cook's Voyage around the World in HMS Endeavour with Determinations by James Britten.* London: Longmans and Co.

Barlow, N., ed. 1934. *Charles Darwin's Diary of the Voyage of H. M. S. "Beagle."* Cambridge: At the University Press.

_____. 1945. *Charles Darwin and the Voyage of the Beagle.* London: Pilot Press.

_____. 1958. *The Autobiography of Charles Darwin.* New York: Harcourt, Brace and Co.

Bates, H. W. 1892. *The Naturalist on the River Amazon.* New York: D. Appleton and Co.

Beaglehole, J. C., ed. 1955. *The Journals of Captain James Cook on his Voyages of Discovery.* (2 vol.) Cambridge: At the University Press.

Beard, W. 1958. *Navigators Immortal.* Glebe, Australia: J. Bell and Co.

Bowman, R. I. 1961. *Morphological Differentiation and Adaptation in the Galapagos Finches.* Berkeley: University of California Press. (vol. 58, University of California Publications in Zoology).

Bryant, J. 1928. *Captain Matthew Flinders, R. N., his Voyages, Discoveries and Fortunes.* London: J. Alfred Sharp.

Burroughs, R. D., ed. 1961. *The Natural History of the Lewis and Clark Expedition.* East Lansing, Mich.: Michigan State University Press.

Cameron, H. C. 1952. *Sir Joseph Banks.* London: The Batchworth Press.

Conway, A. and F. Conway. 1947. *The Enchanted Islands.* New York: G. P. Putnam's Sons.

Cooper, H. M. 1953. *The Unknown Coast, Being the Exploration of Captain Matthew Flinders, R. N. along the Shores of South Australia, 1802.* Adelaide, Australia. (Printed for the author; limited to 500 copies.)

Cutright, P. R. 1940. *The Great Naturalists Explore South America.* New York: Macmillan Co.

Darling, L. 1960. "The 'Beagle'—A search for a lost ship." *Nat. Hist.* 69: (5) 48-58.

Darwin, C. R. 1901. *The Structure and Distribution of Coral Reefs.* New York: D. Appleton and Co.

_____. 1896. *Geological Observations of the Volcanic Islands and Parts of South America Visited During the Voyage of H. M. S. "Beagle."* New York: D. Appleton and Co.

_____. 1845. *Journal of Researches into Geology and Natural History of the Various Countries Visited by H. M. S. "Beagle."* New ed. London: H. Colburn, 1939. (Later ed. of vol. 3 of *Narrative of the Surveying Voyages of His Majesty's Ships "Adventurer" and "Beagle,"* describing only the "Beagle" voyage.)

_____. 1930. *A Naturalist's Voyage Around the World in H. M. S. "Beagle."* London: Oxford University Press.

_____. 1958. (reprinted) *The Voyage of the "Beagle."* New York: Bantam Books.

_____. 1896. *Origin of Species.* New York: D. Appleton and Co. (Original date of publication 1859. Several other copies prepared by different publishers are available.)

Darwin, F., ed. 1896. *The Life and Letters of Charles Darwin.* 2 vol. New York: D. Appleton and Co.

_____. 1950. *Charles Darwin's Autobiography.* New York: Henry Schuman.

Duncan, A. 1821. *Life of Sir Joseph Banks.* Edinburgh: P. Neill.

Eiseley, L. C. 1954. "Alfred Russel Wallace." *Sci. Amer.* 200:(2) 70-84.

Flinders, M. 1814. *A Voyage to Terre Australis; Undertaken for the Purpose of*

Completing the Discoveries of that Vast Continent. London: G. and W. Nicol.

_____. 1814. *Charts of Terre Australis, Commander of Investigator 1799-1802.* London: G. and W. Nicol.

George, W. 1964. *Biologist Philosopher, A Study of the Life and Writing of Alfred Russel Wallace.* New York: Abelard-Schuman.

Goodspeed, T. H. 1961. *Plant Hunters in the Andes.* Berkeley: University of California Press.

Hagen, V. W. von. 1945. *South America Called Them.* New York: A. A. Knopf.

_____. 1949. *Ecuador and the Galapagos Islands.* Norman, Okla.: University of Oklahoma Press.

Hawkesworth, J. K. 1773. *An Account of the Voyages Undertaken by the Order of his Present Majesty for Making Discoveries in the Southern Hemisphere and Successfully Performed by Commodore Byron, Captain Wallis, Captain Carteret, and Captain Cook in the Dolphin, the Swallow, and the Endeavour.* 3 vols. London: W. Strahan and T. Cadell.

Herber, E. C. 1963. *Correspondence Between Spencer Fullerton Baird and Louis Agassiz—Two Pioneer American Naturalists.* Washington, D. C.: Smithson-Institution.

Hermannsson, H. 1928. *Sir Joseph Banks and Iceland.* London: Oxford University Press.

Hill, E. 1941. *My Love Must Wait; The Story of Matthew Flinders.* Sydney: Angus and Robertson.

Hoff, R. and H. de Terra. 1959. *They Explored!* New York: Henry Z. Walck. (Chapter 1, Von Humboldt)

Holmyard, E. J. 1951. *British Scientists.* London: Dent.

Lack, D. L. 1947. *Darwin's Finches.* Cambridge: At the University Press.

Lloyd, C., ed. 1949. *The Voyages of Captain James Cook Round the World.* New York: Chanticleer Press.

Mackaness, G. 1936. *Sir Joseph Banks, His Relations with Australia.* Sydney: Angus and Robertson.

Maiden, J. H. 1909. *Sir Joseph Banks, the Father of Australia.* London: Kegan, Paul, Trench, Trubner and Co.

Moore, R. E. 1955. *Charles Darwin, A Great Life in Brief.* New York: Alfred A. Knopf.

Niles, B. 1946. *Journeys in Time.* New York: Coward-McCann.

Parker, J. W. 1844. *Sir Joseph Banks and the Royal Society.* London: John W. Parker. (A popular biography with historical introduction and sequel. Appendix includes: Presidents of Royal Society and Abstract of Status of Royal Society, 1840.)

Pekin, L. B. 1938. *Darwin.* New York: Stackpole Sons.

Rawson, G. 1946. *Matthew Flinders, Narrative of his Voyage in the Schooner Francis: 1798.* London: The Golden Cockerel Press.

Scott, E. 1914. *The Life of Captain Matthew Flinders R. N.* Sydney: Angus and Robertson.

Sears, P. B. 1950. *Charles Darwin, The Naturalist as a Cultural Force.* New York: C. Scribner's Sons.

Suttor, G. 1855. *Memoirs, Historical and Scientific, of the Rt. Hon. Sir Joseph Banks.* Parramatta, Australia: E. Mason.

Wallace, A. R. 1876. *The Geographical Distribution of Animals.* New York: Harper and Bros.

West, G. 1938. *Charles Darwin: A Portrait.* New Haven: Yale University Press.

CHAPTER 13

EXPLAINERS OF CHANGES IN POPULATIONS

Evolution, like many theses of modern science, was originally formulated by philosophers. With the coming of the Ionian Greek philosophers (Chapter 3), the concept of change in nature began to take form. Empedocles brought together the ideas of his predecessors and developed a number of concepts that might be interpreted now as consistent with the modern theory of evolution. He believed in spontaneous generation of lower forms, which could explain the origin of elementary life, but his spontaneous generation concept did not include the origin of higher forms; he considered them to develop from lower forms by a slow and gradual process.

Empedocles visualized plants as being the first living things to appear on the earth. According to this view, they were pushed up from beneath the surface of the earth and through the earth slime by internal fire. From the plants, animals were produced. This transition, like other major changes among living things, was not rapid but was in progress for ages and came about through the fortuitous acts of the two great forces, love and hate, acting upon the four elements fire, air, water, and earth. Thus higher animals were not produced directly as such, but were at first strange and awkward monsters, most of which could not adapt to their environment and therefore perished. After innumerable trials, nature found some combinations that resulted in animals capable of living and reproducing their kind. Although Empedocles' concepts may seem crude and primitive, he hit upon a number of truths that have been verified by present-day biologists: (1) higher forms developed gradually, (2) plants evolved before animals, (3) poorly adapted forms were replaced by forms better adapted to the prevailing environment; that is, poorly adapted forms became extinct while well-adapted forms lived and reproduced their kind.

Aristotle also believed in the slime origin for lower forms of life, but, like Empedocles, he thought that higher forms represented a progressive series. According to Aristotle, position in the series depended on the degree of *Psyche* that each form contained. Plants and plantlike animals such as sponges had only a vegetable soul, lower animals had an animal soul, higher animals a rational soul, and the highest type of human beings had an intellectual soul. A broad phylogenetic sequence was thus postulated. Aristotle believed in a great plan that governed the universe and all living things. The plan was so designed, however, that change and progress could occur. Necessity brought about change in nature. This idea led to a belief in prenatal influence and in the inheritance of acquired characteristics. Aristotle was more interested in experiment and induction than were his predecessors, contemporaries, or most of those who followed. He did not, however, use these methods in the broad study of evolution but arrived at his ideas through philosophical reasoning. In spite of the weaknesses of Aristotle's work on evolution, it was not materially challenged for some 2,000 years. Although his general concept of evolution is valid today, most of Aristotle's specific theories bearing on evolution have been either discarded or revised.

Some theologians of the Christian era commented on the subject of evolution. St. Augustine (A.D. 354-430), for example, favored an allegorical interpretation of the book of Genesis in the *Bible* and openly promoted an evolutionary concept as opposed to special creation. The six-day creation plan was interpreted as a "series of causes" long enough for those things that were in the mind of God to be brought to their completion. St. Augustine took a position halfway between abiogenesis and biogenesis in his discussions on creation. From the first, he said, there had been two types of germs present on earth: (1) the visible, associated with the bodies of plants and animals, and (2) the invisible, which under proper conditions would grow into living organisms without the cooperation of other living organisms. Not only did he apply this idea to all living things but to inorganic matter as well; thus, the moon, sun, and earth were, at first, germs. Because this process, once established, was self-perpetuating, continued change was inevitable.

Pre-Darwinian Evolutionists

From the time of the early Greek philosophers until the first part of the Christian era, evolution was discussed freely and without prejudice by at least a nucleus of scholars. The more liberal church fathers were not shocked by it, and some even considered it a valid alternative theory to special creation and not inconsistent with the scriptures. During the Middle Ages there was little interest in such subjects. The Renaissance was marked by sporadic developments in biology. Few contributors of that period were deeply enough involved in philosophical subjects to consider evolution seriously. The seventeenth and eighteenth centuries were dominated by the systematists who attempted to classify natural objects, including plants and animals, into fixed and rigid systems (Chapter 9). The fixity of species became firmly established by Linnaeus and his followers. In that atmosphere, evolution was crowded out of the scene, and few biologists gave any attention to the subject until the nineteenth century. Other aspects of biology were developed prior to the nineteenth century, but little was done to establish the biological principle of evolution.

Buffon

George Louis Leclerc, comte de Buffon (1707-88; Fig. 9.4) was a French naturalist, politician, and writer who discussed evolution extensively and took different positions on the subject at different periods of his life. During his early life, he took an extreme view in favor of special creation, much like that of his contemporary Linnaeus. He did not, however, agree with Linnaeus in other respects. Linnean classification he depicted as trifling and artificial. Buffon considered nature as a whole and looked for large likenesses rather than trivial differences. This obviously led to a consideration of broad natural relations and evolution.

Later in life Buffon developed an extreme view on the inheritance of acquired characters. The factors that he visualized as influencing evolutionary change were: (1) direct influences of the environment, (2) migration, (3) geographical isolation, and (4) overcrowding and struggle for existence. These factors would be expected to result in a gradual development of new forms of life rather than abrupt changes.

At a still later stage in his career, when he was more involved in politics, Buffon adopted a more liberal, middle-of-the-road position and wavered between the extremes of his earlier views. He compromised the position of special creation and evolution with a number of wild speculations. The pig, for example, was described as a compound of other animals, the ass was a degenerate horse, and the ape a degenerate man. Buffon was a prolific writer and an interpreter of contemporary thought, but he was not an original investigator. He had broad interests in mathematics, physical science, and biology, and wrote a *Natural History* that was intended to embrace all scientific knowledge. Some 44 volumes were actually published, some by an assistant after his death.

E. Darwin

Erasmus Darwin (1731-1802), English philosopher and free thinker, was somewhat clearer than Buffon on the subject of evolution. The name of his best-known book, *Zoonomia* (1794), was coined to represent the laws of organic life. In this book Darwin developed the theme of the inheritance of acquired characters. The age of the earth was described in millions of years and life was considered to have originated from a primordial protoplasmic mass. The struggle for existence that was elaborated by Charles Darwin, grandson of Erasmus, was suggested in *Zoonomia*.

In the preface to *Zoonomia*, Erasmus Darwin wrote, "The Great Creator of all things has infinitely diversified the works of his hands but has at the same time stamped a certain similitude of the features of nature, that demonstrate to us that the whole is one family of one parent. On this similitude is founded all rational analogy." Erasmus Darwin independently concluded that species descend from common ancestors. He speculated on the variations in animals, the reproduction of the strongest, and the struggle for existence. It has been recognized that Charles Darwin received more from his grandfather than was originally supposed. For every volume written by Charles,

Figure 13.1
Jean Baptiste Lamarck,
French biologist and
pre-Darwinian evolutionist.
Reprinted with permission
from Gardner, *Principles
of Genetics.* Copyright ©
1960, 1964, 1968 by
John Wiley & Sons, Inc.

there is a corresponding chapter by Erasmus. Charles, in one of his written treatises expressed his disappointment in finding *Zoonomia* "more speculative than scientific."

Lamarck

The most important of the eighteenth century evolutionists was Jean Baptiste Lamarck (1744-1829; Fig. 13.1). When his father died, he was 16 years old (the youngest of 11 children) and was enrolled in a Jesuit college. Since he could no longer remain in school, he joined the French Army in the Seven Years' War and was stationed in Monaco. He injured his neck while scuffling with a friend and was returned to France for treatment. A lymphatic gland was affected and Lamarck never fully recovered, even after surgery. When he became sufficiently strong, he entered upon the study of medicine and botany in Paris and supported himself with part-time work in a banker's office. While in Paris he came under the influence of the well-established botanist, Jussieu. In 1781 and 1782 Lamarck traveled across Europe studying plants and taking notes for his *Dictionary of Botany*. Later he wrote *Flora of France,* published by the French government. Through the assistance of Buffon, he obtained a job in the Museum of Natural History where he contributed further to botanical literature.

While in the midst of becoming a good botanist, Lamarck changed his field to invertebrate zoology, and over a period of years he prepared a seven-volume systematic study of the invertebrates. His best-known written work was *Philosophical Zoology* (1809). He was married four times, had many children, was always poor, and in his later life he was blind. When death came to him at the age of 85, he was alone and unremembered.

Lamarck's contributions to systematics, botany, and comparative anatomy have already been mentioned. His contribution to evolution was a substantial one, even though it has not always been fully appreciated. Lamarck described the animal kingdom as a series that graded from simple to complex forms. In his view, no group became extinct through abrupt catastrophes as Cuvier contended, but one form changed into another. Lamarck was not as well trained and disciplined as Cuvier and Buffon, but he gave the first detailed defense of evolution. When he explained the mechanics of evolution, he made use of the theory of inheritance of acquired characters that had come from the Greeks and had been developed by Erasmus Darwin and Buffon. Lamarck made bold speculations and considered the subject in great detail, and even though he did not originate the view, his name has become associated with the concept of inheritance of acquired characteristics. Briefly, he held that: (1) the environment modifies plants and animals, (2) new needs modify old organs and bring new ones into use, (3) use and disuse modify development, and (4) these modifications are inherited.

Lamarck's other contributions to evolution were more sound but, ironically, are not as well remembered as his speculation on the inheritance of acquired traits. He appreciated the function of isolation in the formation of new species, recognized the influence of proximity in destroying differences between varieties within species, saw unity existing in nature, provided the first diagram of an evolutionary tree, and understood, at least in a general way, the physiological balance maintained in nature. Lamarck was first to use the word "species" to describe a natural unit of related animals or plants. Aristotle had used the term, and in the years that followed the word was used extensively in logic. It had been applied to groups of animals and plants by Ray and other systematists, but Lamarck gave the word its modern interpretation.

Cuvier and Saint-Hilaire

Georges Cuvier has already been cited for his contributions in systematics (Chapter 9), comparative anatomy, and paleontology (Chapter 10). Although he materially aided in perfecting the geological timetable and possessed the fundamental knowledge on which evolution is now based, his personal influence on evolution was negative. He openly opposed evolution as a theory and supported the alternative view of fixity of species. His theory of catastrophism and successive new creations was popularized at the expense of the alternative theory of gradual change supported by Lamarck. Cuvier belittled Lamarck and was largely responsible for the lack of recognition accorded Lamarck in later life.

Geoffroy Saint-Hilaire (1772-1844), a contemporary and colleague of Cuvier, opposed Cuvier's views and defended the evolution concept. He believed, with Buffon and Lamarck, that the direct effects of environment explained small variations. Saint-Hilaire recognized the evolutionary effects of isolation and visualized physiological as well as geographical isolation as a factor in species formation.

A well-known incident arose from a controversy between Cuvier and Saint-Hilaire concerning the origin of the squid. Instead of discussing the matter privately and objectively, a public debate was announced and widely publicized, making the issue more emotional than scientific. Saint-Hilaire, who had an idea of evolution, but only meager and crude observations, was right in principle but poorly prepared and perhaps wrong in detail. Cuvier explained the origin of the squid on the basis of special creation. He was wrong in general principle but well prepared with supporting details and won the debate by a better argument and a dramatic presentation. The effect was to retard the study of evolution for several generations.

Charles Darwin and the Evolution Concept

During his five years on the *Beagle,* Charles Darwin observed and collected plants, animals, and fossils in different parts of the world (Chapter 12). On the Galapagos Islands he was impressed with the gigantic tortoises and large crabs that were not like those on the shores of the mainland some 600 miles away. Darwin noted the diversity of the finches. Freed from competition on isolated islands, these birds had become specialized to occupy ecological niches they could not have filled on the mainland. Associated with their ecological specialization were adaptive structural changes, for example, in the size and shape of the beak. To Darwin this indicated that a single ancestor or a few ancestral forms had arrived on these islands, the descendants of which had moved into different habitats and formed new species. These observations were of great significance in the formation of his concept of evolution by natural selection.

Beginnings of a Theory

Darwin observed that different islands sometimes had entirely different species of armored animals, while transitional forms might or might not occur. He kept notes on all these observations and pondered over the strange relations that were indicated. As his data accumulated, his faith in the fixity of species was shaken. In 1838, following his return to England, he published the *Journal of Researches* and took care of other matters of immediate importance. When these tasks were completed, he returned to his notes on species formation and took time to reflect on the significance of his observations.

The problem to be solved was: are species fixed and unchangeable or do groups of living creatures become modified over periods of time in nature, and if so, how? The general belief of the day was that plant and animal species were originally created essentially as they are at any given time. A few observers had postulated that existing species were descended from other species and had gradually become modified in the process. But why should modification occur? It was common knowledge that man could "select" certain types of domestic animals and alter the characteristics of a breed. But how selection could be applied to organisms living in nature was a mystery.

At first Darwin was merely classifying observed facts. As a theory took shape, however, he began working with a purpose. The key to the problem apparently came to him in 1838, when he read an essay by Thomas Robert Malthus entitled *An Essay on the Principle of Population; or a view of its Past and Present Effects on Human Happiness.* Briefly, the theme of the essay was that man multiplies more readily than does his supply of food; therefore, competition occurs for the requirements of existence. The prize for which the competitors struggle is life itself, and the success of one means the failure of others. Darwin decided that if he could demonstrate that favorable variations tend to be preserved in living populations and unfavorable variations are destroyed, he could show how new species come into being. He was trying to find out what actually occurred in living population units in nature. The harder he worked, the more fascinated he became.

Darwin spent some 20 years working over his theory. In the early part of this period (1842), he prepared a small paper outlining the theory, which he sent to his biologist friends for criticism. With the help they provided, he strengthened each point and improved the theory. He read, discussed, observed, and spared no effort in making the case as clear and well documented as possible. A book was planned in which all of his data would be presented. In the meantime, while the great book was taking form, he undertook the writing of an abstract. This was about half written when Darwin suffered a shattering blow. He received for review a short manuscript written by the explorer-naturalist, Alfred Russel Wallace, entitled *On the Tendency of Varieties to Depart from the Original Type.* So closely did it agree with his own theory, he commented later, that it might have been an abstract of his own work.

The two authors of the evolution theory had arrived at their conclusions through entirely different paths. Darwin had pondered the matter for some 20 years and collected volumes of data. Although Wallace had given the general subject considerable thought during the preceding three or four years, he apparently had arrived at the conclusion in a single flash of insight. In February, 1858, during an attack of yellow fever, Wallace was confined to his bed and he had time and inclination to think about the problem of how living populations have arrived at their present status. He remembered the thesis of Malthus on population which he had read many years before and hit upon the idea of "survival of the fittest" as it applied to animals and plants. He thought out the theory in a few hours and by the evening of the same day he prepared a rough outline of the idea. Two days later he had written the paper which was later sent to Darwin for review.

Darwin generously recognized the contribution of his then obscure, young colleague and suggested that Wallace's paper be published immediately. Friends of Darwin, Charles Lyell, and J. D. Hooker, who knew of Darwin's long and persistent work on the same problem, intervened and suggested a joint publication by the two pioneers, summarizing the new theory of evolution. A joint paper entitled, "On the Tendency of Species to Form Varieties; and on the Perpetuation of Varieties and Species by Natural Means of Selection," was read by Lyell and Hooker before the Linnean Society on July 1, 1858. The joint paper by Darwin and Wallace was published in the *Journal of the Proceedings of the Linnean Society* for August 20, 1858. When the paper was read there was interest but little discussion. The reaction that followed was tremendous. Wallace discussed aspects of evolution in later years (Chapter 12) and became an authority in the related area of geographical distribution of animals, but Darwin took the main responsibility of developing the theory that he and Wallace had presented jointly.

Figure 13.2
Charles Darwin as an
elderly man. (From
Gardner, *Principles of
Genetics.* Copyright©
1960, 1964, 1968 by John
Wiley & Sons, Inc.)

Origin of Species

The great book which Darwin had planned was never written. He moved into high gear on his "abstract" (of some 350 printed pages) and completed it in 13 months. It was published on November 24, 1859, under the title, *On the Origin of Species by Means of Natural Selection, or the Preservation of Favored Races in the Struggle for Life.* Every copy of the first edition was sold on the day of publication.

Darwin (Fig. 13.2) was not the first evolutionist, as shown in the first part of this chapter. Much of his theory may have come from his grandfather, Erasmus Darwin. Some recent reviewers have contended that Charles Darwin has received too much credit, but on the other hand, even though his work may not have been entirely original, he deserves much credit for refusing to accept the unproductive hints, guesses, and hypotheses that had gone before him. When the *Beagle* sailed from Plymouth in 1831, no leading scientist was actively interested in evolution. When Darwin puzzled over the facts collected in his notebooks, he was the only man in the world, so far as is now known, who was seriously considering the possibility that members of one species could be modified in such a way that they could ultimately give rise to another species.

Darwinism, or the theory of natural selection, was Darwin's own creation although it was perhaps based on the work of Erasmus Darwin. The theory may be paraphrased: All living creatures multiply so fast that unless the greater part of each generation perished without leaving offspring, the world would rapidly become overpopulated; but the facts prove that the size of animal and plant populations remains roughly stationary. Competition must exist between species and within species for a means of life, with the penalty for defeat being death. It follows that if any member of a species differs from its fellows in a way that gives it an advantage in competition, this member is more likely to survive than are other members of the same population. Such a favored individual is more likely to have offspring and the offspring would be expected to perpetuate the advantage which enabled the parent to survive and reproduce in its environment. Darwin considered all variation to be inherited.

Source of Variation

Darwin was puzzled concerning the ultimate mechanism for variation and he recognized this problem as one of the important but unsolved problems of biology at the time the *Origin of Species* was published in 1859. When the book, *The Variation of Animals and Plants under Domestication,* was completed in 1868, Darwin had arrived at an explanation (i.e., pangenesis) of why variation occurs. His pangenesis theory was based on a modified interpretation of the theory of inheritance of acquired traits. It was merely an attempt to explain the source of hereditary variation and had no bearing on his basic theme of natural selection. Natural selection was based on inherent properties in the organisms concerned.

A much discussed example of evolutionary change was the development of the giraffe's long neck. Lamarck thought that the extended neck was the direct result of each generation of giraffes stretching more and more for the top branches of the trees. The Darwinian theory stated that the giraffes that had inherited genetic determiners for long necks had less chance of starving than those that inherited factors for shorter necks. If there were too many giraffes for the food supply, those with shorter necks must starve when all the lower branches of the trees were stripped of leaves. Those with the longest necks would be most likely to grow to sexual maturity and leave long-necked descendants. Thus giraffes would develop longer necks, generation after generation, until there was no further advantage in this trend. Darwin remained puzzled about the mechanics of variation on a generation through generation basis, but considered particular traits to be inherited although influenced in some way by the environment.

In proving that natural selection was operable, Darwin showed that: (1) new forms can be developed by man's initiation of artificial selection; (2) conditions in nature are sufficient to exert a similar selection in the natural world; and (3) the wide variety of forms existing over the world can be explained by this means. The suggestion was that all animals are related and have a common ancestor. A great controversy developed in this regard, particularly among laymen not acquainted with the basic principles of biology.

Darwin's Later Writings

While the controversy over evolution raged, Darwin worked quietly, gathering support for his theory and resolving other biological problems. He spent much time in botanical research. In 1862, he published his observations on the fertilization of orchids under the title, *On the Various Contrivances by which British and Foreign Orchids are Fertilized by Insects, and on the Good Effects of Intercrossing.* This title seemed innocent enough on the surface, but the content of the book was related to the earlier work on the origin of species. In commenting on this book Darwin said, "My chief interest in the orchid has been that it was a 'flank movement' on the enemy." During this period, the orchid represented an example repeatedly used by anti-Darwinians. Of all of nature's creations, these delicately beautiful flowers were held to be a conspicuous instance of "botanic art for art's sake." Darwin quietly undermined the old argument of special creation by showing that the apparently meaningless ridges and horns of the orchid attract insects that pollinate the flowers. The intricate structures of trapdoors and spring mechanisms had evolved from simpler forms to serve a vital function in the life of the plant.

Two years later (1864) Darwin produced *The Movements and Habits of Climbing Plants.* A preview of this book was published first in the *Journal of the Linnean Society,* and the material was expanded into a book in 1875. Darwin's first enlargement of the evolution theory came in 1868, with *The Variation of Animals and Plants under Domestication.* Next he went to the *Descent of Man and Selection in Relation to Sex* (1871) which included mankind in the animal kingdom, to which the evolution theory could be applied. Darwin also developed theories of sexual selection to explain the mechanics of evolutionary change. In 1872, Darwin made a further contribution in *The Expression of the Emotions in Man and Animals*. His later writings were devoted mainly to botanical subjects.

Darwin's impact is attributable to something that cannot be understood by simply reading an account of his life or reviewing his contributions. He was an upright man, an indefatigable naturalist, and an honest thinker. But integrity, tirelessness, and honesty are qualities that often come together but fail to produce the effect that this man's life produced. Darwin amounted to much more than the sum of his parts. In his mature years, he proved to be a man of extraordinary capacity and theoretical insight. These qualities enabled him to build on the knowledge he gained from his unusual and most valuable experience on the *Beagle* voyage. He was the pioneer who catalyzed a revolution in human thought. His influence upon his own and succeeding ages was enormous.

In some of his last writings, Darwin said modestly that he regretted that he had not done more direct good to his fellow creatures. His kindly character is revealed in his comment on his favorite mode of relaxation, which was novel reading, in which he "blessed all novelists," provided they wrote books with happy endings. His own ending came on April 19, 1882, in his 74th year, and science mourned the loss of the thinker who had changed man's concept of the world. Charles Darwin was buried in Westminster Abby, a short distance from the grave of Isaac Newton.

T. H. Huxley

Thomas H. Huxley (1825-95) was the foremost of Darwin's defenders. At an early age he showed a talent for art which served him well throughout his life. He had an inquiring mind and a tendency to indulge in metaphysical speculations. By the time he was 12 years of age, he had read most of the books in his father's large library, but his formal instruction was scanty because of the then poor quality of the public schools in the vicinity of his home in England. His first introduction to higher education was at Sydenham College, where he did preparatory work for the university. In 1845, he received the M.B. degree at the London University. Two years later, he completed the required medical studies but was too young to qualify for the College of Surgeons.

He joined the medical service of the British Navy and was assigned to the ship, *Victory*. After seven months he transferred to the *Rattlesnake*, which was leaving England on a four-year cruise, mostly in Australian waters. During this voyage Huxley industriously took notes

and included detailed accounts of observations on biological subjects. When he returned to England his records were so voluminous
that no private publisher would accept the work. A choice had to
be made between summarizing the data briefly enough to be acceptable to a commercial publisher or financing the production by some
other means. The manuscript was presented to the British government
for publication, but the Admiralty was also reluctant to publish such a
large work. The Royal Society of London had sufficient funds from a
grant and undertook the publication in 1854, under the title *Oceanic
Hydrozoa.* This momentous work established Huxley in the scientific
world, and he was soon in the front ranks as a naturalist.

Although evolution was being discussed by several scholars during
this period, Huxley was not impressed with the evidence for the
theory. Not only did he consider the evidence insufficient, but the
current explanation of the causes of transmutation of species did not
seem adequate to explain the phenomena. He had studied Lamarck's
work but was not convinced. In 1852, he became acquainted with
Herbert Spencer, a leading proponent of evolution, who tried to
convert him to the evolution concept. Huxley resisted because he
saw no mechanical explanation of its mode of origin. Shortly after
his first interview with Darwin, he expressed his view of rigid lines of
demarcation between species. As he became better acquainted with
Darwin's reasoning, however, he found it to be an intelligible hypothesis that he considered sufficient as a working basis.

Darwin's work, which culminated in the *Origin of Species,* was in
progress and Huxley acted as adviser and agent. He talked with influential men and persuaded them toward the view of natural selection.
In November, 1859, when the *Origin of Species* was published, Huxley
began a vigorous campaign in support of the work. In 1860, a meeting
was held at Oxford in which the controversial issue of special creation versus evolution was debated. Bishop Wilberforce, a teacher of
mathematics, took the side of special creation and Huxley defended
evolution. It was in this manner that Huxley obtained the nickname
"Darwin's bulldog." He was called upon many times to champion the
cause of evolution, and he capitalized upon opportunities afforded by
public debates.

In his later life Huxley was recognized as one of the great biologists of the day. He published many papers on different biological
subjects. Much of his time was devoted to preparing and presenting
public addresses. A broad knowledge of biological subjects was
developed through travel and study. In the course of his 70 years, he
lectured and published on virtually all areas of biological science.

Some of his best known works were textbooks on physiology, comparative anatomy, and paleontology. His grandsons, Julian and Aldous, have carried on the tradition of scholarly accomplishment.

Evolution Today

Considerable discussion has been devoted to the theory of evolution since the publication of the *Origin of Species* in 1859. Supporting material for the theory was greatly altered and improved by Darwin himself to meet the difficulties that he recognized and those pointed out by others. Later editions of the *Origin of Species* were so different from the first that questions have arisen as to what Darwin actually said on certain topics. On the whole, the theory that finally emerged has been well received by biologists universally.

The science of genetics and other even more recent developments in biology have provided data that have clarified some of the mechanics involved in the process. In fact, Darwinism is more firmly established now than ever before because of such advances. Four basic concepts have been recognized since Darwin's time that help to explain how evolution operates. They are: (1) mutation, the original source of variation; (2) gene recombination, which is more effective than gene mutation in providing immediate variation in species with sexual reproduction; (3) selection (natural or artificial), the main directing force in evolution; and (4) isolation, which makes speciation possible; that is, physical barriers confine units of a population for a long enough period of time to make it possible for genetic barriers to develop and for new species to originate.

The evolutionary process that produced currently living things has operated over millions of years and obviously can never be reconstructed in detail. In addition, the life span during which a human being may make observations is too short to allow the individual to see much of "evolution in action." Substantial evidence for the fact of evolution, however, has been obtained from fossil records, geographic distribution patterns, comparative anatomy, embryology, and in varying degrees from virtually all of the other biological sciences.

It is now assumed that the same processes that occurred over long periods of time and were responsible for evolutionary change in the past are operating today. Some of these are being explored experimentally. Although many details are yet to be explained, the broad principles are now understood on the basis of the hereditary process and the influence of the physical environment. Problems that have been investigated recently and some now being resolved include: (1) sources of variation in natural populations; (2) establishment of new variations in populations; (3) directing forces of evolution; and (4) isolating mechanisms through which new species can be separated from related breeding populations.

Summary

The evolution concept was approached by early Greek philosophers, discussed by Christian theologians, particularly St. Augustine, and mentioned during the Renaissance. Systematists of the seventeenth and eighteenth centuries established a fixed and rigid system for plants and animals that was useful for purposes of classification but took little cognizance of the natural relations existing among animals and plants.

Biologists of the eighteenth century, particularly Buffon, E. Darwin, and Lamarck, refocused attention on the natural relations among groups of living things. Charles Darwin, reflecting on his observations during the voyage of the *Beagle,* spent some 20 years developing his theory concerning the origin of species. He would undoubtedly have continued methodically gathering support for the theory if Wallace's similar theory had not come to his attention. The papers of Darwin and Wallace were presented jointly in 1858, and Darwin shifted into high gear and prepared the major manuscript for publication in 1859. Following publication of the *Origin of Species,* Darwin continued his scholarly studies and writings while T. H. Huxley and others publicly defended Darwin's work on evolution.

Darwin's basic theory of natural selection has been widely accepted today. The original theory was modified in several ways by Darwin and later scientists. It has now been supplemented by the mutation theory, accounting for the origin of variation; Mendelian recombination, which provides the more significant source of variation in sexually reproducing organisms; and modern explanations of isolating mechanisms that facilitate speciation.

References and Readings

American Philosophical Society Proceedings. 103:(2) April 23, 1959. Commemoration of the Centennial of the Publication of the *Origin of Species.*

Ardrey, R. 1963. *African Genesis.* New York: Dell Publishing Co.

Barnett, S. A., ed. 1958. *A Century of Darwin.* London: Heineman.

Barzun, J. 1941. *Darwin, Marx, Wagner, Critique of a Heritage.* Boston: Little, Brown and Co. (Part I on Darwin and *Origin of Species.*)

Bates, M. and P. S. Humphrey, eds. 1956. *The Darwin Reader.* New York: C. Scribner's Sons.

Blinderman, C. S. 1957. "Thomas Henry Huxley." *Sci. Monthly* 84:171-182.

Boyd, W. C. 1950. *Genetics and the Races of Man.* Boston: Little, Brown and Co.

Cannon, H. G. 1959. *Lamarck and Modern Genetics.* Springfield, Ill.: Charles C. Thomas.

Carter, G. S. 1957. *A Hundred Years of Evolution.* London: Sedgwick and Jackson.

Clark, J. D. 1958. "Early man in Africa." *Sci. Amer.* 199:(1) 76-83.

Coon, C. S. 1962. *The Story of Man.* New York: A. A. Knopf.

Daniel, G. E. 1959. "The idea of man's antiquity." *Sci. Amer.* 201:(5) 167-176.

Daniels, G., ed. 1968. *Darwinism Comes to America.* Toronto: Blaisdell Publishing Co.

Darlington, C. D. 1959. "The origin of Darwinism." *Sci. Amer.* 200:(5) 60-66.

Darwin, C. R. 1859. *On the Origin of Species by Means of Natural Selection, or the Preservation of Favoured Races in the Struggle for Life.* London: John Murray. (Numerous revisions and reprints have been made since original publication. Now available in paperback edition by Mentor Book Co., New York.)

––––––. 1868. *On the Various Contrivances by which British and Foreign Orchids are Fertilized by Insects, and on the Good Effects of Intercrossing.* London: John Murray.

––––––. 1868. *The Variation of Animals and Plants under Domestication.* 2 vol. London: John Murray.

––––––. 1871. *The Descent of Man, and Selection in Relation to Sex.* 2 vol. London: John Murray.

––––––. 1872. *The Expression of the Emotions in Man and Animals.* London: John Murray.

––––––. 1872. *Insectivorous Plants.* London: John Murray.

––––––. 1875. *The Movements and Habits of Climbing Plants.* London: John Murray.

––––––. 1876. *The Effects of Cross and Self Fertilization in the Vegetable Kingdom.* London: John Murray.

––––––. 1877. *The Different Forms of Flowers on Plants of the Same Species.* London: John Murray.

––––––. 1880. *The Power of Movement in Plants.* London: John Murray.

––––––. 1881. *The Formation of Vegetable Mould, Through the Action of Worms, with Observations on their Habits.* London: John Murray.

––––––. 1950. *Charles Darwin's Autobiography.* ed., F. Darwin. New York: Henry Schuman.

Darwin, C. R. and A. R. Wallace. 1858. "On the tendency of species to form varieties; and on the perpetuation of varieties and species by natural means of selection." *Proc. Linnean Soc.* (August 20, 1858).

Darwin, F., ed. 1896. *The Life and Letters of Charles Darwin.* 2 vol. New York: D. Appleton and Co.

de Beer, G. R. 1962. "The origins of Darwin's ideas on evolution and natural selection." *Proc. Royal Soc.* Series B. 155:321-338.

Dobzhansky, T. 1951. *Genetics and the Origin of Species.* 3rd ed. New York: Columbia University Press.

––––––. 1955. *Evolution, Genetics and Man.* New York: John Wiley and Sons.

Drachman, J. M. 1930. *Studies in the Literature of Natural Science.* New York: Macmillan Co.

Dupree, A. H. 1959. *Asa Gray 1810-1888*. Cambridge: Harvard University Press.
_____. 1958. *Darwin's Century*. New York: Doubleday Anchor Books.
Eiseley, L. C. 1959. "Charles Lyell." *Sci. Amer.* 201:(2) 98-106.
_____. 1960. *The Firmament of Time*. New York: Atheneum.
Fothergill, P. G. 1952. *Historical Aspects of Organic Evolution*. London: Hollis and Carter.
Garn, S. M. 1961. *Human Races*. Springfield, Ill.: Charles C Thomas.
Gillispie, C. C. 1958. "Lamarck and Darwin in the history of science." *Amer. Sci.* 46:388-409.
Glass, H. B., O. Temkin, and W. L. Straus, Jr., eds. 1959. *Forerunners of Darwin: 1745-1859*. Baltimore: Johns Hopkins Press.
Gray, A. 1963. *Darwiniana*. A. H. Dupree, ed. Cambridge, Mass.: Harvard University Press.
Greene, J. C. 1959. *The Death of Adam*. Ames, Iowa: Iowa State University Press.
Hooton, E. A. 1946. *Up From the Ape*. New York: Macmillan Co.
Huxley, F. 1959. "Reappraisals of Charles Darwin: Life and Habit." *Amer. Scholar* Autumn, 1959, pp. 489-499; Winter, 1959, pp. 85-93.
Huxley, J. 1943. *Evolution, the Modern Synthesis*. New York: Harper and Bros.
_____. 1953. *Evolution in Action*. New York: Harper and Bros.
_____. 1939. *The Living Thoughts of Darwin*. New York: Longmans. (Also Greenwich, Conn.: Fawcett Publ., 1959.)
Huxley, L. 1900. *The Life and Letters of Thomas H. Huxley*. 2 vol. New York: D. Appleton and Co.
_____. 1863. *Evidence as to Man's Place in Nature*. London: Williams and Norgate. (Reprinted by D. Appleton and Co., New York, 1871.)
_____. 1896. *Evolution and Ethics and Other Essays*. New York: D. Appleton and Co.
Huxley, T. H. 1896. *Darwiniana*. New York: D. Appleton and Co.
_____. 1909. *Autobiography and Selected Essays*. Boston: Houghton Mifflin Co.
Irvine, W. 1955. *Apes, Angels and Victorians*. New York: McGraw-Hill Book Co.
Kettlewell, H. B. D. 1959. "Darwin's missing evidence." *Sci. Amer.* 200:(3) 48-53.
King-Hale, D. 1964. *Erasmus Darwin*. New York: Scribner's.
Krause, E. 1880. *Life of Erasmus Darwin*. New York: D. Appleton and Co.
Lamarck, J. B. de M. 1963. *Zoological Philosophy*. trans. H. Elliot. New York: Hafner Publishing Co.
Loewenberg, B. J. 1959. *Darwin, Wallace, and the Theory of Natural Selection*. Cambridge, Mass.: Arlington Books.
Loewenberg, B. J., ed. 1959. *Evolution and Natural Science*. Selections from the writings of Darwin. Boston: Beacon Press.
Lyell, C. 1892. *Principles of Geology; or the Modern Changes of the Earth and Its Inhabitants Considered as Illustrative of Geology*. London: John Murray.
Malthus, T. R. 1914. *An Essay on the Principle of Population*. London: J. M. Dent and Sons.
Mayr, E. 1957. *The Species Problem*. Washington, D. C.: A. A. A. S. Publ.

Newman, H. H. 1932. *Evolution Yesterday and Today.* Baltimore: Williams and Wilkins.

Olby, R. C. 1967. *Charles Darwin.* London: Oxford University Press.

Oparin, A. I. 1957. *The Origin of Life on the Earth.* 3rd ed. New York: Academic Press.

Osborn, H. F. 1894. *From the Greeks to Darwin.* New York: Macmillan Co.

Petersen, K. 1961. *Prehistoric Life on Earth.* New York: E. P. Dutton and Co.

Poulton, E. B. 1909. *Charles Darwin and the Origin of the Species.* London: Longmans, Green and Co.

Prosser, C. L. 1959. "The 'Origin' after a century: prospects for the future." *Amer. Sci.* 47:536-550.

Radl, E. 1930. *The History of Biological Theories.* trans. from German. E. J. Hatfield. London: Oxford University Press.

Schmidt, O. 1895. *The Doctrine of Descent and Darwinism.* New York: D. Appleton and Co.

Sears, P. B. 1950. *Charles Darwin, the Naturalist as a Cultural Force.* New York: C. Scribner's Sons.

Smith, H. W. 1953. *From Fish to Philosopher.* Boston: Little, Brown and Co.

Stauffer, R. C. 1959. "'On the Origin of Species', an unpublished version." *Science* 130: 1449-1452.

Tax, S., ed. 1960. *Evolution after Darwin.* 3 vol. Chicago: University of Chicago Press.

Wendt, H. 1955. *In Search of Adam.* Boston: Houghton Mifflin Co.

CHAPTER 14

ORGANIZATION IN LIVING SYSTEMS

Biologists soon recognized unprecedented horizons when the microscope came into being. The basic discovery that bodies of animals and plants were organized around small units that were eventually called cells gave biologists new direction. Theorizers, who suggested the broad significance of isolated observations and postulated a unity among all living things with reference to their fundamental organization, came into their own. Then numerous investigators tested such generalizations and extended the theories. Thus the cell eventually came to be recognized not only as a unit of structure but as a unit of function (metabolism), reproduction, and growth and differentiation.

The classical microscopists of the seventeenth century made critical observations and prepared sketches of the minute structural units they saw in the bodies of living things. Malpighi observed tiny sacs of "utricles" in plant bodies. Grew observed similar bags that he called "bladders" and he noted that they were filled with fluid. Leeuwenhoek saw structures that would now be identified as cells, but he did not assign them a particular name. Swammerdam was also aware of such structural units as he made his minute anatomical dissections and critical observations. These men were effective observers, but they were not theorizers. Not until the nineteenth century were the fragments of information put together and the cell theory developed.

Observation 18 in Hooke's *Micrographia* (1665) was an illustrated study of the minute structure of cork (Fig. 8.7). Open spaces surrounded by heavy walls were observed and described as little boxes or "cellulae." The open space and the honeycomb appearance of the boxes impressed Hooke, and therefore he chose the Latin root *cella* meaning "a little room." The term had been used to describe confined spaces such as prisoners' cells and compartments occupied by monks in monasteries. This word gained acceptance as a term to describe the structural unit of living things. Now that the "cell" has come to represent more than the dead, empty box that Hooke observed, the term is generally considered to be a poor one. The cell's living content is now known to be the important element, and many people have suggested that the name should be changed. The term "protoplast" has been suggested as a replacement, but the word "cell" is firmly interwoven in the literature of biology and a change in the established and time-honored terminology is unlikely.

Developing Background for a Cell Theory

Between the seventeenth and nineteenth centuries there were speculations and indirect suggestions that might now be considered as having anticipated the cell theory. Some of these were associated with developments in embryology and physiology, in which the ideas about cells were incidental to the main objectives of the investigations. Other propositions were not supported by observation.

Lamarck, in his *Philosophical Zoology* (1809), called attention to little masses of gelatinous matter from which living bodies were organized. He did not, however, consider the cell as an individual unit. In the same year, the French botanist, C. F. Mirbel, concluded that plants were composed of a continuous cell membrane. Lorenz Oken (1779-1851), as early as 1805, had stated that "all organic beings come from vesicles or cells." He used the word "Urschleim" or primitive slime to describe the living substance, that is, protoplasm, which composes the cells of living things. Oken was a philosopher and not an observer, but he had voiced an important generalization that influenced later contributors. In the early part of the nineteenth century many observations were made suggesting that minute structures were constantly present in animal and plant bodies. Henri J. Dutrochet (1776-1847), in 1824, expressed the idea of individuality of cells. Their universal distribution and significance as units of structure, however, were not appreciated until later.

The two main structural parts of the cell (Table 14.1), the nucleus and cytoplasm, were described before the cell theory was formulated. In 1831, Robert Brown (1773-1858; Chapter 12), while working with plants that he had collected in Australia, observed constant structures in the cells of the epidermal layers. He illustrated these structures and used the word "nucleus" to identify them. Later he found similar structures in pollen grains and in the ovules and stigmas of different plants. The cytoplasmic part of the cell had been described previously by several investigators. Felix Dujardin (1801-62), in 1836, described

Table 14.1   Contributors to structural aspects of cell theory

Contributor	Dates	Contribution
Hooke	1665	Plant cell wall, name "cell."
Brown	1831	Nucleus.
Dujardin	1836	Cytoplasm.
Schleiden	1838	Cell theory, plants.
Schwann	1839	Cell theory, animals.
von Mohl	1846	Protoplasm.
Pringsheim and Sachs	1868	Plastids.
Strasburger	1875	Mitosis in plants.
Flemming	1882	Mitosis in animals.
van Beneden	1883	Reduction division.
Waldeyer	1888	Named "chromosomes."
Boveri	1888	Mitotic spindle, named "centrosomes."
Benda	1897	Mitochondria.
Golgi	1898	Golgi apparatus.

cytoplasm, which he called "sarcode," in living amebae. He found the living substance to be homogeneous, elastic, contractile, and transparent. It refracted light a little more than water and much less than oil. In the living substance he could not distinguish any organization. This description referred to the basic living substance without suspended elements.

Robert Brown had also described the cytoplasmic parts of the cells in his paper entitled, "Microscopic Observations of the Pollen of Plants" (1828). In his studies of cytoplasm, he had observed the dancing movement of particles in the protoplasm which is now called, in his honor, "Brownian movement." At first he thought the particles were alive and that the movement was a fundamental property of living things. Later he learned that it was a physical process that could occur in nonliving suspensions as well as in living materials. One important contribution of these early observers was the demonstration that the living substance was basically a fluid.

Statement of Cell Theory

Two German biologists, M. J. Schleiden (1804-81) and Theodor Schwann (1810-82; Fig. 14.1), finally stated the generalization that animals and plants are composed of cells. Even though these two men are listed as coauthors of the theory, a current evaluation would not give them equal credit. Schleiden was first to arrive at the conclusion from studies on plant cells published in 1838. Schwann published his results, based on observations of animal cells, a year later. His work was much more comprehensive and had a better biological foundation than did that of Schleiden. Schwann entered into the broader considerations of cells, provided better support for the theory than had his colleague, and thus earned a larger share of credit for the accomplishment.

Schleiden

Trained as a lawyer, Schleiden deserted his law practice at the age of 27 and returned to the university with the initial intention of studying medicine. He specialized in botany, which at that time represented a major aspect of the medical program, and later became professor of botany at the University of Jena. In 1837, he began a series of studies on plant organization and development. These studies extended the work of Robert Brown in showing that the cell nucleus must be important in plant development. Schleiden's background in basic science was weak, however, and he made serious errors in his observations and conclusions. His most conspicuous mistake came from his erroneous idea that cell propagation was based on physical processes of crystalization. He visualized a nucleolus as the starting point from which the cell developed by continuously forming new layers. A form of crystalization was thus described in the formation of a nucleus around the nucleolus. Bubbles observed on the sides of the nucleus were interpreted to represent a sort of budding process which was believed a part of cell propagation. Cell reproduction was thus supposed to be initiated by the nucleolus. A nucleus was next crystalized around the nucleolus. Finally the cytoplasmic part was thought to develop around the nucleus and the entire mass to break out of the parent cell.

Schleiden made few if any original observations in support of the cell theory. By capitalizing on the work of others he arrived at a sweeping generalization that placed cells in a basic position as units of organization in living things. "Each cell leads a double life," he wrote, "one independent, pertaining to its own development alone; the other incidental, as an integral part of a plant."

Schwann

Schwann was a thoughtful and persistent investigator. For many years at the University of Würzburg and the University of Berlin he had done microscopic research on plant and animal cells and had classified tissues. He published (in 1839) a report of his observations in a paper entitled, "Microscopical Researches on the Similarity in Structure and Growth of Animals and Plants." The minute anatomy of cells and tissues from animal and plant materials was described and comparisons were made. Cells and cell parts were illustrated in full detail. Schwann not only substantiated the generalization stated by Schleiden, that living things are composed of cells, but supported it with data from his own observations. Shortly after the publication of his work on the cell theory, Schwann received a professorship at the University of Louvain in Belgium.

In studying connective tissues in animals, Schwann found it necessary to add a phrase to the original statement of the cell theory. Some parts of living organisms were not composed of cells. Analysis showed that the matrix of bone and connective tissue was acellular but represented a cell product. Therefore, he amended the theory to read, "all living things are composed of cells and cell products." Like Schleiden, he described cells as having an independent life, but he also observed that cells are subjected to the control of the

Figure 14.1
Theodor Schwann who
developed the cell theory.

organism. Schwann used the word "cytoblastema" to describe the living material which has since been called protoplasm. He also coined the word "metabolism" to encompass all chemical processes carried on in the cell. The cell was described as the structural unit of living organisms. Indeed, Schwann referred to the cell as the unit of life.

Corrections and Extensions of Cell Theory

Both Schleiden and Schwann emphasized the structural unit or box when they spoke about cells. They attached importance to the living processes that must go on in cells, particularly during development of the organism; but the variety and complexity of cell contents remained unknown or at least unappreciated. The next step was to identify cell parts and to determine their functional significance in the activities of cells.

In this process, Schleiden's misconception about the behavior of the cell and its nucleus in cell reproduction, a point which Schwann had passively accepted, had to be corrected. In 1841, Robert Remak (1815-65) studied the frog egg and followed the early developmental stages. He discovered that cells came from preexisting cells. Karl Nägeli (1817-91) followed this work in 1844, with a general description of cell division. He did not observe the details of mitosis but showed that Schleiden's idea was incorrect. Nägeli made chemical analyses and found that different cell structures were made of different chemical materials. Some cell parts, such as the nucleus, contained nitrogenous material, whereas the cell wall was made of carbohydrates. Some bodies (grains) found in the cytoplasm were entirely made up of starch.

Hugo von Mohl (1805-72) coined the word "protoplasm" (in 1846) to describe the living substance in cells. Working with living plant cells, he observed the streaming of cytoplasm around the central vacuole. Soon a group of investigators working with protozoa found that ciliary movement reflected a fundamental property of living things. Instead of movement being confined to the inside of the cell, as in streaming protoplasm, a special organization of structures made it possible for the entire cell to move. The complex activities involving muscular action in higher animals eventually were shown to be refinements of the streaming movement observed in generalized cells.

Albrecht Kölliker (1817-1905), a student of Johannes Müller, applied the cell theory to embryology and histology. He showed that the egg is a single cell, that the nucleus is the most constant cell part, and that the nucleus has great significance in cell reproduction. Nuclear division was found to precede cell reproduction. Remak and other cytologists supported and clarified this view, and the cell theory was expanded to consider the cell as a unit of development. Kölliker also visualized the cell as carrying hereditary determiners long before definitive data on this point were available. He conducted investigations on complex nerve fibers in the central nervous system and found them to be cellular structures with greatly extended processes. The involuntary muscles he examined were found to be composed of cells. He showed that a large group of animal structures not specifically related to cells in previous studies were composed of cells. Kölliker later authored valuable textbooks in embryology and histology.

Max Schultze (1825-74) recognized the prevailing trend in 1865, and emphasized the contents of the cell. The cell was now known to be a living, functioning unit rather than a dead, empty box. Schultze defined the cell as a "mass of cytoplasm with a nucleus." This definition is more concise than some others, recognizes the essential living unit, and serves well as a brief definition of a cell.

The next major contribution toward an understanding of the cell

Figure 14.2
Theodor Boveri, German
cytologist and embryologist
who made significant
contributions to the
chromosome theory of
inheritance. (From Gardner,
Principles of Genetics,
Copyright © 1960, 1964,
1968 by John Wiley &
Sons, Inc.)

was the detailed description of cell division. Eduard Strasburger (1844-1912), Professor of Botany at Bonn, observed (in 1875) the mitotic process in plant material and described and illustrated the details of mitotic cell division. Four years later he demonstrated that nuclei arise only from preexisting nuclei. He used living materials and there was no doubt that the structures observed existed in live cells.

Walther Flemming (1843-1915), Professor at Prague and later at Kiel, named and described the process of mitosis in animal cells. His reports appeared in 1879, 1882, and periodically during the next 10 years. Flemming used modern methods of fixation and staining on tissues from amphibian larvae, and worked with killed instead of living material. In this way more precise stages were observed and a detailed and accurate account of mitosis was published. As a drawback, though, there was always some question as to whether the structures observed were present in the living cells or were artifacts resulting from the treatment. Flemming coined terms that are still used to describe structures and stages in the process including such useful designations as: mitosis, aster, chromatin, prophase, metaphase, anaphase, and telophase. He observed the longitudinal division of structures that originated in the nucleus during cell division and disappeared when the division was completed. These structures, visible only during the division stages of the cell, were named (in 1888) chromosomes by W. Waldeyer (1836-1921).

Also in 1888, Theodor Boveri (1862-1915; Fig. 14.2), Professor at Würzburg, described the cytoplasmic structures associated with spindle formation in the mitotic division of animal cells and named them centrosomes. It soon was observed that plant cells did not have such structures. The cellular cytoplasm can apparently accomplish the same function in plant cell division that the centrosomes accomplish in the division of animal cells. Plastids in the cytoplasm of plant cells were described by Nathaniel Pringsheim (1823-94) and Julius Sachs (1832-97). Mitochondria, cytoplasmic cell organelles in both plant and animals, were described by Benda in 1897.

The work of the great investigators of the last quarter of the nineteenth century developed the study of cells to a point where it justified recognition as the distinct biological science of cytology. The book by Oscar Hertwig (1849-1922) entitled *Cell and Tissue,* which was published in 1893, constituted a landmark in the development of the then new science of cytology.

Modern techniques have facilitated more extensive and intensive studies of cells. Ross Harrison (1870-1959), at Johns Hopkins University, developed a tissue culture technique that made it possible to raise embryonic cells outside of the body and watch their activities under the microscope. Alexis Carrel of the Rockefeller Institute and Charles Lindberg improved the technique. It has since been adapted to permit the study of growth and differentiation in normal cells and the study of abnormal growth in cancer cells. Tissue culture rapidly became an important tool, useful in many aspects of experimental biology. The phase contrast microscope, developed in 1941, revitalized interest in studies of living cells. This instrument allows the observer to distinguish otherwise transparent cells and parts of cells and to watch cell division and other cell processes without killing or staining the material. The electron microscope has been useful in facilitating the description of minute structural parts of killed cells. More recently this instrument has been used for photographing the internal structures of chromosomes.

Figure 14.3
Rudolf Virchow,
German pathologist.

Cellular Pathology

A further extension in the cell theory, that the cell is the locus of disease as well as life, was made by the German pathologist, Rudolph Virchow (1821-1902; Fig. 14.3). Virchow followed his predecessors in the observation that all cells come from other preexisting cells, and in his early life he established the cell's character as an independent life unit. He made detailed microscopic observations of bone and connective tissue in healthy and diseased animals. This led to his discovery that abnormal or pathological events in the body were connected with unusual cellular activities. Virchow's first paper dealing with cells was entitled "On the Evolution of Cancer," and a long series of studies on malignant tumors followed.

Virchow took his medical degree in Berlin in 1843, joined the Charité Hospital in that city immediately, and in 1847 qualified as a teacher at the University of Berlin. The same year, he co-founded the *Archiv für pathologische Anatomie und Physiologie,* of which he was sole editor through 170 volumes. He achieved early renown by his lectures at the Charité. This history-making series of lectures, first published in Germany in 1858, was reissued in edition after edition, to become a classic of medical history.

In 1856 he became professor of pathology and director of the Berlin Institute of Pathology. He was largely responsible for the classification of body components into epithelial organs, connective tissues, and the more specialized muscle and nerve; he proved the presence of neuroglia in the brain and spinal cord, discovered crystalline haematoidine, and made out the basic structure of the umbilical cord. He made the first distinction between "fatty infiltration" and "fatty degeneration," and initiated the histological study of necrosis. Virchow recognized and named leukemia, and first used the terms thrombosis, embolus, osteoid tissue, and parenchymatous inflammation. He is the author of numerous works, of which *Cellular Pathology* is regarded as his particularly great contribution.

In *Cellular Pathology* (1858), Virchow directly applied theory to pathology and a new concept of "cellular pathology" was initiated. Before Virchow, humoral pathology was the main theme; after Virchow, cellular pathology took precedence. With keen insight, Virchow developed the concept of genetic continuity from cell to cell. To quote from Virchow's *Cellular Pathology*:

> Where a cell exists there must have been a preexisting cell, just as the animal arises only from an animal and the plant only from a plant. The principle is thus established, even though the strict proof has not yet been produced for every detail, that throughout the whole series of living forms, whether entire animal or plant organisms, or their component parts, there rules an eternal law of continuous development.

In a very real sense, modern cell biology started with this statement from Virchow. As a direct result of the insights of Virchow and others of this period, by the end of the nineteenth century a comprehensive cell biology was developed. It could then be stated categorically that all cells carried chromosomes which in turn carried determinants. By 1900, it was further suggested that the significant material in chromosomes, the actual hereditary material, was the nucleinic acid discovered in 1869 by Friedrich Miescher.

Despite his great interest and early accomplishment in cellular pathology, Virchow shifted in later life to a consideration of the body as a whole and the influence of infectious agents. In addition to his investigations and technical writings, accomplished independently and in connection with his professorship at the University of Berlin, he was an anthropologist, sanitarian, and statesman of considerable importance. His work in public health resulted in a great saving of human life. For some 55 years he was editor of *Archives for Pathology*.

Problems and Criticisms of the Cell Theory

Many problems arose as the cell theory was applied to various groups of organisms. When bacteria were found to occupy an important place in the living world, a question arose concerning their relation to the cell theory. At first bacteria did not seem to fit the definition of Schultze because no formed nucleus could be identified, but bacteria were known to contain nuclear material. Current evidence suggests a nuclear organization much like that in higher forms. Some bacteria (e.g., *Escherichia coli* K 12) are known to utilize a type of sexual reproduction.

Then came the discovery of viruses. Are they cells? Certainly they are not cells in the usual sense. Although they lack some characteristics of cells, such as semipermeable membranes to control the diffusion of materials in and out of the unit, they do exhibit other cell characteristics. The same chemical materials found in the nuclei of higher organisms are present in viruses. They evidence mutation and recombination, activities that are associated with genes in higher organisms.

Some structures of certain organisms and some whole organisms are not divided into cells in the usual way. Molds of the genus Mucor are coenocytes; that is, their nuclei are suspended in the hyphae without being partitioned off into compartments. Some protozoans that were originally considered to be single-celled organisms were later found to have more than one nucleus. Likewise, some muscles such as heart muscle are not divided into discrete cells. In such cases, however, even though no partitions establish specific units, there are nuclei surrounded by cytoplasm. Schultze's definition applies satisfactorily in these cases since he did not emphasize the boundary around the "unit of protoplasm containing a nucleus." It must be recognized, however, that the separateness of cells remains a general rule, even though many exceptions occur. Presence or absence of a surrounding barrier is considered to be only a superficial characteristic which does not weaken the basic cell theory.

Are complex protozoans one-celled animals as they are often identified in elementary texts? The malaria organism is one-celled at one stage of its cycle, but it goes through a process of sporulation, and at one stage, 16 cells are formed. The 16-celled organism then breaks into 16 independent merozoites. Therefore the answer to the question depends on the stage in the cycle of the organism. Higher animals, including man, are one-celled animals during one stage of their cycle. Some protozoans, such as Euplotes, are much more complex and highly organized than some metazoans. The distinction between protozoa and metazoa as one-celled and many-celled organisms is obviously subject to criticism.

Finally, is the cell or the organism the unit of life? Schwann said that the functions of life reside not in the organism but in separate elementary units or cells. This was interpreted to indicate that cells are independent units of life. Indeed, some cells of higher animals do remain alive for hours or days after the organism as a whole is dead. The status of the organism as a whole, however, is generally considered the more important measure of life.

Cells and Heredity

Greek philosophers considered inherited traits of individuals to be acquired through direct contact with the environment. This idea, although not precisely stated until the eighteenth century, was widely accepted. The French scientist, Lamarck, formulated a common view of eighteenth century biologists into a theory which bears his name (Lamarckism) and is known as the theory of inheritance of acquired characteristics. This theory emphasized use and disuse over long or short periods of time as the significant factor in determining the characteristics of the individual. Direct influence of the environment on the parents was considered to be represented in the germinal material (eggs and sperm) and therefore to be transmittable in inheritance.

The mechanism proposed by Lamarck was never demonstrated experimentally. Present views concerning the nature of the germinal material hold that direct hereditary change by environmental modification is most unlikely. The alternative concepts based on mutation, genetic recombination (Chapter 17), natural selection, and other directing forces in nature (Chapter 13) adequately explain most of the things that seemed to the last century biologists to require a Lamarckian explanation.

Germ Plasm Theory

An alternative explanation to Lamarckism was provided, in the latter part of the nineteenth century, by August Weismann (1834-1914) of

Freiburg, a prominent German biologist. This theory emphasized the remarkable stability of the germ plasm. The environment was considered to have little, if any, effect on the hereditary material, even though environmentally induced modifications of external characteristics occurred. The germ plasm was distinguished from the somatoplasm (body cells) and was described as the hereditary material carried from generation to generation in the reproductive process. According to Weismann, reproduction was accomplished not by somatoplasm but by germ plasm, which is transmitted essentially unchanged from generation to generation. Although some details of the germ plasm theory have been modified, the fundamental premise is well established. Germ plasm is remarkably stable, but it occasionally undergoes change by spontaneous mutation. The mutation rate can be increased experimentally by irradiation and by some other chemical and physical factors supplied in the environment, but the genetic changes occur at random.

Weismann studied under Rudolph Leuckart and became proficient in the microscopic study of cells. In middle life, however, he began to suffer from eye trouble which made observation with the microscope impossible. He subsequently developed great skill as a theoretician and formulated the conception of continuity of germ plasm from parent to offspring. According to Weismann, the offspring resembles the parent because it is derived from the same substance, the germ plasm, the essential element of germ cells. The body, with its other cells, he described as a vehicle for the transmission of germ cells. Body cells nourish, protect, and transmit to succeeding generations germ cells received from past generations. This was visualized as a continuous stream of life. Samuel Butler (1835-1902) described the continuity of germ plasm when he said, "a hen is only an egg's way of producing another egg." It remained for the geneticists and cytologists of the twentieth century to determine the mechanisms through which stability and continuity are maintained through the germ plasm while, at the same time, genetic modifications produced through influences in the environment are also transmitted through gametes and zygotes.

Chromosomes

Cytologists have contributed to the understanding of inheritance mainly through their studies of the chromosomes, which were shown by Roux (1883) and Boveri (1888) to be hereditary factors.

Chromosomes soon became associated with the determination of sex, the most obvious characteristic of animals and plants. The first investigations relating chromosomes to sex determination were carried out late in the nineteenth century. H. Henking, a German biologist, discovered in 1891 that a particular nuclear structure could be traced in the spermatogenesis process (through which sperm are produced) of certain insects. Half the sperm received this structure and half did not. Henking did not speculate on the significance of this body but merely identified it as the "X" body and showed that some sperm were different from others because of its presence or absence. In 1902 these observations were verified and extended by C. E. McClung, who made cytological observations on many different species of grasshoppers and demonstrated that somatic cells in a female grasshopper carry a different chromosome number than do corresponding cells in a male. He followed the X body in spermatogenesis but did not succeed in tracing the oogenesis (through which eggs are produced) of the female grasshopper, which is more difficult to study. McClung associated the X body with sex determination, but erroneously considered it to be peculiar to males. Had he been able to follow oogenesis, his interpretation would undoubtedly have been different.

Valuable contributions to basic knowledge about sex determination were made in the early part of the present century by the distinguished American cytologist E. B. Wilson (1856-1939; Fig. 14.4). Wilson and Miss N. M. Stevens, beginning in 1905, reported extensive cytological investigations on several different insects, notably the genus Protenor, an uncommon group of insects closely related to the boxelder bug. In these insects, different numbers of chromosomes were observed in the germ cells of the two sexes. Wilson and Stevens succeeded in following oogenesis as well as spermatogenesis. The unreduced cells of the male carried 13 chromosomes, and those of the female carried 14. Some male gametes, that is, sperm, were found by microscopic observations during the later stages of spermatogenesis to carry 6 chromosomes whereas others from the same individual carried 7. The female gametes, eggs, all had 7. Eggs fertilized with 6-chromosome sperm produced males and those fertilized with 7-chromosome sperm produced females. The "X" body of Henking was thus found to be a chromosome that influenced sex determination. It was identified in several insects and became known as the sex or X chromosome. All the eggs of these insects carried the X chromosome, but it was included in only half of the sperm. This was called the XO mechanism of sex determination.

During the same year (1905) that Wilson and Stevens reported on XO sex determination, they observed another chromosome arrangement in the milkweed bug, *Lygaeus turcicus*. In this insect the same number of chromosomes was present in the cells of both sexes. The one identified as the mate to the X, however, was distinctly smaller and was called the Y chromosome. Sex determination based on equal chromosome numbers in the two sexes but with different kinds of chromosomes making up one pair was called the XY type. The XY type is now considered characteristic in most of the higher animals and occurs in at least some plants (e.g., *Melandrium album*).

Variations in Chromosome Number

Classifications of chromosome changes are arbitrary and superficial because such changes are necessarily interpreted in terms of obvious additions or eliminations of parts of chromosomes, whole chromosomes, or whole chromosome sets. The presently accepted classification system, therefore, is merely a working tool. Two main classes are euploidy and aneuploidy (*ploid*, Greek for unit; *eu*, true or even; and *aneu*, uneven). Euploids have chromosome complements consisting of whole sets or genomes. The chromosome number of euploid organisms is basically represented by the monoploid or haploid (n). Euploids with chromosome numbers above the monoploid level may be diploid (2n), triploid (3n), tetraploid (4n), or have some other "polyploid" number.

The first critical study of aneuploid plants was made by A. F. Blakeslee and J. Belling using the common Jimson weed *Datura stramonium*, which normally has 12 pairs of chromosomes in the somatic cells. These investigators announced in 1924 the discovery of a "mutant type" having 25 rather than 24 chromosomes. At the meiotic metaphase one of the 12 pairs was found to have an extra member; that is, one trisome was present along with 11 disomes. This originally discovered trisomic plant differed from wild-type plants in several specific ways. Conspicuous deviations were observed in shape and spine characteristics of seed capsules. The chromosome complement with one extra member in addition to the regular set was illustrated by the formula 2n + 1. Theoretically, because the complement was composed of 12 chromosome pairs differing in the genes they carried, 12 distinguishable trisomics were possible in Jimson weeds. Through experimental breeding Blakeslee and his associates succeeded in producing all 12 possible trisomics. These were grown in Blakeslee's garden and each was found to have a distinguishable phenotype

Figure 14.4
E. B. Wilson, American cytologist, who made extensive contributions including the chromosomal mechanism of sex determination. (From Gardner, *Principles of Genetics*. Copyright ©, 1960, 1964, 1968 by John Wiley & Sons, Inc.)

which was attributed to an extra set of the genes contained in one of the 12 chromosomes.

Polyploidy

In contrast to aneuploids, which differ from standard $2n$ chromosome complements in single chromosome, euploids differ in multiples of n. Monoploids (n) carry one genome; that is, one each of the normally present chromosomes. The n chromosome number is usual for gametes of diploid animals, but unusual for somatic cells. Monoploidy is seldom observed in animals except in the male honey bee and other insects in which male haploids occur. Normal chromosome behavior in animals and plants is based on diploids which are used in the following examples as standards for comparison. Organisms with three or more genomes are polyploids. Polyploidy is common in the plant world. Fully one-half of all known plant genera contain poly-ploids, and about two-thirds of all the grasses are polyploids but poly-ploids are rarely seen in animals.

Polyploidy combined with interspecific hybridization provides a mechanism by which new species may arise suddenly in nature. E. Anderson has shown this process to be important in the evolution of many plant groups. In one of his investigations, the blueflag, *Iris versicolor* with 108 chromosomes, was shown to have the doubled chromosome complement of a hybrid derived from a 72-chromosome iris of the Mississippi Valley and a 36-chromosome arctic iris from Alaska. Evidently these species, now separated geographically, had grown near each other at some time in their ancestral history. E. B. Babcock concluded, from his elaborate analysis of 196 species of Crepis, that polyploidy was involved in the formation of about 8 per cent of the species in that genus.

Chromosome Structural Modifications

In 1917 C. B. Bridges observed that a sex-linked recessive gene in Drosophila came to expression when it was presumed to be in hetero-zygous condition. He postulated that a section of the homologous chromosome containing the dominant allele was missing; that is, a deficiency had occurred in a chromosome. When a recessive gene presumed to be homozygous did not come to expression, Bridges postulated that a dominant allele was present in another place in the chromosome set; that is, a duplication of a chromosome section had occurred. Rearrangements of sections within chromosomes (inver-sions, 1926) and exchanges of parts between entirely different chro-mosomes (translocations, 1919) were postulated by A. H. Sturtevant and Bridges to explain various genetic irregularities. Many years elapsed, however, before the predicted structural changes could actually be observed through the microscope. The first cytological demonstration of plant chromosome rearrangements was made in maize by Barbara McClintock in 1930.

Giant Polytene Chromosomes in Diptera

Large coiled bodies about 150 to 200 times as large as gonad cell chro-mosomes were observed in the nuclei of glandular tissues of dipterous larvae as early as 1881 by E. G. Balbiani. He described banded struc-tures in the nuclei of cells of larval midges in the genus Chironomus but did not attach any particular significance to the observation. Three years later J. B. Carnoy made further morphological observations, and in 1912 F. Alverdes traced the development of these structures from the early embryo to a late larval stage. In 1930, D. Kostoff suggested a relation between the bands of these structures and the linear sequence known to occur among genes. The anatomical significance of the nuclear bodies was further studied by E. Heitz and H. Bauer in 1933, in the genus Bibio, a group of March flies whose larvae feed on roots of grasses. These authors identified the bodies as giant chromosomes occurring in pairs. They described the morphology in detail and discovered the relation between the salivary gland chromosomes and other somatic and germ cell chromosomes. They also demonstrated that comparable elements occurred in the giant chromosomes and in the chromosomes of other cells of the same organism.

It is largely through the work of T. S. Painter that Drosophila salivary gland chromosomes were first used for cytological verification of genetic data. Painter related the bands on the giant chromosomes to genes, but he was more interested in the morphology of the chromosomes and implications concerning speciation than the association of chromosome sections with particular genes. Bridges, beginning in 1934, made extensive and detailed investigation of the salivary gland chromosomes and in the course of his investigations developed a tool of practical usefulness in relating genes to chromosomes. In applying this method to *Drosophila melanogaster* he prepared a series of cytological (chromosome) maps to correspond with the linkage maps already available. This project of constructing maps of all four chromosomes of *D. melanogaster* was in progress at the time of his death in 1938.

Chronology of Events

1661	M. Malpighi, "utricles."
1665	R. Hooke, *Micrographia,* used the word "cell."
1674-77	A. van Leeuwenhoek observed blood cells, sperm, protozoa and bacteria.
1682	N. Grew, "bladders."
1759	C. F. Wolff, "little globules."
1805	L. Oken suggested cell theory.
1809	J. B. Lamarck *Philosophical Zoology;* importance of cells in living organisms.
1824	R. J. H. Dutrochet, individuality of cells.
1828	R. Brown, cytoplasm in plant; "Brownian movement."
1831	Brown, nucleus in plant cells.
1835-46	H. von Mohl, general aspects of cell division and protoplasm.
1836	F. Dujardin, cytoplasm.
1838-39	M. J. Schleiden and T. Schwann, cell theory.
1841	A. Kölliker, sperm and eggs are sex cells; applied cell theory to embryology and histology.
	R. Remak, cells come from preexisting cells.

1844	K. Nägeli, general aspects of cell division and chemistry of cells.
1846	H. von Mohl, protoplasm.
1855-58	R. Virchow, *Cellular Pathology*.
1865	M. Schultze, cell, "a mass of cytoplasm with a nucleus."
1865-92	N. Pringsheim and J. Sachs described plastids.
1875	E. Strasburger, mitosis in plant cells.
1879-82	W. Flemming, mitosis in animal cells.
1881	E. G. Balbiani, giant salivary chromosomes in Chironomous larvae.
1883	E. van Beneden, reduction in *Ascaris*.
1888-92	T. Boveri, centriole and mitosis in *Ascaris*. W. Waldeyer named "chromosomes."
1891	H. Henking, "X" body in some sperm of some insects.
1892	A. Weismann, germ plasm.
1893	O. Hertwig, *Cell and Tissue*.
1897-99	C. Benda, mitochondria in sperm and other cells.
1898	C. Golgi, Golgi apparatus in nerve cells.
1902	C. E. McClung, "X" body identified as a chromosome.
1905	E. B. Wilson and N. M. Stevens, XO and XY sex determination.
1907	R. Harrison, tissue culture technique.
1917-19	C. B. Bridges, structural and numerical changes in chromosome.
1924	A. F. Blakeslee, extra chromosome (trisomic) in Datura.
1926	A. H. Sturtevant, chromosome inversion in Drosophila.
1930	B. McClintock, chromosome rearrangement in maize.
1933	E. Heitz and H. Bauer, related giant salivary chromosomes to somatic-cell chromosomes. T. S. Painter used giant salivary chromosomes for cytological verification of genetic data.
1934	C. B. Bridges, salivary chromosome maps of *Drosophila melanogaster*.

References and Readings

Ackerknecht, E. H. 1953. *Rudolph Virchow, Doctor, Statesman, Anthropologist*. Madison: University of Wisconsin Press.

Brown, R. 1833. "Observations on the organs and mode of fecundation in Orchideae." *Trans. Linn. Soc.* 16: 685-745.

Clark-Kennedy, A. E. 1945. *The Art of Medicine in Relation to the Progress of Thought*. Cambridge: At the University Press.

Conklin, E. G. 1939. "Predecessors of Schleiden and Schwann." *Amer. Nat.* 73: 538-546.

Espinasse, M. 1956. *Robert Hooke*. Berkeley: University of California Press.

Gerould, J. H. 1922. "The dawn of the cell theory." *Sci. Monthly* 14: 268-277.

Goldschmidt, R. B. 1956. *Portraits from Memory*. Seattle: University of Washington Press.

Hertwig, O. 1895. *The Cell*. London: Swan Sonnenschein and Co.

Hooke, R. 1961. *Micrographia*. (reprinted) New York: Dover Publications.

Hughes, A. 1959. *A History of Cytology.* New York: Abelard- Schuman.

Lamarck, J. B. 1963. *Zoological Philosophy, An Exposition with Regard to The Natural History of Animals.* trans. H. Eliot. New York: Hafner Publishing Co.

Moulton, F. R., ed. 1940. *The Cell and Protoplasm.* Publ. of A. A. A. S. No. 14, Washington, D. C.: The Science Press.

Rudnick, D., ed. 1958. *Cell, Organism and Milieu.* (Society for Study of Development and Growth.) New York: Ronald Press.

Sachs, J. von. 1909. *A History of Botany.* English trans. Oxford: Clarendon Press.

Suner, A. P. 1955. *Classics of Biology.* New York: Philosophical Library.

Virchow, R. 1855. "Cellular-Pathologie." *Virchow's Archiv.* 8:1.

Voeller, B. R. ed. 1968. *The Chromosome Theory of Inheritance.* New York: Appleton-Century-Crofts.

Wilson, E. B. 1928. *The Cell in Development and Heredity.* New York: Macmillan Co.

CHAPTER 15

BIOGENESIS AND MICROBIOLOGY

THE ORIGIN of life has been a subject of interest throughout the history of biology. Among the theories advanced, the most widely discussed among early biologists was that of spontaneous generation. This theory implied an origin of living things by processes not involving parent organisms. A favorable combination of nonliving materials was considered sufficient to account for the origin of living things.

The theory of spontaneous generation was commonly accepted from the earliest periods of biological history until the middle of the last century. Aristotle, like many of his contemporaries and followers, considered flies and many other small creatures to be formed spontaneously from the mud in the bottom of streams and pools. Aristotle was able to observe reproductive mechanisms in some animals. Cuttle-fish and octopus for example, were observed while engaged in the reproductive process. These animals he recognized as being formed from eggs produced by the female of the species and activated by "milt" from the male. Higher organisms were considered by Aristotle to develop from eggs even though the eggs of many had not been observed, but lower organisms whose reproduction was unobserved were believed to arise spontaneously (Table 15.1).

Table 15.1   Investigators for and against spontaneous generation

Investigator	Year	FOR Method or argument	Investigator	Year	AGAINST Method or argument
Aristotle	B.C. 384-322	Observed small creatures but not reproductive process.	Redi	1668	Flies screened out.
Needham	A.D. 1745	Hay infusion heated.	Joblot	1710	Hay infusion heated.
Needham	1769	Germ of life destroyed by excess heat.	Spallanzani	1765	Hay infusion heated.
Pouchet	1859	Hay infusion heated.	Schwann	1836	Heated air introduced.
Bastian	1872	Oxygen removed by treatments.	Schulze	1836	Air treated with chemicals.
			Schröder, von Dusch	1854	Cotton plug for filter.
			Pasteur	1857-75	Sealed flasks, Pasteur tubes.
			Tyndall	1876	Optically pure air introduced.

Redi's Experiments on Flies

One of the first men to question the spontaneous origin of living things was the Italian physician and naturalist, Francesco Redi (1621-97; Fig. 15.1). After studying medicine at the University of Pisa, he became court physician to Ferdinand Medici, Grand Duke of Tuscany. Redi was a member of the famous Academy of Experiments (Chapter 7) that was organized in 1657, at Florence, Italy.

Redi set out to determine by experimentation whether flies could be produced spontaneously. His procedure followed a precise pattern by killing three snakes and placing them in an open box where they were allowed to decay. Maggots appeared on the decaying flesh, thrived on the meat, and grew rapidly. During the growth process the maggots were observed critically day after day. Following the period of rapid and continuous growth the maggots became dormant (pupated) and after a few days they emerged as flies. Several varieties of pupae were identified by size, shape, and color and each was observed to give rise to a particular type of adult fly.

Redi continued his experiments, testing different kinds of flesh both raw and cooked. He used the meat of oxen, deer, buffalo, lion, tiger, duck, lamb, kid, rabbit, goose, chicken, and swallow, and he also studied several kinds of fish, including swordfish, tuna, eel, and sole. Particular types of maggots developed on the meat, emerged in due course, and seemed to give rise to particular types of adult flies. Sometimes the maggots were all of the same type and Redi observed that only one type of fly emerged, whereas on other specimens several different types were identified. Different types of maggots were later isolated and each was clearly demonstrated to give rise only to a particular kind of fly. Adult flies of the same kinds that emerged from the maggots were observed to hover over the decaying meat and Redi noticed that the flies dropped tiny objects on the meat. Some flies would remain quiet on the meat and deposit several units in one place but others would deposit single objects while hovering above the decaying meat. Redi theorized that maggots might be developing from the objects dropped by the adult flies on the putrefying meat.

An experiment was carefully designed to test this hypothesis. Portions of fish and eel were placed in flasks. The openings of some flasks were completely sealed off and the meat was observed through the glass as it underwent decay; comparable flasks prepared in the same way were left uncovered as controls. Flies were soon attracted to the opened flasks and in a few days maggots appeared on the meat. Similar flies were also observed inside. Occasionally maggots appeared on the tops of the sealed flasks. They would wriggle on the surface and appear to be trying to get through the glass to the putrefying meat inside. This indicated that the maggots were developed from the elements dropped from the adult flies and were not derived spontaneously on the decaying meat.

Even though the results seemed conclusive, Redi was not content. He tried variations of different kinds to see if the results could be repeated. Experiments were carried out at different seasons of the

Figure 15.1
Francesco Redi, Italian
physician who demonstrated
that flies were not
produced by spontaneous
generation.

year with various kinds of vessels and different kinds of meat. He even buried meat underground and observed that no maggots were developed in the covered meat, but maggots did appear and eventually emerged into flies on similar meat exposed to the air. One difficulty was realized to be inherent in the experiments; the sealed containers and the soil covering some pieces of meat could exclude some vital force necessary for the spontaneous generation of life. To obviate this uncertainty, Redi designed a further experiment in which air was permitted to enter but flies were excluded.

For this experiment Redi covered glass containers with a fine veil that allowed air to enter but through which flies were unable to reach the decaying meat. He found that the covered meat would not produce maggots, but the unprotected containers provided as controls were well populated with maggots and in due time with adult flies. Redi observed carefully the activities of flies during the different phases of the experiment.

Flies were attracted to the meat as soon as it showed the first signs of decay and they laid eggs on the outside surface of the unprotected meat. Some eggs were deposited on the veil and the larvae that emerged would have wriggled their way through the meshwork and entered the containers but Redi removed them as fast as they appeared. He watched closely the method of egg deposition and noted that in a few cases active young were deposited by adult flies. The eggs had apparently hatched in the body of the mother. Some adults would remain quiet on the surface and deposit several eggs at one time whereas others would drop single eggs or larvae from the air without lighting.

Redi, through these simple but ingenious experiments, was able to demonstrate that flies do not develop spontaneously on putrefying meat but that they must come from other flies through the medium of eggs. In his book, *Experiments on the Generation of Insects,* published in 1668, Redi recorded the results of his experiments that disproved spontaneous generation of organisms as complex as flies. With the advent of microbiology, however, the whole controversy flared up again and more critical experiments were designed to resolve the problem. In dealing with minute organisms, more elaborate tools and refined techniques were required.

Microorganisms and Spontaneous Generation

One of the earliest experiments on microorganisms that produced evidence bearing on the problem of spontaneous generation was performed by the French microscopist, Louis Joblot (1645-1723). He observed in 1710, as Leeuwenhoek had done in an earlier period, that when hay was infused in water and allowed to stand for a few days it gave rise to countless microorganisms called "Infusoria." According to present-day taxonomic arrangements the organisms included in a hay infusion would be mostly bacteria and protozoa. Only a small portion would represent the specific class of protozoa now known as infusoria.

Joblot's contemporaries, and many who followed, considered the presence of microorganisms in a hay infusion to be conclusive evidence for spontaneous generation. Joblot was critical of this interpretation, however, and carried out an experiment to test the prevalent idea. He boiled the fluid to be used for a hay infusion and divided the boiled material into two parts. One was placed in a container that was sealed off completely and thus protected from air, whereas the other was left open. The fluid in the open container soon was cloudy and had numerous microorganisms, but the material in the closed container was clear and free from all living things. By this experiment, Joblot showed that the boiled infusion alone was not capable of producing life. Something in the air was required for organisms to become established in the originally sterile infusion.

Needham, For Spontaneous Generation

In 1745, a report on a similar study, but with different results, was published by John T. Needham (1713-81), an English Catholic priest who was interested in science. In his experiments organisms developed in the heated and closed hay infusions as well as in those left open, supporting spontaneous generation. Obviously, there were technical differences in the way the two experiments were conducted by the two investigators. Hay is now known to carry resistant spores that are not killed by ordinary boiling. Needham apparently did not heat his cultures to a temperature sufficiently high to kill the spores.

On the strength of his contribution, Needham was elected a member of the Royal Society of London and later he became one of eight foreign associates of the French Academy of Science. One of the main reasons for the wide recognition accorded Needham's results was the support and systematic treatment provided to him by Buffon. In his enthusiasm for Needham's work, Buffon gave it considerable space in his own publication and added his own comments favoring spontaneous generation. Needham was invited to Paris as the guest of Buffon and collaborated with Buffon on the second volume of Buffon's encyclopedia of scientific knowledge. The distinguished London churchman and the famous French encyclopedist thus formed a strong team favoring spontaneous generation.

Bonnet, Preexisting Germs

At this same time, a slightly different idea concerning the origin of life was presented by another distinguished French naturalist, Charles Bonnet (1720-93) who spoke of a preexistence of germs. This idea was speculative, like many others of that period, but it supported the existence of microorganisms and the stability of the living processes. The preexistence of organisms was never demonstrated and the idea suffered for want of experimental support. A heated discussion, however, followed the presentation of Bonnet's speculation. It has since been shown that neither the view of Bonnet nor that of Needham and Buffon was correct.

Spallanzani, Against Spontaneous Generation

Another scientist entered the controversy at this point and added an impressive chapter to the history of spontaneous generation. This was the Italian physiologist, Lazaro Spallanzani (1729-99; Fig. 15.2). Spallanzani was an experimenter who recognized the problem of spontaneous generation as one lending itself to the experimental procedure.

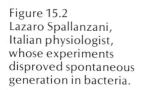

Figure 15.2
Lazaro Spallanzani,
Italian physiologist,
whose experiments
disproved spontaneous
generation in bacteria.

The actual experiments of Spallanzani were similar to those of Needham, but he designed and conducted them with greater care and he prolonged the period of sterilization. Several different media were used, including urine and beef broth as well as hay infusion. In one experiment Spallanzani boiled the medium for an hour and hermetically sealed the containers while the medium was hot, thus completely excluding all organisms. Spallanzani considered the matter settled when in 1765 he published his results in Italian.

In 1769, the work of Spallanzani was translated into French and published with accompanying notes written by Needham, who found a different explanation for most of Spallanzani's results and conclusions. The result that seemed to Spallanzani most conclusive, that is, the continuous absence of microorganisms in boiled and sealed cultures, was attacked by Needham on the basis of the extreme and prolonged heating that Spallanzani had applied to his cultures. High temperatures over long periods of time were considered by Needham to be sufficient to destroy the "vegetative force" that he thought was necessary for the development of any life. He accused Spallanzani of "torturing" the vegetative infusions to the point where the vital material was weakened or destroyed. Furthermore, argued Needham, the air remaining in the empty part of the vessels was completely spoiled by the heat treatment.

Needham then suggested a type of treatment by which supposed foreign organisms could be removed and the vegetative force uninjured. These provisions were considered by Spallanzani not sufficient to remove all the organisms from the culture, but he was challenged to take up his experiments again. He found that, in order to destroy all microorganisms, it was necessary to boil the medium for at least three quarters of an hour. The argument between the two adversaries then faced an insurmountable difficulty. Needham insisted on a treatment that would not destroy the presumed "vegetative force," whereas Spallanzani insisted that the treatment prescribed by Needham was not sufficient to destroy all foreign organisms present in the cultures.

The problem was further confused by an observation of the eminent French chemist, Gay-Lussac (1778-1850). He analyzed the air in the sealed containers in which meat and grape juice had been boiled and found no oxygen. Therefore, he concluded that preserved foods could be kept only in the absence of oxygen, which he considered to be the life-giving property of the medium. Needham's objections then seemed even more readily justified, because oxygen was considered to be removed by the treatment performed by Spallanzani. Even though the Italian experimenter felt secure in his earlier position, some doubts concerning the broader aspects of the problem were raised in his own mind by the objections of Needham and Gay-Lussac.

Schwann and Schulze Against Spontaneous Generation

Another development in the controversy was initiated by the German physician and physiologist, Theodor Schwann (Fig. 14.1). In 1836, Schwann reported an experiment in which a preparation of lean meat in a glass flask was sealed off in a flame and the entire preparation was boiled. The material was allowed to cool in the sealed flask and no organisms appeared. To test the objections made previously concerning this type of experiment, Schwann modified his experimental procedure in such a way that air could be added continuously to the culture by a system of tubes. When the air coming in was heated, no growth occurred in the medium. It was shown further that air could be heated and then cooled before it was introduced, and still no contamination would occur in the meat juice. This experiment reinforced

Spallanzani's view and was contrary to Needham's observation. This experiment contradicted, in part at least, Needham's objection that a vegetative force had been destroyed and it disproved Gay-Lussac's hypothesis that preservation could be accomplished only in the absence of oxygen.

A further problem arose, however, when Schwann prepared four flasks with a solution of cane sugar. Some were opened to permit ordinary air to enter while others were closed off and only heated air was allowed to enter through appropriate tubes. At the end of this experiment, organisms were present in the flasks that had been exposed to ordinary air, but those in which only heated air was allowed to enter were completely free from organisms and no fermentation was detected. When these experiments were repeated, some of the flasks left open to ordinary air did not show fermentation. Schwann concluded that it was not oxygen, or at least not oxygen alone, in the atmosphere that was responsible for alcoholic fermentation but rather a principle sometimes present and sometimes absent in ordinary air, that Schwann considered to be a living organism.

Although Schwann's experiments yielded strong evidences in support of an air-carried living principle involved in alcoholic fermentation, the conclusion was not widely accepted. Uncertainty still persisted in the minds of most investigators as to whether living organisms were introduced with air or whether some chemical alteration of air such as the removal of oxygen or some other vital material was responsible for the origin of living organisms in the medium.

Frantz F. Schulze (1815-73), a Berlin agricultural chemist, added another step to the experimental procedure when he (in 1836) devised an experiment similar to that of Schwann. Instead of calcining (heating) the air, however, he passed it through concentrated solutions of potassium hydroxide or sulfuric acid. His results were similar to those of Schwann. The question of the effect of the heat on the air was resolved by this method but the fundamental objection was not overcome. It was possible, the opponents insisted, that sulfuric acid and potassium hydroxide as well as heat could destroy the vegetative force or the "germs of life."

Schröder and Von Dusch Introduce the Cotton Plug

Two other German scientists, Heinrich G. F. Schröder (1810-85) and Theodor von Dusch (1824-90), devised a method of filtering air through cotton-wool. This did nothing to the air that could be construed to destroy any force it contained or alter in any way its fundamental properties. However, dust particles carrying microorganisms were successfully filtered from the air as the air was passed through the cotton. The original method used by Schröder and von Dusch was an elaborate procedure making use of a complex system of flasks and glass tubes. A plug of cotton-wool was placed in the glass tube that entered the flask containing nutrient medium and a pump was arranged to force the air through the cotton, along the tubes, and into the flask. Every precaution was taken in the preparation and heating procedure to make sure that all of the organisms were removed from the original material and air was introduced only through the cotton plug.

Different kinds of media, such as meat juice, beer "must," and milk were used in the experiments. Some of these were shown to behave differently. Meat juice and beer "must" remained free from organisms and without physical change after the treatment. Milk, however, curdled and spoiled as readily in the tubes through which the air was filtered as in the tubes through which ordinary air was introduced. Different kinds of changes were thus associated with different kinds of material. A putrefaction was occurring in milk that required only oxygen, whereas in meat juice and beer "must," factors from the air were required before alcoholic fermentation could proceed. In their publications of 1854 and 1859, these authors concluded that some agency carried by the air was required for alcoholic fermentation. Furthermore, they demonstrated that this agency could be filtered out of air by ordinary cotton appropriately arranged in the tubes.

The cotton plug has since been developed to simple and practical usefulness in bacteriology. Flasks and test tubes containing sterile media or cultures of microorganisms are commonly stoppered with plugs of sterile nonabsorbent cotton. As long as the plugs remain tight and dry they serve the valuable purpose of excluding the microorganisms carried in dust particles, and provide an efficient means through which the air can circulate in and out of the container.

In spite of the excellent experimental work of the investigators listed above, the controversy concerning spontaneous generation had not been resolved to the satisfaction of everyone. Recognizing the uncertainty existing on this fundamental question, the French Academy of Science offered a prize for the best dissertation on the subject "Attempts by Well-Conceived Experiments to Throw New Light on the Question of Spontaneous Generation." The main competitors for this prize were the French naturalist, Felix A. Pouchet (1800-72), and the French chemist, Louis Pasteur (1822-95; Fig. 15.3).

Pouchet carried out experiments (1859) in which he prepared flasks containing hay infusions, sterilized them by boiling, and then

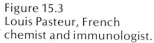

Figure 15.3
Louis Pasteur, French
chemist and immunologist.

sealed off the necks of the flasks. He subsequently observed no difference between the control flasks left open to the air and those
sealed off, following the heat treatment that was presumed to be
sufficient for the removal of the organisms. In the course of his
studies, he obtained air from various places, including the edge of a
glacier at 6,000 feet elevation, and found that all the containers, regardless of the source of air or the presence or absence of air, had
growing organisms.

Pasteur Disproves Spontaneous Generation

Louis Pasteur had been trained in chemistry and, although at this
time in his career he had done some research on fermentation, he
was considered a chemist rather than a biologist. Pasteur's experiments were similar to those of Pouchet, but he used sugared yeast
water for the medium instead of hay infusion. Nineteen of Pasteur's
prepared flasks were opened and then sealed in a lecture hall of the
Museum of Natural History in Paris. Following incubation, four

showed evidence of living organisms. Of the 18 which were opened outdoors under trees, 16 developed organisms. Pasteur was thus able to show that different results were obtained when air was introduced from different sources. This suggested that it was not air itself, but something that was sometimes carried in air that was responsible for the microorganisms.

Insufficient heat treatment was considered by Pasteur as a possible explanation for the presence of organisms in Pouchet's sealed containers. He did not know that hay infusions may carry organisms in a dormant stage (as spores) which are not destroyed by the ordinary boiling. (Spores are carried in the hay itself and they provide an insidious means by which living organisms can remain dormant for long periods and develop when conditions are suitable.) Since yeast water had no resistant spores, the experiments of Pouchet as well as those of Pasteur yielded results that could now be repeated and explained. Unfortunately, the reason for the difference was not understood at that time. Later Pasteur became aware of the presence of resistant spores of certain bacteria in such materials as hay infusions. He also noted the existence of anaerobic organisms that cannot live in the presence of free oxygen.

Pouchet withdrew from the competition and Pasteur won the prize. In his detailed treatise, Pasteur described the experiments in which flasks carrying media were sterilized and sealed off—no life was found in these flasks. Some were opened in various places, incubated, and observed for the presence of organisms. Twenty were opened in the high Alps and all except one were negative, whereas *all* of those opened in a dusty building developed living organisms. Flasks opened on the streets in downtown Paris were all contaminated. A decisive experimental treatment was thus recorded by Pasteur that not only

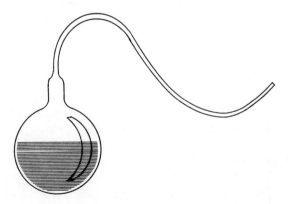

Figure 15.4
Pasteur tube illustrating
a device used by Pasteur
in his experiments on
spontaneous generation.

won the prize offered by the French Academy, but also won the immediate argument about spontaneous generation.

A further controversy developed between a London physician, Henry Charlton Bastian (1837-1915), and Pasteur. It followed the publication in 1872 of Bastian's two-volume work titled *The Beginning of Life*, in which difficulties with Pasteur's experiments were cited. Bastian was particularly concerned with the discrepancies between the experiments of Pasteur and those of Pouchet, and he reemphasized the old theory of Gay-Lussac that oxygen was the only essential factor in the origin of life. Pasteur drew out the neck of a flask which contained a sterile medium which would sustain bacteria and produced a "Pasteur tube" (Fig. 15.4). The end of the tube was completely open to the air but the bend in the tube provided a place for dust particles to settle without entering the flask. No growth of bacteria occurred in properly prepared Pasteur tubes.

Tyndall and Optically Pure Air

John Tyndall (1820-93), an English physicist, studied the optical effects of dust. He devised a method of identifying optically pure air; that is, a method that used optical means to determine when air is free from dust. This approach provided a more decisive method than those used by Pasteur for removing the foreign bodies in the air without altering the air.

In the years 1876 and 1877, Tyndall presented the results of his experiments in *Philosophical Transactions,* describing a method of purifying air by allowing it to settle in closed boxes. Following the settling period he introduced light through a slit in an otherwise closed box in such a way that rays of light could be reflected from suspended particles in the air. The optical test could be used to indicate when the air was completely free from foreign bodies. When pure air was introduced into a medium capable of supporting living organisms, no organisms were produced. It was thus established that organisms carried in the air were responsible for contamination. These results, coupled with those of Pasteur, constituted the final step in the long controversy concerning immediate spontaneous generation of bacteria. It was then concluded, only to be taken up again in the present century in a more modern context at the level of the virus.

A further contribution that prevented a recurrence of the idea of spontaneous generation for organisms as complex as bacteria was made later by Pasteur. He showed that a particular kind of organism kept in pure culture could be introduced into a medium and, if the

Figure 15.5
Joseph Lister, English
surgeon who devised
methods of antiseptic
surgery.

culture were protected from contamination, only the kind of organism in the original culture would be present. Other media inoculated with organisms from other types of pure cultures would produce only those particular types. This evidence was clear and decisive and showed that parent microorganisms generate only their own kind of organism.

With the controversy over spontaneous generation of bacteria closed, life histories and reproductive processes of microorganisms took on great significance. Practical applications were made in food preservation and problems concerning fermentation. It was no longer hopeless to attempt to preserve foods and a canning industry soon developed. Applications were also made to problems concerning the control of disease. Antiseptic methods were applied in hospitals and particularly in surgical procedures. In Paris at this time, about one in every 19 pregnant women who went to hospitals to have babies died of puerperal fever, a disease familiarly called "childbed fever."

The Hungarian physician, Ignaz Phillipp Semmelweis (1818-65), was a pioneer in employing asepsis. While on the staff of the hospital in Vienna, he recognized the infectious nature of puerperal fever and insisted that attendants in obstetrical cases thoroughly cleanse their hands. He thus greatly reduced the mortality rate from infection in childbirth in the Vienna hospital. When doctors became aware of the importance of washing their hands and sterilizing their instruments, the mortality from this disease dropped considerably.

Lister and Antiseptic Surgery

Even more striking were the applications of antiseptic surgery pioneered by the English surgeon, Joseph Lister (1827-1912; Fig. 15.5). He was interested in bacteriology, and as early as 1878 had obtained pure cultures of bacteria. Following the impressive work of Pasteur and Tyndall, Lister devised methods of sterilizing the operating room with all its equipment by spraying 5 per cent carbolic acid over the hands of the doctors before an operation and on the immediate surroundings while the operation was in progress. This was a much needed improvement at a time when most surgical treatments were complicated by infection. The legendary old Doctor Lowry, who preceded Lister in England, was reported to have performed 1,000 operations and only three of his patients survived.

Current View of Origin of Life

Spontaneous generation is now being considered in another setting against a different background. Modern naturalistic discussions concerning the origin of life have centered around the possibility of life developing once in the far distant past by the combination of inorganic materials present in a primordial "soup." An energy source such as lightning has been suggested for promoting the chemical synthesis. Simple amino acids and nucleic acids have been produced in the laboratory by bringing together materials known to have been present in the early stages of the earth's history at a favorable temperature and introducing electrical energy. To be sure, there is some distance between these crude organic materials and the complex proteins and nucleic acids that occur in the bodies of living organisms, but the experiments have been provocative.

Pasteur showed that living things as complex as bacteria could not arise spontaneously in a short period of time under the conditions of his experiments. The possibility is not excluded that much more simple organisms having the power of self-replication could have arisen by natural means in the long period of the distant past. It is not likely that this could occur now because the greater quantity of oxygen in the atmosphere would immediately oxidize an unprotected proto-plasmic mass and the new living organism could not survive. If it did avoid immediate oxidation, it would surely be eaten or absorbed by some other form of life which is now so abundant in any place suitable for life to originate spontaneously.

Problems of Fermentation

In the middle of the last century, fermentation was considered to be a chemical process. The chemist Liebig described the process as de-pending on a chemical ferment, an alterable substance that decom-poses and in so doing excites chemical change, that is, fermentation, in a ground substance. The disintegration of a ferment was thus be-lieved to cause a disturbance among the molecules that brought about the changes associated with fermentation. According to this view, the primary cause of fermentation was a chemical instability. Other chemists of the period, including Berzelius (1779-1848) and Bertholet (1827-1907), held similar views concerning alcoholic fer-mentation. In 1837 Schwann had described alcoholic fermentation as a process dependent on a yeast, but none of the leading chemists of the time considered seriously the possibility of a living organism as the causative agent of fermentation.

Pasteur pursued Schwann's idea further and became a pioneer in experimental studies on fermentation. He confirmed the observation made by Schwann and showed that a living organism, a yeast, is re-quired for the chemical change that transforms sugar into alcohol and carbonic acid. In 1857 Pasteur discovered that a different kind of or-ganism was associated with the process by which sugars were broken down to lactic acid. These observations led Pasteur to a hypothesis that microorganisms of some kind are essential for each kind of fer-mentation. To obtain evidence concerning this hypothesis, Pasteur isolated microorganisms associated with fermentation and raised them in nutrient solutions. These were inoculated into appropriate natural media from which the natural organisms had been removed and fer-mentation was produced experimentally. A temperature of 30 to 50 degrees centigrade was found to be optimum for the process.

Pasteur followed carefully the cycle of fermentation. He repeatedly removed the organisms from fermenting cultures, transferred them to other sterile media, and found that fermentation always resulted when the proper organisms were present and the appropriate conditions were arranged. At the end of the experiment, Pasteur identified the materials that had been produced. He then described the chemical changes which required organisms that were involved in the transformation of sugar to lactic acid. Different organisms were observed under the microscope and their characteristics were carefully described. Lactic acid organisms were found to resemble in some ways those associated with alcoholic fermentation. More recent studies have shown that the two kinds of organisms are not as closely related as Pasteur supposed. The lactic acid organism is a bacterium rather than a yeast. Distinctly different kinds of organisms were thus shown to be responsible for different kinds of fermentation. When brewer's yeast was added to a standard medium containing sugar, alcoholic fermentation would proceed; when lactic acid organisms were added to a similar medium, lactic acid was the product.

During the next 20 years (1857-77), Pasteur studied a number of other fermentable materials and became an authority in the field. In 1858, he discovered that the organism associated with fermentation in ammonium tartrate was a mold. Yeasts, bacteria, and molds had now been identified with fermentation. Pasteur's memoir on alcoholic fermentation, written in 1860, remains to this day a classic on the subject.

Pasteur investigated (1861) butyric acid fermentation and made another important discovery; that fermentation can proceed in the absence of oxygen. The rod-shaped organism associated with butyric fermentation was considered by Pasteur to be animal in character and was named vibrio. While examining a drop of fluid culture containing butyric vibrio under the microscope, Pasteur observed that the organisms in the margin of the drop where they came in contact with the air were not active, but in the center they were actively motile. He developed the hypothesis that oxygen had an inhibiting effect on the growth and activity of these organisms. To test the hypothesis, Pasteur passed a stream of oxygen through an active butyric fermenting culture and observed that all action was inhibited. Two words, aerobic and anaerobic, were used respectively to distinguish between the two kinds of organisms. Also, in 1861 Pasteur published his research on acetic acid fermentation. He showed that this fermentation was accomplished by organisms of the genus Mycoderma, which he studied in detail in 1862. A full scale study of vinegar preparation was undertaken and the results were published in 1864 and 1868.

Pasteur was then asked by his former professor, J. B. Dumas, and personally requested by Napoleon III, to study the problem of souring wine that threatened the important French wine industry. He pursued the investigation with vigor and enthusiasm and in 1866 published an impressive memoir on wine. This 264-page treatise dealt with the so-called diseases of wine, which Pasteur described as being

Figure 15.6
Hieronymus Fracastorius,
Italian physician of the
sixteenth century, who
developed the concept
of contagion.

due to foreign organisms that invaded the wine and altered its chemical and physical properties. Pasteur showed that undesirable organisms could be removed by partially sterilizing, at a temperature below the boiling point, the juice from which wine was to be made. This was not sufficient to destroy the valuable properties of the juice. Other organisms capable of producing desirable qualities could then be introduced from pure cultures. The process, now called pasteurization, was later adapted to milk for the purpose of removing most of the natural organisms and all of the pathogens such as the tuberculosis, brucellosis, and typhoid organisms.

Pasteur returned in 1871 to his investigation of fermentation, this time devoting his attention to the fermentation of beer. The results were published in 1876 in a detailed, nearly 400-page paper which not only discussed the results of investigations on beer, but also summarized Pasteur's general knowledge substantiating his concept that the fermentation process was based on living organisms. The proposed explanation of fermentation was again challenged, this time by Liebig, who had developed a modified version of his older theory in which enzymes were postulated to be responsible for fermentation. No enzymes could be demonstrated, however, and Pasteur's view that living organisms were responsible for the process eventually won support.

Germ Theory of Disease

The germ theory of disease came as an extension of the principle of biogenesis. Living organisms produced by parent organisms were found to be associated with disease in domestic animals and man in much the same way as other organisms were associated with fermentation and decay. One of the first to suggest the relation between living organisms and disease was Hieronymus Fracastorius (1484-1553; Fig. 15.6), who (in 1546) wrote a treatise entitled "On Contagion." This was the earliest well-considered exposition on the concept of contagious infection. In this treatise, he postulated seeds of disease called "seminaria" that were capable of transmitting disease from individual to individual. Fracastorius suggested several modes of transmission such as direct contact, passive transfer on clothing or other articles, and through the air. As an illustration of airborne transmission Fracastorius mentioned the spread in the air of a substance released when onions are peeled. The eyes of people a few feet away smart and release water. This example was considered only as a model demonstrating how disease could be transmitted from person to person even at considerable distance. Fracastorius presented accurate descriptions of several common diseases including typhus fever, plague, rabies, and syphilis. Leeuwenhoek, in 1676, described organisms not visible to the unaided eye as being widespread in nature. These observations added considerable support to the idea of living organisms which might presumably be the causative agents of disease.

Sydenham and Specific Cause for Disease

A seventeenth century physician, Thomas Sydenham (1624-89; Fig. 15.7), also contributed to the idea of specific causes for specific diseases. Sydenham had great influence on his fellow physicians, but he had no immediate successors to carry on what he taught. He insisted that disease had an infectious cause and thus initiated the beginning of the end of the humoral theory of disease. Sydenham gave clear accounts of the differential characteristics of infectious diseases, such as smallpox, dysentery, plague, and scarlet fever. In order to treat a disease properly, he contended, it was necessary to understand the cause. When the cause was understood and appre-

Figure 15.7
Thomas Sydenham,
English physician who
developed a specific
treatment for malaria.

ciated, a cure might be devised. In many cases the cure would be the removal of the cause. At the time this concept was developed, few diseases were known well enough to provide a test for the hypothesis. Even now few diseases are completely understood and entirely curable. Sydenham made one observation that resulted in a practical treatment for an important disease when he observed that cinchona bark, now known to contain quinine, was useful in the treatment of malaria. When he tried this remedy on other fevers, he found it had no value. It was thus not a "cure-all," but a specific cure for a specific disease. Later it was found that quinine only arrested but did not actually cure malaria.

Jenner and Smallpox Vaccination

Another practical accomplishment was made in 1796 by Edward Jenner (Fig. 10.5), a brilliant London physician, who had lived and studied with John Hunter (Chapter 10). In his medical practice he had observed that people who had recovered from the comparatively mild disease, cowpox, did not get smallpox. This suggested the possibility of a vaccination against smallpox with the less severe cowpox. Vaccination for smallpox had been used before the time of Jenner, but he developed it to a stage of practical usefulness. Jenner and his contemporaries were not able to explain the reason for the success of the vaccination, however, and it remained for Pasteur to develop the explanation on the basis of his attenuated virus theory.

Davaine Identifies Organisms with Disease

The first recorded demonstration of a specific organism associated with a specific disease was made by the Swiss professor of philosophy, Bénédict Prévost, at an academy in Montauban, France in 1807. He observed germination of spores of the wheat bunt organism and conceived the idea that this organism penetrated the young wheat plant and was the actual cause of the disease. The French pathologist Casimir J. Davaine (1812-82) was first to identify a microorganism with a disease in an animal. He was led to this accomplishment through his interest in symptoms that might prove valuable for diagnostic purposes. In 1850, Davaine and Rayer noticed rod-shaped bodies in the blood of animals that had died with splenic fever or anthrax. At the time, they did not attach any significance to the observation. In 1861, after reading Pasteur's memoir on butyric acid fermentation, Davaine returned to his anthrax study and in 1863 became the first to associate a particular organism with a specific disease. The rod-shaped organism, *Bacillus anthracis*, was identified as the causative agent of anthrax. Robert Koch added the next step in the sequence when he succeeded in tracing the life cycle of the anthrax bacillus.

Koch, Pioneer Bacteriologist

Robert Koch (1843-1910; Fig. 15.8), a German physician, teacher, and bacteriologist, was closely associated with the firm establishment of the germ theory of disease. His education was obtained at the University of Göttingen. Here he studied under the pathologist Jacob Henle, who in 1840 had anticipated the germ theory of diseases. Koch received the doctor of medicine degree in 1866, and during the next few years he practiced medicine, served in the Franco-Prussian War, and began research projects in the field of microbiology. In 1876, he traced the life cycle of *Bacillus anthracis* and developed pure cultures of the anthrax organism.

Using these cultures, Koch observed the multiplication of the organism and discovered avenues through which infection was transmitted from animal to animal. He also explained the delayed action that had puzzled earlier investigators. Cattle, placed in a pasture where diseased animals had been buried months or years before, contracted the disease. Koch showed that organisms buried with diseased animals could form spores and persist in the soil for long periods of time. They were brought to the surface by earthworms and thus became associ-

Figure 15.8
Robert Koch,
German bacteriologist.

ated with the vegetation the animals ate. Once introduced into an animal body, the organisms became active, multiplied, and caused the symptoms of the disease. Through his careful studies of the life cycle of the organism and the effects of the organism when transmitted into new host animals, Koch discovered an immunizing agent in which Pasteur also became interested.

Another valuable contribution of Koch that did much to verify the germ theory of disease was the isolation, in 1882, of the tubercle bacillus, *Mycobacterium tuberculosis.* This represented a milestone in the history of bacteriology and epidemiology. The infectious nature of an important human disease was understood for the first time. Tuberculosis was attributed to a living organism capable of reproducing itself and thus spreading the infection from one individual to another.

In 1883, Koch discovered the cholera vibrio, *Vibrio comma.* Control of another important human disease, cholera, was then accomplished by improving sanitary conditions in the areas where cholera formerly had spread in epidemics.

Koch also helped improve techniques used in bacteriology. One contribution was the introduction of solid media for bacteriological cultures. In the early days of bacteriology, only fluid media were used. These are effective for growing most bacteria, but the individual organisms are difficult to catch and study critically. In addition, colony formation could not be observed when the organisms are dispersed in an excess of fluid. Koch recognized the desirability of isolating organisms in a solid or semisolid material in order to observe their growth and colony formation. He visualized stationary organisms, "stopped like ships frozen in the ice," as ideal objects for study. If division could continue while the organisms were localized, colonies would appear.

Koch first used gelatin as a solid medium. Gelatin can be liquefied by heat and poured while warm into dishes. When cool it is solid, but it retains enough moisture to provide a suitable medium to support bacterial growth. This was a great advance in the technical aspect of bacteriology because colonial formations and characteristics of organisms growing in groups as well as individuals could now be followed. Initially, flat glass plates were covered with the warm gelatin and then protected from air contamination with bell jars. The gelatin ran off the edges, however, and created an untidy situation in the laboratory. To circumvent this difficulty, Koch's assistant, Richard J. Petri (1852-1921), invented a dish with the edges turned up, called a petri dish (Fig. 15.9). A lid, loose enough to allow air to enter, was devised later to protect the plate from contamination.

Another contribution of Koch was the hanging drop or depression slide (Fig. 15.10) method of bacterial culture. This was accomplished by preparing a slide with a depression in the center. A drop of culture containing organisms was then placed on a cover glass that was placed over the depression in such a way that the drop hung freely into the open space. Sufficient air was provided in the surrounding space, and

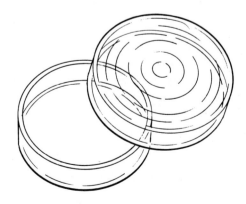

Figure 15.9
A petri dish illustrating
the method used by Koch
for raising bacteria on
a solid medium.

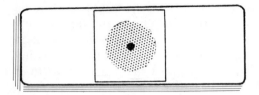

Figure 15.10
A depression slide
illustrating the hanging
drop method used by Koch
for raising certain types
of bacteria.

the drop was protected from contamination. The slide could be placed conveniently in the incubator and held at the appropriate temperature for the development of the organisms. When it was necessary to inspect the culture, the glass depression slide could be placed on the microscope without disturbing the activities of the living organisms.

The hanging drop technique was later adapted and widely used for blood studies. Another valuable application has involved using the depression slide technique for tissue culture. By this means tissue cells taken from the body of an animal are kept alive and may be studied in minute detail. Growth characteristics of such cells have been followed by making frequent observations under the microscope, and valuable contributions have been made in the study of normal as well as abnormal growth. When ordinarily rigid cells, such as connective tissue and bone cells, are raised in tissue culture, they grow in spongy masses without restriction.

Koch also made contributions in perfecting staining techniques. He introduced methyl violet as one of his first bacteriological stains, and through the years it greatly facilitated the identification and study of microorganisms.

Another contribution, which had theoretical as well as technical implications, was the method devised by Koch to associate specific organisms with specific diseases. People were beginning to note organisms that occurred in association with disease symptoms and, without critical evaluations, were considering such organisms to be the cause of the disease. A system was needed, therefore, by which organisms could be properly identified as causative agents of particular diseases. Perhaps it was even more important to keep organisms not causing the disease from being wrongly incriminated through circumstantial evidence.

Koch formulated a set of postulates that required critical analysis of the organism in association with the symptoms. (1) The organism must be associated with the disease. (2) The investigator must isolate that organism from the diseased animal and establish it in a pure culture. (3) Organisms from the pure culture must then be injected into animals free from the disease and in due time the inoculated animals must show symptoms of the disease. (4) The organism must be isolated from the second host. In many cases it is impossible to carry out all of the steps required by these postulates. Some organisms cannot be cultured under usual laboratory conditions. Nevertheless, the standard established by Koch was well considered and did much to eliminate the loosely drawn conclusions that were all too prevalent at that time.

Not the least of Koch's accomplishments was the influence that he exerted on the developing science of microbiology. His own contributions as a pioneer in the field of bacteriology were basic and significant, but the contributions made by his many students and assistants represented a vital body of knowledge adding impetus to the development of the new science. One of the most original of Koch's assistants, Paul Ehrlich, carried out research similar to that of Koch on staining methods. He also pioneered in the field of chemotherapy; that is, the aspect of medicine devoted to the treatment of internal diseases by chemical reagents.

Ehrlich and Chemotherapy

Paul Ehrlich (1854-1915) was born in Strehlen, Upper Silesia, the son of a Jewish innkeeper. His early training was in chemistry, but he had basic preparation in biology. He attended medical school to learn about biological processes in living systems and to search for methods through which diseases might be cured. Paul's cousin, Carl Weigert, who was nine years older than Paul, had pioneered the field of differential staining in tissues and bacterial cultures.

In 1878, Weigert had observed that certain kinds of bacteria were susceptible to different stains. Ehrlich was fascinated by this discovery, learned the art of staining from Weigert, and soon was doing independent research. Eventually he also discovered that certain tissues in the body were susceptible to certain chemical preparations. When methylene blue, for example, was injected into the living animal, it was found to be taken up by the fine structures of the brain, thus providing a differential stain for specific tissues. This concept was developed by anatomists as a tool to pick out tissues that had affinities to certain stains.

A related possibility was also considered by Ehrlich: if different tissues had varying affinities for different stains, they might also have different affinities for drugs. Thus, certain materials might be found which would have an immediate toxic effect upon invading microorganisms without seriously affecting the tissues of the host. His idea of "magic bullets" developed from this line of reasoning. Certain problems were immediately evident. Why does one tissue stain blue and another red when the same stain is employed? Why does the cell nucleus accept one stain and the cytoplasm another? Why was it that different bacteria exposed to the same dye are not all affected in the same way—some remain unchanged while others are deeply stained? Even more immediate and practically applicable questions could be asked on the basis of facts then known about certain important diseases. Diphtheria toxin, for example, could be injected into a pigeon without any harm or injury to the bird. When the same quantity of toxin was placed in a child's body, great disturbance and possible death would follow. There must be a fundamental difference, therefore, between the cells of the pigeon and those of the child, with respect to their affinity to diphtheria toxin.

Ehrlich did more than postulate relations and ask questions, he worked out ingenious methods and tools to test his hypotheses. First he developed diagnostic measures to identify symptoms of diseases to be studied. Laboratory techniques were then invented by which chemicals could be introduced into the cells of experimental animals. Throughout all of this work, Ehrlich was motivated by the idea that chemical affinities represented the keys to life's secrets. The first logical question in his study of diphtheria was, "How does the toxin injure the cell?"

Ehrlich was largely responsible for developing the antitoxin to neutralize the effect of diphtheria toxin, but he was not content with the practical application of an antitoxin. He wanted to know how the antitoxin prevented the injury that the toxin could produce. Furthermore, he wanted to know how the animal body produced an excess of antitoxins and thereby became immune. Since toxins and antitoxins must both be chemicals, he attempted to produce toxins and antitoxins in the laboratory. Ehrlich himself was unsuccessful, but progress in that direction has been made since his time.

Following his brilliant work on diphtheria, which was completed in 1904, Ehrlich devoted his attention to cancer chemotherapy and searched for a chemical agent that would offset abnormal growth and malignant change in different parts of the animal body. This goal was not realized, but the pattern of Ehrlich's experiments has been continued, with some progress.

Ehrlich's crowning achievement was the discovery of salvarsan, an arsenical used for the treatment of syphilis. Shortly after the discovery of the causative agent of syphilis, *Treponema pallidum*, by Fritz Schaudinn in 1905, Ehrlich began experimenting with various arsenicals. His objective was to find a drug that would be toxic to the spirochaete, but not seriously toxic to the individual being treated. In the course of this investigation he and his associates discovered the famous "606" remedy. Members of Ehrlich's research team tried 605 compounds without appreciable success but number 606 (salvarsan) gave positive results. It was found later that salvarsan did not completely rid the individual of the disease-producing organism, but it did largely control the disease symptoms.

Salvarsan was the first great therapeutic agent discovered in modern times. Sydenham, many years before, had discovered a therapeutic agent, quinine, for the control of the malarial symptoms in the human body, but Ehrlich was first to make a systematic approach to broader problems involving chemotherapy. Arsenicals remained the best agents for treating syphilis until the discovery of the sulpha drugs and penicillin in the 1930s.

Pasteur's Contribution to the Germ Theory of Disease

In 1865, Pasteur's study of fermentation was interrupted by another request from Professor Dumas. A silkworm disease called pebrine was threatening the important silk industry in France, and Pasteur was asked to devote his attention to this disease with the hope that he might achieve an understanding and develop controls. Six years were devoted to this investigation.

Pasteur began by observing diseased worms and found that a sporozoan was present in the intestinal canal of silkworms that had the disease. The parasite was found to spread throughout the body of the worm, eventually filling the gland in which the silk is normally spun and making the worms unable to produce silk. Pasteur discovered that the organism was spread from parent worms to their offspring in the egg. He then was able to find a link in the reproductive cycle that could be broken to eradicate the parasite.

Silkworm eggs could be examined under the microscope for detection of parasites. At first, silkworm growers considered this method tedious and impractical, but adequate tests proved it to be both practical and convenient. By examining eggs and destroying those in which the parasite was present, cultures free from the disease could be maintained. In this way, the disease was eradicated and the French silk producers were able to reestablish the industry. In the course of his work, Pasteur discovered that two diseases were involved instead of one. Pebrine was spread by a protozoan, but another disease, flacherie, which had previously been indistinguishable from pebrine, was caused by a bacterium.

Following the completion (1876) of the lengthy paper on beer production, Pasteur became interested in human infectious disease. He had already studied anthrax, a disease that affected human beings as well as animals, and he was interested in making further observations and developing practical applications of the germ theory as it related to man. In 1880, he examined cases of osteomyelitis and puerperal fever. These diseases were attributed to microorganisms and Pasteur believed that pus, transferred from one person to another on doctors' hands or instruments, was a means of spreading the disease. Following preliminary observations, Pasteur became interested in the so-called "virus diseases," such as smallpox, and was particularly impressed with the immunity established following an attack of such diseases. He had great respect for Edward Jenner's work on vaccination.

To study immunity experimentally, he chose a disease common in France known as fowl cholera. This disease had been identified as a bacterial infection, and the causative organisms had been cultured. Pasteur began by testing the organisms in pure culture on appropriate media, and he discovered that organisms transferred from culture to culture would tend to lose their virulence, that is, their ability to infect chickens when inoculated. In 1880, Pasteur described weakened organisms produced in this way as attenuated viruses. The virulence of a culture was found to depend on the time that elapsed between successive transfers. When cultures were allowed to remain for several months without transfer, the disease-producing ability of the organisms became weak.

Quite by chance, he inoculated a group of chickens with a culture of virulent organisms after they had already been inoculated with attenuated organisms. The chickens were found to be immune to the virulent culture, presumably having been immunized from the mild attack of the disease brought on by the weakened virus. The principle of acquired immunity was derived from these results. The attenuated virus was a protective agent that stimulated the body to build up an immunity against the specific bacterium.

Fowl cholera vaccine soon was being produced on a large scale and used widely for injection into flocks of chickens. It protected them in most cases, but unfortunate results were sometimes observed. Once in a while the inoculation itself adversely affected the chickens, and occasionally it killed them. Although criticism came from some sources, the principle of achieving immunity by inoculating with an attenuated virus soon became well established. In 1888 Pasteur recommended a virulent culture of pathogenic organisms for the destruction of rabbits in Australia and New Zealand. Pilot experiments were attempted, but the results indicated that this method was not promising. Pasteur successfully applied the principle of acquired immunity to anthrax, however, and later to swine erysipelas and rabies.

When the anthrax bacillus was cultured in broth at 42 and 43 degrees C for eight days, it lost its virulence and did not produce the symptoms of the disease when injected into experimental animals such as guinea pigs and rabbits. Sheep were also found to develop no symptoms after inoculation with the attenuated virus. Pasteur recommended the inoculation of sheep with the newly found vaccine as a protective measure against anthrax. The Agricultural Society of Melun became interested in the prospects and offered 60 animals to Pasteur for a demonstration. Pasteur accepted the challenge and the test was made in a farmyard a mile from Melun. Thanks to a correspondent of the Paris Times, M. de Blowitz, the publicity was spread throughout the world.

On May 5, 1881, 24 sheep, 1 goat, and 6 cows were inoculated with five drops each of a living but attenuated strain of anthrax bacillus. At the same time, 24 sheep, 1 goat and 4 cows were assembled but were not inoculated. On May 17, all of the animals that had been previously treated were reinoculated with a less attenuated culture. The crucial test was applied on May 31, by inoculating the immunized animals and the unprotected animals with a highly virulent anthrax culture. On June 2, when Pasteur and his assistants arrived to make their observations, all of the vaccinated sheep were well. Twenty-one unprotected sheep and the goat were dead of anthrax. Two others of the unprotected sheep died at the end of the same day. The six vaccinated cows were well and showed no symptoms, whereas the four unprotected cows had swellings at the site of inoculation. On June 3, one of the vaccinated sheep died. It was pregnant and the autopsy indicated that it had died following the death of the foetus. This demonstration won wide publicity and Pasteur's attenuated virus was soon well known and used throughout the world.

At the International Medical Congress, held in London a few weeks after the demonstration, Pasteur delivered an address on, "Vaccination in Relation to Chicken Cholera and Anthrax." In this address he explained the use of vaccination as a prophylactic measure and described for the first time the reason for the success of smallpox vaccination many years before.

Anthrax vaccine was used extensively and soon many animals were inoculated. Results were not always successful, but in general they were highly favorable. In 1882, Pasteur reported that 85,000 animals had been inoculated. In a report covering 79,392 sheep, the mortality from anthrax had fallen from 91.01 per cent among the unprotected to 0.65 per cent among the vaccinated. By 1894, 3.5 million sheep were reported to have been inoculated for anthrax. The disease has since become relatively unimportant because the vaccine proved such an efficient prophylactic measure.

Pasteur's next project, in which he was associated with L. Thuillier, was an investigation of swine erysipelas. The organism was found in the blood of diseased animals, and the investigators succeeded in obtaining a pure culture. They next attempted to attenuate the virus, but were not entirely successful. Different effects of the immunization were observed on different animals and varying susceptibilities were encountered in different races of pigs. This confused the study, and made it difficult to tell the extent of immunity that was being induced by the vaccination. In the course of the experiments, it was found that the virus could be attenuated more uniformly by passage through the body of a rabbit. Large scale production began, and between 1886 and 1892, over 100,000 pigs were inoculated in France alone and many in other countries.

The last of Pasteur's great achievements was concerned with the disease rabies. This disease was not common in either animals or human beings, but it was spectacular when it occurred because the symptoms aroused the imagination of people and caused excitement and publicity whenever a case was reported. Pasteur and his collaborators, Chamberland, Roux, and Thuillier, began a series of investigations involving the animal nervous system. After three years of work, it was found that the rabies virus could be attenuated for the dog, rabbit, and guinea pig by passage through a series of monkeys. Protective inoculations for the dog were then devised by the use of virus attenuated in a monkey. An incubation period was established for the virus and Pasteur reported in 1885 that dogs could be protected even after having received the virus from the bite of a mad dog. The method developed consisted of drying the spinal cords of rabbits that had died from the disease after having been injected with the virulent organisms. In the course of about two weeks of drying, the materials taken from the cord became highly attenuated. By inoculating the attenuated virus from the dried cord, it was possible to immunize a healthy dog against the effects of the virus. The results on dogs suggested that the method might have practical application for human beings.

A case was brought to Pasteur before he was ready to test the material on human beings. On the 6th day of July 1885, a nine year old boy, Joseph Meister, was brought from Alsace suffering from bites on the hands and legs from a mad dog. The next day, 60 hours after the bites had been inflicted, the boy was given his first injection of the attenuated rabbit material which was 14 days old. In the 12 inoculations that followed, the virus was made stronger and stronger and the boy was saved from the symptoms of the disease. By October of 1886, some 2,490 persons had been inoculated. A few deaths occurred following inoculation, but for a disease that had previously been nearly 100 per cent fatal, these few failures were not especially disturbing. It was estimated in 1905 that over 100,000 persons in the world had received treatment. The records of the Pasteur Institute showed that up to 1935, 51,057 persons had been treated in that Institute in Paris. About 150 deaths had occurred among the inoculated people, a mortality of 0.29 per cent.

The number of persons who came to Pasteur's laboratory for treatment soon exceeded the capacity of the limited facilities available, and it was suggested by a commission of the Academy of Sciences that a scientific institute be erected for providing inoculations and further scientific research. Public subscriptions were solicited, and a total of 2,586,680 francs were contributed. A plot of land was purchased in Paris, and the Pasteur Institute was erected in 1888. The inauguration on November 14, 1888, was attended by a large group of distinguished people from France and elsewhere in the world.

Pasteur died seven years later on September 28, 1895. A memorial was erected to him in the main entrance of the Pasteur Institute in Paris. On the walls and ceiling of the sanctuary are represented in picture form his major accomplishments.

Arthropod Vectors of Disease

Further support of the germ theory of disease came from another line of investigation. Following the example set by Koch in his study on anthrax life history, studies were designed to determine the life cycle of various pathogenic organisms and their relation with their animal hosts. In some cases, intermediate hosts were found to transmit the organism from one animal to another. In 1880, Alphonse Laveran (1845-1922) observed an organism in the blood of human beings suffering from malaria. This was a triumph in the study of one of the most important diseases of mankind. A protozoan, Plasmodium, was identified as the causative organism for malaria. Ronald Ross, in 1898, showed that the bird malaria organism, *Plasmodium cathemerium*, was carried by a mosquito of the genus Culex. In the same year, Grassi and Bignami discovered that the human malarial organism was transmitted by a mosquito of the genus Anopheles.

Figure 15.11
Theobald Smith, United
States microbiologist.

During the same period, insect transmission of other diseases was being considered. In 1893, David Bruce (1855-1931), found that African sleeping sickness was transmitted by a tsetse fly. Walter Reed, in 1900, showed that the yellow fever virus was transmitted by a mosquito *Aedes aegypti*. Theobald Smith (1859-1934; Fig. 15.11) and his assistant, Kilborne, delineated the cycle of Texas cattle fever. Smith was the first American to distinguish himself in the field of microbiology.

In 1889, Smith and Kilborne performed a series of controlled experiments and established the fact that ticks, *Boophilus annulatus,* were carriers of cattle fever but the nature of the disease and its causative agent still were unknown. When Smith examined the blood of infected cattle, he found little bodies within the red corpuscles that he identified in 1891 as protozoa (Babesia). He next demonstrated that the tick functioned in the cycle by carrying these organisms from infected animals to healthy animals. His most striking discovery was that the protozoa could be introduced into cattle by the bites of young ticks that had received the infection from the mother ticks through the eggs. The report of the investigation of Texas cattle fever, in 1893, was one of the classics in the literature of bacteriology. For the first time in history, a complete cycle of transmission of an infectious disease by an arthropod carrier was established.

Sulfa Drugs and Antibiotics

A new day dawned in the control of microorganisms and the treatment of disease with the discovery of sulfa drugs and antibiotics. In 1932 Gerhard Domagk (1895-1964) discovered that the azo dye Prontosil had a curative action in experimental streptococcus infections. The active principle in Prontosil was found to be sulfanilamide. The sulfa drugs are a group of sulfanilamide and related organic compounds that are especially effective against streptococcus, gonococcus, meningococci, certain types of pneumococcus, staphylococcus, brucella and clostridium infections, and as a urinary antiseptic.

Antibiotics are substances produced by microorganisms that kill or prevent the growth of other microorganisms. Thousands of soil microorganisms have been found to produce substances that have antibiotic properties. Some such substances were recognized by their ability to diffuse in the soil and to prevent the growth of certain other species in their vicinity. Penicillin, produced by a mold, *Penicillum notatum,* was known for many years as a contaminant of bacterial cultures. Clear spaces devoid of growth had been seen around certain colonies on culture plates containing both bacterial and mold growth. Many bacteriologists had observed this phenomenon and some had indicated that the clear zone was due to some bacteriotoxic substance given off by the mold.

Alexander Fleming (1881-1955), British bacteriologist, recognized the significance of this observation and investigated the inhibitory action of the mold. In 1928, while engaged in research on influenza, he observed that mold had developed accidentally on a staphylococcus culture plate and had created bacteria-free circles around the units of mold. Results of his investigations were published in the *Journal of Experimental Pathology* in 1929. He investigated the growing mycelium on culture plates and also in broth medium. By passing broth cultures through filters he removed the mold filaments and was able to study the growth products mixed in the broth which Fleming called "penicillin." A highly toxic effect was demonstrated when this penicillin was brought into contact with sensitive bacteria. It was effective even when diluted 800 times.

The active principle in the broth (now called penicillin) was eventually purified and since 1940 has been produced and marketed on a large scale. It is used in treatment of infections by staphylococcus, streptococcus, pneumococcus, gonococcus, meningococcus, and clostridium and has been employed for treatment of syphilis and actinomycosis. A biosynthetic preparation is now available that causes fewer side effects than the original preparation.

Among the thousands of antibiotic substances that have been found by extensive screening of microorganisms in soil, seawater, and other materials, only about 30 have been commercially marketed. To be produced on a large scale, each antibiotic must be effective, safe, and, in the United States, approved by the Food and Drug Administration. Most of the better ones are now the older ones such as actinomycin D, chloramphenicol, cyclohexamide, erythromycin, neomycin, novobiosin, puromycin, streptomycin, and tetracycline, which have been established by long usage. Large scale screening for new antibiotics and the refining and testing of products that was carried on extensively in the period 1940-60 has since been curtailed or abandoned. When an individual becomes sensitized against an antibiotic and is unable to tolerate it, or when the organism which the antibiotic is intended to control mutates and the antibiotic is no longer effective, the patient may be given a different one that will take care of his needs. Combinations of antibiotics are frequently prescribed to guard against failures that may occur because mutations make a single one ineffective.

In the golden age for antibiotics (1940-1959) virtually all infectious diseases decreased markedly in incidence and mortality. Not much was known about the chemical structure and clinical action of antibiotics, but they had been shown empirically to be effective against infectious diseases and were used more or less indiscriminately for many different infections. Widespread use of antibiotics in hospitals resulted in the selection of mutant strains of pathogens that became a problem in hospitals and they were carried by patients to their home communities where they became dispersed. Sensitivity of patients that have been treated repeatedly with antibiotics also became a problem. More knowledge of the antibiotics and their reactions in patients became necessary to keep up with the mounting problems.

Later research on antibiotics was centered around a correlation of chemical structure. Antibiotics tend to be related compounds that the organisms can produce. Among the thousands of antibiotics that have been discovered, patterns have been detected that permit their classification into chemical groupings. When chemical structures became known, attention was given to biosynthesis through which alterations were accomplished and new properties were observed. Many antibiotics in nature have been synthesized and new ones have been built by developing different chemical combinations. Mechanisms of resistance of organisms to antibiotics have been shown to depend on mutations and an R factor through which cyclic genetic material that is not a part of chromosomes can be transferred from one microorganism to another. This is a frightening prospect because harmless bacteria can transfer resistance to pathogenic bacteria. The mode of action of the antibiotic (the opposite to resistance) is another approach. Antibiotics have been found to act by preventing the synthesis of protein.

Chronology of Events

B.C.
384-322	Aristotle, spontaneous generation for lower forms, bisexual reproduction for higher forms.

A.D.
1546	H. Fracastorius, concept of contagious infection.
1668	F. Redi, disproved spontaneous generation of flies.
1674-76	A. van Leeuwenhoek described bacteria, protozoa.
1680	T. Sydenham, quinine, a specific drug for malaria.
1710	J. Joblot, particles in air and not spontaneous generation cause life in hay infusion.
1745	J. T. Needham, spontaneous generation in hay infusion.
1749	C. Buffon supported Needham on spontaneous generation.
1764	C. Bonnet, preexisting germs.
1765	L. Spallanzani, no spontaneous generation in hay infusion.
1769	J. L. Needham contended Spallanzani had destroyed "vegetative force."
1796	E. Jenner, smallpox vaccination.
1807	B. Prévost, a specific organism causes wheat bunt.
1836	T. Schwann allowed only heated air to enter media flask and no growth occurred; showed alcoholic fermentation to depend on yeast.
1836	F. F. Schulze, passed air through acids and bases and removed organisms.
1854	H. G. F. Schröder and T. von Dusch filtered air with cotton plug.
1857-80	L. Pasteur, particular organisms cause fermentation; disproved spontaneous generation; showed that organisms are carried in air; cure for silk worm disease; attenuated viruses used for resistance to fowl cholera; immunization for anthrax; inoculation for rabies.
1863	C. Davaine, a specific organism *Bacillus anthracis* causes anthrax.

1876-82	R. Koch, life cycle and pure culture of anthrax bacillus; isolated tubercle bacillus; discovered cholera vibrio.
1876	J. Tyndall, optically pure air used to disprove spontaneous generation.
1878	J. Lister, antiseptic surgery.
1878	C. Weigert, certain bacteria susceptible to particular strains.
1880	A. Laveran found malaria organism in human blood.
1893	D. Bruce, African sleeping sickness transmitted by tsetse fly.
1898	R. Ross, bird malaria organism carried by a mosquito.
1900	W. Reed, yellow fever virus transmitted by a mosquito.
1904	P. Ehrlich, chemical treatment for syphilis.
1905	F. Schaudinn, discovered causative agent for syphilis.
1909	P. Ehrlich, chemical cure for syphilis.
1929	A. Fleming discovered penicillin.
1932	G. Domagk discovered sulfa drugs.

References and Readings

Brock, T. D., ed. 1961. *Milestones in Microbiology.* Englewood Cliffs, N. J.: Prentice-Hall.

Burrows, W. 1959. *Textbook of Microbiology.* 17th ed. Philadelphia: W. B. Saunders Co.

Cohen, B. A. 1937. *The Leeuwenhoek Letter.* (Original Leeuwenhoek Letter translated and reprinted by Society of American Bacteriologists, Baltimore, Maryland.)

Conant, J. B., ed. 1952. *Pasteur's Study on Fermentation.* Cambridge, Mass.: Harvard University Press. (Harvard Case Histories in Experimental Science, Case 6. Translation of portions of Pasteur's memoir on lactic acid fermentation and excerpts from Tyndall's lecture on the same subject arranged and supplemented by Dr. Conant.)

Curtis, H. 1965. *The Viruses: Their Role as Agents of Disease and as Probes into the Nature of Life.* Garden City, N. Y.: American Museum of Natural History.

De Kruif, P. H. 1932. *Microbe Hunters.* New York: Harcourt, Brace and Co.

_____. 1950. *Men Against Death.* New York: Harcourt, Brace and Co.

Doetsch, R. N., ed. 1960. *Microbiology: Historical Contributions from 1776-1908.* New Brunswick, N. J.: Rutgers University Press.

Drewitt, F. D. 1931. *The Life of Edward Jenner.* London: Longmans, Green and Company.

Dubos, R. J. 1950. *Louis Pasteur, Free-lance of Science.* Boston: Little, Brown and Co.

Eberly, J. C. 1963. *Edward Jenner and Small Pox Vaccination.* London: Chalto and Winters.

Ford, W. W. 1939. *Bacteriology.* New York: Harper and Bros.

Frobisher, M., Jr. 1944. *Fundamentals of Bacteriology.* Philadelphia: W. B. Saunders Co. Chapter 1 on history.

Gage, S. H. 1936. "Theobald Smith, investigator and man." *Science* 84:117-122.

Gladston, I. 1954. "Erlich, biologist of deep and inspiring vision." *Sci. Monthly* 79:395-399.

Green, J. R. 1909. *A History of Botany.* Oxford: Clarendon Press.

Guthrie, D. 1955. *From Witchcraft to Antisepsis.* Lawrence, Kansas: University of Kansas Press.

Lechevalier, H. and M. Solotorovsky. 1965. *Three Centuries of Microbiology.* New York: McGraw-Hill Book Co.

LeFanu, W. R. 1951. *A Biobibliography of Edward Jenner.* London: Harvey and Blythe.

Levine, I. E. 1960. *Edward Jenner: Conquerer of Small Pox.* London: Blackie.

Metchnikoff, E. 1939. *The Founders of Modern Medicine.* New York: Walden Publishers.

Paget, S. 1914. *Pasteur and After Pasteur.* London: A. and C. Black.

Reddish, G. F., ed. 1954. *Antiseptics, Disinfectants, Fungicides and Chemical and Physical Sterilization.* Philadelphia: Lea and Febiger.

Sexton, A. N. 1951. "Theobald Smith, first chairman of the Laboratory Section, 1900." *Amer. J. of Public Health,* Part I, 41:125-131.

Strode, G. K., ed. 1951. *Yellow Fever.* New York: McGraw-Hill Book Co.

Taylor, B. 1950. *Edward Jenner, Conquerer of Small Pox.* London: Macmillan.

Thompson, C. J. S. 1934. *Lord Lister.* London: John Bale, Sons and Danielsson.

Vallery-Radot, R. 1923. *The Life of Pasteur.* trans. R. L. Devonshire. Garden City, N. Y.: Doubleday, Page and Co.

CHAPTER 16

EXPLAINERS OF LIFE PROCESSES

THROUGHOUT the history of biology, physiological data have been developed more slowly than have the morphological aspects of plants and animals. One reason is that the human mind more easily comprehends observable structural characteristics than intangible functional relations which must be approached by experimental procedures. A second reason is that physiology is closely allied with the physical sciences, particularly chemistry and physics. Not until the last century did these sciences make sufficient progress to be effectively adapted to the explanation of biological phenomena.

Mechanistic Approach to Vital Activity

The French philosopher and physiologist, René Descartes (1596-1650;

Figure 16.1
René Descartes, French
philosopher and
physiologist.

Fig. 16.1) was one of the first individuals known to have an interest in vital activity, the complex of processes through which the organism functions as a coordinated whole. He regarded the phenomena associated with living things as the effects of mechanical forces and, therefore, subject to explanation by the laws of chemistry and physics. This early mechanistic view concerning the nature of life and of living functions resembles the one that has prevailed among physiologists of the nineteenth and twentieth centuries and has been productively used as the basis for an experimental science. Although Descartes was mainly a philosopher, he did much to set the stage for a mechanistic approach to modern physiology.

Descartes began his intellectual activities as a young boy. At ten years of age he was sent by his father to a Jesuit college where he proved to be a good student but was soon labeled as an independent thinker and a nonconformist. He was disturbed by the discrepancies between materials presented by the teacher and those in various books. He also detected occasional inconsistencies in material presented by different teachers on similar subjects. Descartes enjoyed mathematics because of its stability and exactness. In later life he classified the subjects that he had taken, other than mathematics, as being either merely entertaining or worthless.

At 17 years of age Descartes went to Paris to continue his education at the University. After a short introduction to the social life of Paris, he chose to spend most of his time in solitude, devoting himself mainly to study. When he was 21, he entered the military service and was stationed for two years at a garrison in Breda. It was while in the military service that he wrote his first treatise entitled "On Music."

When he was 33 years of age (in 1629) he divorced himself entirely from society and concentrated on independent study. In France he had been subjected to persecution, particularly by church people because of his unorthodox thinking and writing. He was now anxious to find a retreat where he could develop his ideas concerning man's relation with nature, even though these ideas might be contrary to accepted doctrines of the church. To avoid possible conflicts with religious leaders in his home community, he traveled to Holland, where he devoted ten years to conscientious study. His main interests were in mathematics, optics, mechanics, and acoustics, but he made some contributions to the field of anatomy. He was particularly interested in Harvey's theory on the circulation of the blood which had recently been published.

Descartes's great book, *Discourse on Method* (1637), was only a small book consisting of six essays, but it has had great influence through the ages. A seventh essay was prepared but was not published with the other six because it was too unorthodox to be acceptable at that time.

The first essay recounts the story of Descartes's school days and of his resolution to abandon school and study independently. The second essay describes his experiences during a winter while he served with the army. The rules of his study methods are set forth in this essay: (1) Never accept any information that is not clearly and objectively shown to be true. (2) Divide each of the difficulties under examination into as many parts as possible and attempt to solve one part at a time (i.e., "divide and conquer"). (3) Conduct the thinking processes in such order that the simplest and easiest parts are approached first (i.e., a step-by-step buildup will resolve even the more complex problems). (4) Make enumerations so complete and reviews so general that no significant part of the analysis can be omitted.

The third essay is a code of morals, the fourth an account of the mental steps leading to the concept that is translated as "I think, therefore I am." This topic is expanded in his later work entitled *Meditations*. The fifth essay sketches his theory concerning the essence and mutual relations of mind and body. The sixth essay includes a statement of his reasons for not publishing his work at an earlier date.

The unorthodox seventh essay consists of a thoroughly materialistic interpretation of a variety of philosophical and physiological issues of the mid-seventeenth century period. Descartes used scientific facts as far as they were available to him and then proceeded to speculate about the mechanisms involved in various functions of the body. As he wrote, fact and philosophy became intermixed, and it is difficult to determine which of his comments are based on fact. Scientifically, the essay is significant in being the first frank attempt to explain organic functions on a mechanistic basis. Philosophically, it is significant for having established the famous Cartesian (Cartesius is Latin for Descartes) conflict of matter versus mind. Its chief philosophical value lies in what it did to man's view of man. It dared man to consider himself as something to be understood in terms of natural rather than transcendental philosophy. This essay conflicted with generally accepted religious doctrine, and Descartes considered it judicious to delay its publication. It was published 12 years after his death (1662) in Latin and two years later (1664) in French.

Descartes carried on a correspondence with Princess Elizabeth in England over a period of years and won her admiration for his intellectual accomplishments. Queen Christina, of Sweden, heard of Descartes and his scholarly work and, in 1647, she sent for him to come to her court. He considered this invitation for a long time, hesitating mainly because of a chest condition that he feared would be more serious in the cold climate of Sweden. Finally he was persuaded to go to Sweden by the offer of an appointment to the Swedish court and a title with an estate. He remained in Sweden for three years until his death, which probably was caused by tuberculosis, in 1650.

In retrospect, Descartes's supreme achievement was his perception of mathematics as necessary in science. He believed nature was a machine that operated according to the laws of mathematics. His error was made in attempting to interpret nature according to mental concepts, as a mathematician does his theorems, without due attention to observation. Descartes was one of the inventors of the analytical geometry that led to the calculus of Newton. He was successful in theoretical optics. By modern standards, some of his work in physics is quite sound, while some is absurd.

Among his nonphilosophical endeavors, Descartes was most successful in physiology, but even here he was guilty of many factual errors because he did not make original observations. He developed, for example, a mechanistic explanation for the operation of the nervous system to explain how single sensations influence bilateral organs. A central controlling unit was hypothesized as coordinating the machine. To make his ideal plan, Descartes envisioned the pineal body as an organizing and coordinating center. Nerves were imagined to be hollow tubes through which fluids from the pineal body could flow. Valves in the pineal body were visualized as regulators of the flow. The valves were believed to be controlled directly by stimuli entering from external sources. This was an efficient mechanical device; its defect lay in its not being true. The erroneous Cartesian views of biology were soon replaced with other theories that had more substantial foundations.

Mechanics of Nerve and Muscle

G. A. Borelli (1608-79; Fig. 8.2), a student of Galileo at Pisa, followed the mechanistic approach of Descartes, but applied it primarily to muscular action. He attempted to explain the action of muscles according to mathematical laws. His formulas involving statics and dynamics of muscles were sound in theory, but, like the work of Descartes, they were not based on observation of activities in real living systems. Borelli used weights, fulcrums, and energy sources in developing models of muscular activity. On the basis of mechanical models, he tried to explain the flight of birds, swimming of fish, and walking of land animals. Borelli was far ahead of his time in suggesting that the contraction of muscles would eventually be explained on the basis of chemical and physiological processes.

Albrecht van Haller (1708-77; Fig. 11.1), a Swiss physiologist, described the action of the nervous system in terms of living material being sensitive or irritable. Some parts of the body, he found, were specialized to carry impulses and some were capable of contracting, thus moving arms and legs and other structures. He believed that the properties of irritability and contractability were inherent in all living materials, but some parts had become especially efficient in these functions and were thus enabled to serve the entire organism. Nerve channels, according to van Haller, carried messages from receptors to appropriate centers, which were associated with sensations in the brain. In response to impulses sent out from the brain, muscles were activated. Haller had a modern and rational view of nerve and muscle action, but the scientific background available to him was not sufficient to permit his formulating an explanation of the detailed mechanics involved.

Animal electricity was discovered in 1786 by the Italian investigator,

Figure 16.2
Hermann Helmholtz,
German physiologist.

Luigi Galvani (1737-98), who showed that a nerve could be stimulated by an electric current. When a particular nerve was stimulated in an experimental animal, a particular muscle would contract. Sensitive instruments (galvanometers) built by Galvani eventually were used to show that active nerves always carry a small current of electricity. Emil du Bois-Reymond (1818-96), at the University of Berlin, showed that muscles and nerves produce currents during their active state called action currents. This observation has been incorporated into the modern theories of transmission of the nerve impulse.

Hermann Helmholtz (1821-94; Fig. 16.2), German physiologist, expanded the work of Galvani by investigating further the electric current associated with nervous activity. He found that the rate of conduction in the leg of a frog is only about 30 meters per second, which is much slower than the rate at which electricity travels in a copper wire. Helmholtz demonstrated that chemical as well as electrical activity was involved in the transmission of the nerve impulse. Since 1915, theories about nervous transmission have been based on this discovery of Helmholtz.

Reflex Action

The mechanism(s) responsible for involuntary responses of the body was considered and speculated upon by several early physiologists. The term "reflex action," was first used in 1833 by the English physiologist, Marshall Hall (1790-1857), to designate this type of response. Lloyd Morgan (1852-1936), in attempting to simplify the concepts connected with the nervous response, wrote about the nervous system in terms of its smallest possible functional unit. Such a unit, consisting of as few as three nerve cells, was called a "reflex arc." One neuron was required to carry an impulse into the central nervous system, one to make the transfer in the spinal cord, and one to carry the impulse to a muscle and activate a response.

H. S. Jennings (1868-1949) studied protozoa and lower plants and showed that they follow simple physio-chemical rules in responding to stimuli. The word "tropism," used earlier by T. A. Knight (1759-1838) to identify responses of plants to external stimuli, was applied to the responses of protozoa. Names were chosen to indicate the type of stimulus. Response to chemicals, for example, was called chemotropism, response to light, phototropism. Jacques Loeb (1859-1924) extended the work of Jennings to higher invertebrates and showed that many of their actions are controlled by physical conditions. The considerations that involve sensations, emotions, thought, and motives are now included in the fields of modern physiology and psychology.

Life Processes in Plants

Greek philosopher-scientists considered plants to be the lowest forms of life, capable of arising by spontaneous generation. The view that such organisms could arise spontaneously minimized incentive for critical observations concerning their reproduction and nutrition.

Plant Nutrition

Aristotle considered the soil to be a reservoir of preformed food which could be taken into a plant where a sort of digestive process took place (Table 16.1). Cesalpino (1519-1603) tried to homologize plants and animals with respect to their nutrition; he imagined a circulatory system in plants and included a heart, blood vessels, and other vital organs. Cesalpino did not, however, follow the Aristotelian

Table 16.1   Progress in understanding plant nutrition

Investigator	Date	Contribution
Aristotle	384-322 B.C.	Preformed food in soil.
Cesalpino	1583	Elementary materials taken from soil.
Jung	1638	Openings in roots select materials for food.
van Helmont	1648	Water uptake of a tree measured.
Mariotte	1676	Air and water required for nutrition.
Hales	1727	Something in air (CO_2) required.
Priestley	1772	Plants release oxygen.
Ingenhousz	1779	Carbon cycle, respiration, sunlight (energy).
Dutrochet	1837	Chlorophyll, enzyme for photosynthesis.
Sachs	1862	Sunlight is energy source for enzyme action.

view that preformed food was provided in the soil. Rather, he visualized more elementary materials in the soil from which the plant made its food. In the heart of the plant, presumably the crown or the place where the root system meets the part above the ground, simple materials were supposed to be changed to usable food. Cesalpino also attempted to explain, by physical laws, how water was taken up in a plant. A blotterlike material was postulated in the plant as the means for absorbing the water.

J. Jung (1587-1657) thought that tiny openings must be present in the roots which could select the materials from the soil that were needed in the plant for making food. These openings were presumed to have the ability not only to select needed materials but also to reject the soil constituents that were not useful to the plant. Jung also suggested that waste materials resulting from chemical activities in the plant might be released into the air. The importance of air to the basic nutrition of the plant remained unrecognized.

J. B. van Helmont (1577-1644) introduced the experimental technique into plant physiology. He raised a tree in an earthen vessel and measured the water that was added during the period of growth. The weight of the grown tree was comparable with that of the water added. He concluded, therefore, that water was the only requirement for growth. His measurements were obviously not critical enough to detect some two ounces of material from the soil that must have been involved in nutrition of the plant. Furthermore, the functions of air were still unaccounted for.

Edmé Mariotte (1620-84), a French physicist, performed significant experiments on plant nutrition and in 1676 read a paper entitled, "On the Vegetation of Plants," before the French Academy of Sciences. He had observed the upward movement of sap under pressure in trees. Therefore, he reasoned, there must be something in plants that permits the entrance but prevents the immediate exit of fluids. His experiments were crude, but from the results he postulated a chemical process through which a plant could manufacture its own food from common materials available to all plants, particularly air and water. Mariotte concluded that plants do not take preformed food directly from the soil but take up only elementary materials that they can build into food.

The mechanism through which air could be taken into the plant was suggested by the Italian microscopist, Malpighi, who observed minute openings, called stomata, in the undersurface of plant leaves. Malpighi also identified and traced the plant vessels that carry fluids. In addition, he speculated on the functions of leaves and suggested that the leaf elaborates food.

Stephen Hales (1677-1761; Fig. 16.3), English cleric and plant physiologist, followed the example of his predecessors and utilized experimental procedures. Hales had received his training in mathematics and physical science at Cambridge and was well prepared to conduct objective and critical experiments. He became interested in plants through his review of Ray's work, particularly the descriptions of the plants in the vicinity of Cambridge. As Hales's interest in plants developed, he performed many simple but ingenious experiments, the results of which were published in 1727, under the title *Vegetable Staticks*. In one series of studies, he measured the amount of water taken up by the roots and compared this quantity with the amount given off by the leaves. In another series of experiments, calculations were made of the relation between the moisture in the earth and that taken up by the plant. He also studied the rate at which water rises in the plant. Physical models making use of capillary tubes were constructed to illustrate the absorption of water through fine pores in the plant. Processes in living plants were thus explained on the basis of physical laws.

Among the most interesting experiments of Hales were those that supported Mariotte's idea that the air supplies something material to

Figure 16.3
Stephen Hales, English
plant physiologist.

the substance of plants. In one series of experiments Hales planted a sunflower in a pot which was enclosed and covered with a metal plate. Air and water were admitted by a tube extending through the cover. By adding a fixed amount of water to the previously weighed pot containing soil, plant, and equipment, and by making daily measurements during the period of the experiment and at the conclusion, he discovered that some substance was unaccounted for. This he attributed to the air. It is now known that plants get their essential carbon dioxide from the air, but it is doubtful that Hales's experiments were measuring this substance. Perhaps his discovery was more a coincidence than a forerunner of the identification of carbon dioxide as a factor in plant nutrition. Details of plant metabolism had to await classification of the separate gases in the air.

In the seventeenth century, the constituent that made fire burn was considered to be a substance in the material that was burned. Georg Ernst Stahl (1660-1734), a German chemist, had named this substance "phlogiston," and he theorized that every combustible substance contains phlogiston. When a substance burned, phlogiston was sup-

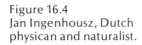

Figure 16.4
Jan Ingenhousz, Dutch
physican and naturalist.

posed to leave and go into the air as smoke while the nonphlogiston constituents remained behind as ash. Coal was considered to be almost pure phlogiston, while sulfur was supposed to contain a high proportion. Not until the eighteenth century was the phlogiston theory subjected to critical review.

Joseph Priestley (1733-1804), English chemist, demonstrated (in 1772) that plants immersed in water give off a gas (oxygen) and that this gas is required by animals. The French chemist, Antoine Lavoisier (1743-94), studied animal respiration and demonstrated the significance of oxygen, carbon dioxide, and water in breathing. The phlogiston theory was then replaced by the demonstration of oxidation. Combustion was defined as the combining of oxygen with the substance being burned.

Jan Ingenhousz (1730-99; Fig. 16.4), Dutch physician and naturalist, discovered the essential aspects of the carbon cycle in nature, a cycle that accounts for the buildup of starches in the bodies of green plants. He developed his "balance of life" concept in a publication with the long name, "Experiments Upon Vegetables, Discovering Their

Figure 16.5
J. E. Purkinje,
Bohemian naturalist and
plant physiologist.

Great Power of Purifying the Common Air in the Sunshine and of Injuring it in the Shade and at Night" (1779). Ingenhousz demonstrated that green plants take in carbon dioxide from the atmosphere in the daylight and give off some carbon dioxide at night. The carbon dioxide taken in during the day is used as the carbon source for nutrition. That given off at night is the product of respiration. Sunlight was established as the energy for photosynthesis.

Animals were not assigned a place in the carbon cycle in nature until later. We now know that they depend on plants for food, and they release carbon dioxide into the air through respiration as well as through the decay of their bodies. The ability to incorporate carbon dioxide into compounds, however, is a fundamental property only of plants. And this particular phenomenon constitutes an important phase of the carbon cycle that is significant in the conservation of nature.

The next major step in the understanding of plant nutrition came with discoveries about the enzyme system that functions in photosynthesis. The green substance in plants had been described and named "chlorophyll" in 1817. Henri J. Dutrochet (1776-1847), a French experimenter, found (in 1837) that only cells carrying the green substance could combine or fix carbon dioxide with other substances to form nutrient material. Chlorophyll was established as the enzyme required in photosynthesis.

By that time, experiments investigating plants had reached the stage of complexity in which special laboratory facilities were required. The first laboratory for the study of plant physiology was established in 1824 by J. E. Purkinje* (1787-1869; Fig. 16.5), in a room of his own home. After studying philosophy and medicine at Prague, he went to Germany and became established at Breslau. When he was appointed Professor of Physiology by King Friedrich III, his broad background and interest in biology helped him make the new area of physiology fascinating to students. His lectures became popular and, as a result, the total number of medical students increased rapidly. In 1844, the Prussian government furnished Purkinje with a physiology institute. Purkinje was a great and dedicated physiologist who had broad interests in the field of animal cytology, animal physiology, and medicine as well as plant physiology.

In his research he was far ahead of his time. According to his biographer, H. J. John (*Jan Evangelista Purkyne, Czech Scientist and Patriot*), he made the following significant contributions in advance of those usually credited with the discovery. (1) In 1835 he observed that the skins of animals were made of cells, which he likened to the "cells" described by Hooke. This observation preceded the announcement of the cell theory by Schleiden and Schwann in 1838 and 1839. (2) Purkinje used the word "protoplasm" in 1839 to describe the living substance. The same term was introduced by von Mohl in 1846. (3) Ciliary movement in vertebrates was described by Purkinje in 1841. Twenty years later (1861) this same observation was made by von Siebold. (4) In 1835 Purkinje showed that the white matter in the nervous system contains only masses of fibers, but the gray matter also includes numerous cell bodies. This fact was formally set forth by Henle in 1841.

Julius Sachs (1832-97), a student of Purkinje, carried on the tradition of experimental plant physiology and became one of the most influential teachers of botany during this period. Sachs performed an experiment (in 1862) by coating parts of live plant leaves with wax and exposing the coated leaves to sunlight. He found that starch was produced only in the uncoated parts. This provided confirmation

* Often shown as Purkyne.

that sunlight is required in the process of photosynthesis. Sunlight was shown to be the energy source for the reactions in which the enzyme chlorophyll is involved.

Sachs and Nathanael Pringsheim (1823-94) continued studies on photosynthesis and described the chloroplastid as the structural carrier of chlorophyll in the cells of green plants (Chapter 14). Chlorophyll was thus shown not to be distributed freely in plant cells and tissues as earlier investigators had supposed, but to be restricted to cytoplasmic structures in cells, the plastids. Indeed, the chlorophyll is synthesized within the chloroplastids.

The major steps in plant nutrition were thus discovered in the last century. More recent developments have been in the direction of working out the chemical steps involved in photosynthesis as well as technical aspects of starch production and storage.

Nitrogen Cycle

Chemists of the nineteenth century, particularly Justus von Liebig (1803-73), Professor of Chemistry at Heidelberg, and Karl Nägeli (1817-91), German chemist and botanist, found that nitrogenous materials were necessary for the continued life and growth of plants. Liebig, in his textbook entitled *Organic Chemistry Applied to Agriculture and Physiology* (1840), showed that ammonia as well as carbon dioxide and water was needed for plant growth. It was further observed that the same three materials required by the growing plant, that is, carbon dioxide, water, and ammonia, are released by decomposition when the plant is dead.

The French chemist, Jean B. Boussingault (1802-87) demonstrated that plants take up nitrogen from the soil in the form of nitrates. This led him to the conclusion that no carbon-containing material is necessary in the soil to promote plant growth if nitrates are present. Thus in his experiments, the carbon was supplied entirely from the carbon dioxide in the air. Further data on nitrogen nutrition came from the work of Mercellin Berthelot (1827-1907). In 1886, he demonstrated that certain bacteria in clay soils "fix" nitrogen into forms (nitrates) which the plant can use. It was soon recognized that nitrogen fixation had great practical significance because of its relation with soil fertility. The combined efforts of many investigators resulted in the understanding of the nitrogen cycle (Fig. 16.6).

Since ancient times, crop rotation has been recognized as a good agricultural practice. A particularly favorable rotation has been that utilizing leguminous crops such as peas, beans, and alfalfa and grain crops such as wheat, barley, and maize. Empirical data showed this rotation to be beneficial but no one knew why. As early as 1686, Malpighi had studied the germination of leguminous plants and found minute nodules on the roots of the seedlings. Several later investigators also observed the nodules but it was not until the latter part of the nineteenth century, while Berthelot's work on nitrogen fixation by soil bacteria was in progress, that other bacteria (Rhizobium) were found in these nodules. Hermann Helriegal and H. Wilforth (in 1888) proved that leguminous plants can absorb atmospheric

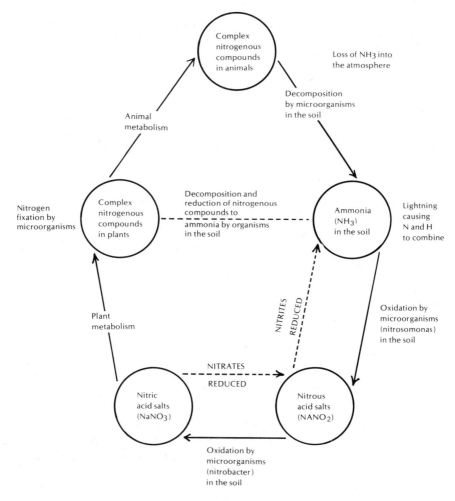

Figure 16.6.   The nitrogen cycle.

nitrogen and that the bacteria in the root nodules are directly involved in this process. It was further demonstrated that leguminous plants can live and grow independently of either ammonia or nitrates in the soil. Bacteria living in the nodules as symbionts, that is, having a mutual living relation with the plants, fix atmospheric nitrogen.

Leading Animal Physiologists of the Nineteenth Century

Müller

Johannes Müller (1801-58; Fig. 16.7) was one of the great German physiologists of the nineteenth century. He studied medicine in his early life, became professor of anatomy and physiology at the University of Bonn, and later was widely known for his *Handbook of Physiology* (1840), which was used extensively as a text and reference. While physiology was emerging as a science, Müller was a leader in his field. He made significant contributions not only through his book, but also through effective teaching and original investigations. In the course of his extensive work, he correlated comparative anatomy, chemistry, and physics, and showed their relation to physiological problems. This was an essential development that had much to do with shaping the immediate future of the science. His main interest concerned the mechanisms of the sense organs and the production of sensations. Color sensations were investigated by appropriate experimental procedures, some of which were original with Müller. The anatomy of the voice organs was studied and relations between structures and various sounds were established. He also gave attention to inner ear structures and explored the mechanism by which impulses are received in the ear and transmitted to the brain.

Müller is widely known for his "doctrine of specific nerve energies." This doctrine states that impulses originate in particular sense organs and are carried over certain tracts to the central nervous system. He showed experimentally that the kind of stimulus is not as important in determining the response as the particular tract over which the stimulus is carried. External events are now known to initiate specific action of certain nerves, which in turn carry the impulses to the brain, where sensations are interpreted.

In his later life, Müller became interested in the broader aspects of morphology and embryology as well as of physiology. Many accomplishments in these fields of biology such as the microscopic anatomy of glandular and cartilaginous tissue are now attributed to the

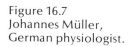

Figure 16.7
Johannes Müller,
German physiologist.

independent work of Müller. Not the least of his accomplishments was the training of several excellent students who carried on in various aspects of anatomy and physiology. Among his more famous students was Theodor Schwann (Fig. 14.1), the zoologist who was responsible for formulating and supporting the cell theory. Another of Müller's prize students was Rudolf Virchow (Fig. 14.3), who carried microscopic work further into the field of pathology and became the father of cellular pathology. Herman Helmholtz (Fig. 16.2), who followed Galvani in the study of animal electricity, was also a student of Müller.

A colleague of Müller and another great teacher of physiology was Karl Ludwig (1816-95). In addition to his skill as a teacher and investigator, he invented several ingenious devices for studying physiology. The mechanically rotating drum or kymograph used to record the heart beat, muscular activity, and movements involved in breathing was worked out by Ludwig. He studied secretions and provided physical and chemical explanations for activities going on within the animal body.

Bernard

Claude Bernard (1813-78; Fig. 16.8) was the leading French physiologist of the last century. He made more basic contributions to the science of physiology than did any other one person. He began his education when he was eight years old, under the supervision of a parish priest. Claude mastered Latin and other classical subjects in his early years. Later, while attending a Jesuit school, he taught language and mathematics in his free time to private students who came to his home.

When he was 18 years of age (in 1832), his family could no longer support him in school, and he went to a neighboring town where he became an apprentice to a druggist, M. Millet. His main responsibilities were the routine and menial tasks necessary in the drugstore. Among other duties he swept the floors, washed bottles, made paper corks, and delivered prescriptions. A veterinary school was located a short distance from the drugstore and Claude frequently carried drugs to the school. He sometimes lingered and watched the operations that were being performed on live animals. Vivisections made a deep impression on him and he frequently discussed his observations with his employer. As Claude gained more experience, he was given the opportunity to make some simple preparations. He took great pride in his first product, shoe polish, and said later, "Now I could make something; I was a man."

Claude was interested in the theater, and on his night off from the drugstore he often attended a comedy or vaudeville. After observing numerous stage performances, he thought he could write as well as some authors whose plays were being presented, and decided to try it. His first attempt was a small play entitled "La Rose du Rhone." It was a farce-comedy based on a popular medicine of that day called "theriac." This was a complex mixture of about 60 different drugs and included opium, squills, spikenard, honey, and wine. In addition to all the regular constituents that were in the recipe, druggists made a practice of adding all the spoiled products and uncalled-for prescriptions that were left on their shelves. No one knew exactly what it contained after all the extras had been added for the formula. Obviously no two batches of theriac were the same, and it was impossible to duplicate a particular batch if it should be found to give unusually good results for some particular illness. Bernard's play was an imaginary story of strange occurrences that followed the use of theriac. The play was sold for a hundred francs and the author was now a playwright. The play had some success at a small theater in Lyons.

The success of this play made Claude overconfident as an author. He next attempted a five-act play. This time he chose a historical theme based on a twelfth century incident concerning King John and his nephew who was heir to the throne. It was a greater undertaking than his time and ability would permit and constituted a tragedy in more ways than one. So much time was devoted to the play that Claude was unable to give proper attention to the responsibilities of the drugstore. M. Millet wrote to his family explaining that after a year and a half as an apprentice, Claude evidently had lost interest and could no longer remain at the drugstore. Claude left the drugstore and went to Paris, carrying with him his unfinished play. He was discouraged considerably by a critic, Saint-Marc Girardin at the Sorbonne, who advised him to learn a profession and write plays only in his spare time. Claude followed the good advice and in 1832 entered medical school at the College of France.

The results of his examination in 1839 for internship were not impressive. Claude ranked 26 in a total of 29 students taking the examinations. This may have been partly due to his indifference concerning his work. But despite Claude's low examination scores and an apparent lack of enthusiasm, Magendie, the French physiologist, took him into his laboratory as an assistant. Magendie, one of the great physiologists of the time, saw that Claude had considerable skill with his hands. Magendie was well recognized in his field, but he was skeptical, intolerant, and sarcastic, with little patience for students and assistants. He was, however, interested in original research, and he made it a practice to teach by making actual dissections along with his students. Claude Bernard thrived in this environment and stayed with Magendie as an assistant for many years. At 35 he was still an assistant. He had tried several times to obtain a university appointment but had not been successful. Although he was a good thinker and an excellent technician, he was characterized as a poor speaker. Finally in 1854, he became a professor of general physiology at the Sorbonne, and in 1858, he was appointed professor of medicine at the College of France, where he replaced Magendie. Only two years after the latter appointment, when Bernard was 47 years of age, he became ill with tuberculosis and retired to his birthplace in Southern France to regain his health. He never practiced medicine but continued his research and writing when his health would permit.

Through his persistence in research, Bernard discovered many facts in the field of physiology. His work provided an impressive example of the effective use of the experimental method in physiology. Bernard's approach to research reflected his great respect for natural law as it operates in the organic as well as in the inorganic world. Not the least of his accomplishments came from his influence on other physiologists of his own time and on the contemporary teaching of medicine and physiology. He wrote one important book while he was ill, *Introduction to Experimental Medicine* (1865). This book was an excellent first attempt to formulate basic laws of experimental physiology and medicine. The thesis is summarized in the following quotation taken from the end of the book.

I believe, in a word, that the true scientific method confines the mind without suffocating it, leaves it as far as possible face to face with itself, and guides it, while respecting the creative originality and spontaneity which are its most precious qualities. Science goes forward only through new ideas and through creative or original power of thought. In education, we must, therefore, take care that knowledge which should arm the mind does not overwhelm it by its weight, and that rules, intended to support weak parts of the mind, do not atrophy the strong and fertile parts. I need not enter into further explanations here; I have had to limit myself by forewarning biological science and experimental medicine against exaggerating the importance of erudition and against invasion and domination by systems; because sciences submitting to these would lose their fertility and would abandon the independence and freedom of mind essential to the progress of humanity.

In his personal research he made significant discoveries. His first publication reported a study of the chorda tympani nerve that supplies the taste buds of the anterior two-thirds of the tongue and the submaxillary gland. This nerve was traced to the seventh cranial nerve and found to be a branch of that nerve. Bernard demonstrated in this early study remarkable facility for dissection as well as the patience and skill necessary in tracing a very tiny nerve over a long distance. He followed this auspicious beginning with many other research papers. He wrote a thesis (in 1843) on gastric juice and its role in digestion. In this study, he traced the digestion of different kinds of foods as it occurs in the various parts of the alimentary canal. His next paper described the spinal accessory nerve and demonstrated its relation to the vagus nerve. Later papers concerned the pancreas and the sympathetic nervous system. His most significant discoveries may be classified under the following subjects:

1. *Pancreatic digestion.* This investigation was initiated when he noticed that the urine of rabbits was clear when the animals were fed meat, but cloudy when they were fed herbs. Therefore, he reasoned, different chemical processes must be involved in the digestion of animal products, compared with that of plant products. He followed the course of digestion and discovered the enzyme steapsin. Furthermore, he found that pancreatic juice contained more than a single enzyme and identified those that act on proteins, carbohydrates, and fats.

2. *Glycogenic function of the liver.* Bernard performed careful experiments on animal starch formation over a period of 12 years and showed that carbohydrates are stored as glycogen in the liver. This work has become important in understanding and treating diabetes. The investigations were more significant to basic physiology, however, in demonstrating that the animal body can build up as well as break down complex chemical materials; that is, it synthesizes as well as analyzes. Bernard showed by chemical methods that the liver could release sugar directly into the blood and he called this process "internal secretion."

3. *The vasomotor nerves.* Bernard observed vasoconstrictors and vasodilators that control the size of the small blood vessels in the ear of a rabbit and found them to be innervated by the sympathetic system. Various chemicals in the blood were identified as

activating the vasomotor nerves and thus influencing blood pressure.

4. *The action of drugs or poisons on the body.* Curare, used by Indians in America to poison arrows many years before its physiological properties were known, was one of the drugs studied by Bernard. He carefully investigated the effects of curare and found that it worked at the point where the nerve entered the muscle. Now it is widely used in physiology laboratories to block nerves of experimental animals while vivisections are being performed. Bernard also studied carbon monoxide (CO) poisoning and found that CO displaces oxygen in the blood by forming a more stable combination than oxyhemoglobin. The oxygen-carrying capacity of the blood thus is destroyed in the presence of CO. Carbon monoxide poisoning is due, therefore, to cell suffocation for want of oxygen.

Endocrinology

Endocrinology, as a distinct area of biology, began with the experiments of W. M. Bayliss and E. H. Starling (1902-05) that demonstrated the existence and manner of action of a chemical substance that came to be known as the hormone, secretin. Starling first used the word "hormone" (Greek, to excite) in 1905, with reference to secretin. It now designates organic secretions of endocrine glands which are released into the blood stream and exercise a wide variety of regulatory roles over physiological activities. The hormone secretin is secreted by the mucosa of the upper part of the small intestine when the acid chyme enters the intestine. Secretin is carried by the blood and stimulates the secretion of pancreatic juice and bile. Bayliss and Starling were the first to prove unequivocally that chemical coordination of bodily functions can occur without assistance from the nervous system. Eventually it was shown that the endocrine glands elaborate hormones and release them into the blood. Hormones thus exert regulatory effects (that is, inhibit as well as excite) upon distant target organs and tissues.

With continually improving techniques, it was possible to isolate and chemically characterize many hormones. First to be crystallized was epinephrine (adrenalin) from the adrenal glands, an accomplishment of J. Takamine and T. B. Aldrich in 1901. E. C. Kendall obtained pure thyroxine from the thyroid glands of swine in 1919; but not until 1926 did C. R. Harington show that thyroxine is related chemically to the amino acid tryosine and thus to be a proteinaceous substance. Insulin was difficult to isolate from the pancreas (islet tissue) because proteolytic enzymes from the acinar portion of the pancreas destroyed the hormone's activity during the extraction procedures. F. G. Banting and C. H. Best (1920) finally succeeded in isolating insulin by using pancreatic tissue from dogs whose acinar tissue had been caused to degenerate by ligation of the pancreatic duct. Finally in 1926, J. J. Abel prepared insulin in crystalline form and demonstrated that this hormone, too, was a protein.

S. Ascheim and B. Zondek (1927) discovered that two previously unidentified hormones, estrone and chorionic gonadotrophin, occurred in the urine of many pregnant animals. Estrone was estrus-producing in action, whereas chorionic gonadotrophin could produce marked development of ovarian follicles. Various estrus-producing hormones, identified collectively by the generic term "estrogens," were isolated directly from ovarian tissue by E. A. Doisy in 1935. K. David and his associates, also in 1935, isolated a pure crystalline hormone from testicular material and named it testosterone. It was then assumed (and later proved) that androsterone, which had previously been found in urine, was a degradation product of testosterone.

Investigations in the United States and Europe between 1936 and 1942 resulted in the isolation of approximately 30 different steroids, saturated hydrocarbons which include hormones of the gonads and adrenal cortex, from the adrenal cortices of slaughterhouse animals. Six of these compounds were biologically potent. Of all the endocrine glands, the pituitary (hypophysis) was the most intriguing and simultaneously the most difficult to study. It was suspicioned that the hypophysis was the "master" gland, but the investigators could do little but hypothesize for many years because the gland was so difficult to remove without injuring the brain. Techniques for performing hypophysectomies were finally developed by various investigators, however, and the operation is now standardized. Pituitary hormones are all proteinaceous, and their chemical elucidation has therefore been understandably slow. Seven hormones have been obtained from the adenohypophysis (anterior lobe): somatotrophin (STH), corticotrophin (ACTH), thyrotrophin (TSH), prolactin (lactogenic or luteotrophin), follicle stimulating hormone (MSH). In contrast, only two hormones are associated with the neurohypophysis (posterior lobe), oxytocin and vasopressin. The antidiuretic effect attributed to neurohypophyseal extracts is due to vasopressin.

The 1940s ushered in a period of unprecedented biochemical expansion and inquiry in the field of endocrinology. Clinical interests were the driving force behind much of the work, but substantial progress depended materially upon the introduction of new tools and methods. Such developments as phase and electron microscopes, improved tissue culture techniques, radioactive isotopes (as tracers), microanalytic methods, and chromatography, to name but a few, have been vital in widening our understanding of biochemical processes. The use of radioactive isotopes has been particularly instrumental in aiding endocrinologists. Perfusing C_{14}-labeled cholesterol or radioacetate has allowed scientists to elucidate the synthesis of many of the steroid hormones. Such use of radioactive iodine is directly responsible for the great progress made in our understanding of thyroid metabolism.

Another technique that provided valuable information is the tagging of "end-groups" of hormones with such compounds as dinitrofluorobenzene. Tagging, in essence, binds the terminal amino group on a protein chain. Using this technique, and subsequent hydrolysis and partition chromatography, F. Sanger (1954) defined the full chemical structure of insulin. In a similar way, V. du Vigneuad and his associates had already (in 1953) determined the structure of oxytocin and vasopressin. In 1957, W. W. Bromer determined the amino acid sequence of glucagon, the hyperglycemic factor that is elaborated by the alpha cells of the islet tissue of the pancreas.

Enzymes

Enzymes are proteins that control chemical reactions in living organisms without being used up in the reaction. Results of enzyme action were observed long before their nature or their mechanisms were known. In 1765, Spallanzani (Chapter 15) observed that pieces of meat were dissolved by gastric juice. William Irvine (in 1785) found that extracts from sprouted barley would cause starch to liquify. A. Payen and J. F. Persoz studied the process of starch solubilization and called the responsible agent diastase. They found that diastase was inactivated by boiling. Without undergoing permanent change itself, a small amount of diastase could cause the conversion of a large amount of starch to sugar. Furthermore, diastase could be concentrated and purified by precipitation with alcohol.

"Ferments" were considered by many investigators of the eighteenth and nineteenth centuries to cause changes in organic materials. When "fermentation" was found to result from the action of living organisms, interest in ferments (or enzymes) lagged. The word enzyme (meaning "in yeast") was introduced by the physiologist W. Kuhne (1878). Studies on fermentation led to the discovery that agents extracted from living organisms could also control reactions. Intra-cellular enzymes could then be studied experimentally and were thus associated with vital processes. When yeast cells cause sugar to ferment, that is, convert it to alcohol, the yeast cells absorb sugar from the liquid medium and return the products, alcohol and carbon dioxide, back into the medium. It was shown by Edward Büchner (1887), German chemist who won the 1907 Nobel Prize in Chemistry, that fermentation can occur in the presence of yeast cell extracts as well as in the presence of living yeast cells. With the tools of cell extracts available, many enzymes were found and studied. Alcohol and lactic acid fermentation were shown to be due to the action of several different enzymes, each acting in sequence on the succes-sive products of the reaction.

J. B. Sumner (1926) purified and crystallized urease, an enzyme that causes a hydrolysis of urea to ammonia and carbon dioxide. The enzyme was shown to be a protein. John H. Northrop and M. Kunitz (1930) prepared pure crystalline pepsin and trypsin and showed that these enzymes were also proteins. In 1946 Sumner and Northrop received the Nobel Prize in Chemistry for their work on enzymes. By this time many enzymes had been studied and it was generally assumed that all enzymes were proteins.

A great many different kinds of enzymes have subsequently been identified. They are elaborated only by living cells and in minute quantities. Each activates only a single chemical reaction or a very limited group of reactions. Enzymes are typically composed of a protein part (apoenzyme) which confers specificity and a nonprotein part (coenzyme) necessary for activity. An enzyme produces its effect on its substrate by temporarily combining with the substrate and activating it, so that the substrate undergoes a chemical change, at which time it loses its combination with the enzyme. Most enzymes exert their effect within cells, but a few, such as the digestive enzymes, are normally secreted to the outside of the cell.

Cellular respiration, that is, the oxidation of molecules such as glucose within the cell, was found to involve a sequence of enzyme reactions. The oxidoreductases necessary for these processes were characterized chemically by the German physiologist, Otto War-burg, in the 1930s. A series of enzymes called dehydrogenases have since been shown to catalyze oxidation reactions by the removal of hydrogen from the substrate.

In 1967, David Harker, United States biochemist, and his associates determined the structural arrangement of one enzyme. These investigators discovered the sequence of the 124 amino acid units in the ribonuclease molecule. B. Gutte and R. B. Merrifield (1969) demonstrated complete chemical synthesis of a ribonuclease with functional and physical properties identical to a naturally occuring ribonuclease. This is the enzyme that breaks down ribonucleic acid (RNA) in the cell. Enzymes are also associated with the synthesis of both RNA and DNA (deoxyribonucleic acid). Thousands of enzymes operate within a single living cell.

Chronology of Events

384-322 B.C.	Aristotle, plant food preformed in soil.
A.D.	
1583	A. Cesalpino (Caesalpinus), theory of circulation; plant food made from elementary materials.
1637	R. Descartes, *Discourse on Method*.
1638	J. Jung, plant roots select and reject soil constituents.
1648	J. B. van Helmont, measured water taken up by a tree.
1676	E. Mariotte, plant makes food from air and water.
1680	G. A. Borelli, mechanistic studies of nerve and muscle.
1707	G. E. Stahl, phlogiston theory related to plant nutrition.
1727	S. Hales measured water taken up by plant roots and released by leaves; something in air (CO_2) required for food.
1759-66	A. van Haller, irritability and contractability in living systems.
1772	J. Priestley, plants give off oxygen.
1779	J. Ingenhousz, sunlight is energy source for photosynthesis; carbon dioxide taken in by plants in daytime and given off at night.
1786	L. Galvani, animal electricity.
1789	A. Lavoisier, oxidation in respiration.
1799	T. A. Knight, trophisms in protozoa.
1833	M. Hall, reflex action.
1837	H. J. Dutrochet, chlorophyll essential in photosynthesis.
1840	J. Müller, *Handbook of Physiology*.
	J. von Liebig, ammonia (nitrogen) needed for plant growth.
1841-44	J. E. Purkinje, ciliary movement and other aspects of physiology.
1844	K. Nägeli, nitrogen needed in plant nutrition.
1850	H. Helmholtz, chemical and electrical activity in nerve impulse.
1852-56	K. Ludwig studied heartbeat and muscular activity with kymograph.
1860	J. B. Boussingault, plants absorb nitrates from soil.
1862-65	J. Sachs, sunlight necessary for photosynthesis; chloroplastids.
1865	C. Bernard, *Introduction to Experimental Medicine*.
1886	M. Berthelot, nitrogen fixation by bacteria.
1887	E. Buchner, fermentation can occur in presence of yeast extracts.
1888	H. Helriegal and H. Wilforth, nitrogen fixation in leguminous plants.

1893	L. Morgan, reflex arc; action controlled by physical conditions.
1901	T. Takamine and T. B. Aldrich, crystallized adrenalin.
1902-05	W. M. Bayliss and E. H. Starling, hormone secretion.
1906	H. S. Jennings, effects of physical and chemical stimuli on protozoa.
1912	J. Loeb, actions in animals controlled by physical conditions.
1919	E. C. Kendall, pure thyroxine.
1920	F. G. Banting and C. H. Best, isolated insulin.
1926	J. B. Sumner, purified urease.
1927	S. Ascheim and B. Zondek, estrone and chorionic gonadotrophin.
1930	J. H. Northrop and M. Kunitz, purified pepsin and trypsin.
1935	E. A. Doisy, estrogens.
1935	K. David, testosterone.
1954	F. Sanger, determined the sequence of amino acids in insulin.
1957	W. W. Bromer determined the amino acid sequence of glucagon.
1967	D. Harker, discovered the amino acid sequence of ribonuclease.
1969	B. Gutte and R. B. Merrifield, complete chemical synthesis of ribonuclease.

References and Readings

Bernard, C. 1949. *An Introduction to the Study of Experimental Medicine.* New York: Henry Schuman.

Clark-Kennedy, A. E. 1929. *Stephen Hales.* Cambridge: At the University Press.

Crombie, A. C. 1959. "Descartes." *Sci. Amer.* 201: (4) 160-173.

Foster, M. 1901. *Lectures on the History of Physiology.* Cambridge: At the University Press.

Green, J. R. 1909. *A History of Botany.* Oxford: Clarendon Press.

Gruber, H. and V. Gruber. 1956. "Hermann von Helmholtz; nineteenth-century polymorph." *Sci. Monthly* 83:92-99.

Hawkes, E. 1928. *The Pioneers of Plant Science.* London: Sheldon Press.

John, H. J. 1959. *Jan Evangelista Purkyně, Czech Scientist and Patriot.* Philadelphia: American Philosophical Society.

Lowndes, R. 1878. *René Descartes; His Life and Meditations.* London: F. Norgate.

Mahaffy, J. P. 1881. *Descartes.* Philadelphia: J. B. Lippincott and Co.

Olmsted, J. M. D. 1938. *Claude Bernard, Physiologist.* New York: Harper and Bros.

_____. 1944. *Francois Magendie, Pioneer in Experimental Physiology.* New York: Henry Schuman.

Olmsted, J. M. D. and E. Harris. 1952. *Claude Bernard and the Experimental Method of Medicine.* New York: Henry Schuman.

Roth, L. 1937. *Descartes' Discourse on Method.* Oxford: Clarendon Press.

Sachs, J. von. 1909. *A History of Botany.* English trans., Oxford: Clarendon Press.

Scott, J. F. 1952. *The Scientific Work of René Descartes.* London: Taylor and Francis.

Stevenson, L. G. 1947. *Sir Frederick Banting.* Toronto: The Ryerson Press.
Suner, A. P. 1955. *Classics of Biology.* New York: Philosophical Library.
Valery, P. 1947. *The Living Thoughts of Descartes.* Philadelphia: David McKay Co.

CHAPTER 17

GENETICS, A UNIFYING SCIENCE

GENETICS developed as a science during the twentieth century, after Mendel's work was discovered in 1900. The roots of present-day genetics, however, can be traced into antiquity. Selection and hybridization were undoubtedly employed by prehistoric plant growers and stock raisers, even though they were not aware of the genetic principles involved. Tablets of stone prepared by the Babylonians some 6,000 years ago have been interpreted as showing pedigrees of several successive generations of horses, thus suggesting a conscious effort toward improvement. Other stone carvings of the same period illustrate artificial cross-pollination of the date palm as practiced by the early Babylonians. The early Chinese, many years before the Christian era, improved varieties of rice. Maize was cultivated and improved in the western hemisphere by the American Indians, beginning at an early period in their history.

Figure 17.1
Gregor Mendel, Austrian
investigator who laid the
foundation for the science
of genetics. (From Gardner,
Principles of Genetics.
Copyright ©, 1960,
1964, 1968 by John Wiley
& Sons, Inc.)

In another era Hippocrates, Aristotle, and other Greek philosophers made observations and speculations suggesting genetic principles. Elements of truth, however, were vaguely stated and mixed
with error. Stories of unusual hybrids were invented by the Greeks
and repeated with additional imaginative flourishes by Pliny, Gesner, and other popular writers of their times. The giraffe was supposed
to be a hybrid between the camel and the leopard. The two-humped
camel was thought to have resulted from a cross between a camel
and a boar. An ostrich was imagined to appear when the camel mated
with the sparrow. Plants were also considered capable of remarkable hybridizations. The acacia tree, for example, crossed with the
palm was said to produce the banana tree. Fantastic explanations of
the mechanism of reproduction and sex determination were associated with these stories. Although many such tales have persisted,
little information that materially contributed to the science of genetics was available before the seventeenth century.

During the seventeenth, eighteenth, and nineteenth centuries sporadic observations were made that have since been associated with
genetic mechanisms. In the twentieth century basic principles of
inheritance and variation were discovered and found to apply to all
living things. Genetic material was defined chemically and a universal
code was discovered. Genetics has thus become a unifying area of
biology.

Mendel and His Experiments

Gregor Mendel (1822-84; Fig. 17.1) is appropriately called the "Father of Genetics." He made revolutionary contributions with his experiments on garden peas *(Pisum sativum)* conducted in a monastery garden (Fig. 17.2) and thus laid the foundation for the science of genetics.

In one experiment, Mendel raised two varieties of garden peas, tall and dwarf. When the flowers of the tall strain were allowed to be fertilized with their own pollen, the offspring were all tall. The other variety produced only dwarfs. Mendel observed that weather, soil, and moisture conditions had an effect on the relative growth of the peas, but the size of these two varieties was controlled largely by inheritance. At least heredity was the main limiting factor under the conditions of his experiments. Tall plants were 6 to 7 feet high whereas dwarfs raised in the same environment measured from 9 to 18 inches. Under his observations, no dwarfs ever turned into tall plants and no tall plants, even under unfavorable environmental conditions, turned into dwarfs. The vines were smaller when the plants from a tall variety were raised under less favorable conditions, but they were long and stringy and easily classified as hereditarily tall.

Mendel proceeded with the hybridization by crossing tall with dwarf plants. From this cross all of the offspring in the first genera-

Figure 17.2
Monastery garden where Mendel conducted his hybridization experiments in Brno, Czechoslovakia, as it appears today. (Courtesy Dr. Ing. Jaroslav Krizenecky.)

tion (F_1) were tall. The dwarf character had disappeared from the F_1 progeny. When the tall hybrid plants were fertilized by their own pollen (selfed) and the progeny (second generation of F_2) were classified, the missing trait turned up again. Some progeny were tall and some were dwarfs. Careful classification of the plants showed that on the average about three tall plants were produced to one dwarf.

In other crosses, different pairs of contrasting traits were studied. Mendel recognized from his studies that all seven pairs evidenced distinct inherited patterns. He observed, for example, only 3:1 ratios in the second generation (F_2) from all monohybrid crosses. One member of each pair seemed to dominate the other, and this one was identified as dominant in contrast to the other which was recessive. Mendel's conclusions from his crosses were based on unit traits, dominance of one member of each pair over the other, and segregation of genetic elements associated with pairs of contrasting traits. Mendel interpreted what he had observed by assuming that living organisms transmit heredity traits through the reproductive mechanism by means of some kind of physical particles which operate as independent units. He visualized with remarkable clearness physical elements which could be involved in inheritance and used a German word, *Anlage* (pl. *Anlagen*), to identify these units. Mendel reasoned that since traits expressed by one parent could be hidden in the first generation and yet appear in the second generation, the anlagen, later called genes, must be paired. He believed that the contrasting expressions in seeds or plants were produced by different arrangements of the two members of a pair.

The most important conclusion that Mendel drew from his first experiments involved the principle of segregation, which describes the separation of pairs of genes. When crosses were made between parents that differed in two gene pairs, all F_1 progeny expressed the traits determined by the dominant genes. In the F_2 a multiple of the 3:1 $[(3 + 1)^2 = 9 + 3 + 3 + 1]$ or 9:3:3:1 was obtained from which Mendel determined that the two pairs of genes were independent with respect to each other. This led to his second principle, that of independent combinations or independent assortment.

In 1900, Mendel's paper was discovered by three men simultaneously: Hugo de Vries, a Dutch botanist known for his mutation theory (see below) and studies on the evening primrose and maize; Carl Correns, a German botanist who studied maize, peas, and beans; and Erich von Tschermak-Seysenegg, an Austrian botanist who worked

with several plants including garden peas. All three of these investigators approached Mendel's principles independently from their own
studies, recognized the significance of this type of investigation, and
found Mendel's work and cited it in their own publications. Correns
and Tschermak-Seysenegg were mainly interested in applications,
particularly of hybrid vigor (heterosis), and they did not promote
Mendel's work. Instead, de Vries, who was studying mutations and
other biological problems related to basic genetics, made the real
theoretical discovery of Mendel's work.

De Vries and the Mutation Theory

The Swiss anatomist and physiologist, Albrecht Kölliker (1817-1905)
had criticized Darwin's theory of evolution by natural selection and
had presented an alternative hypothesis that called for evolution by
sudden change. This idea appealed to the Dutch botanist, Hugo de
Vries (1848-1935; Fig. 17.3). De Vries received his university training at
Leyden, Heidelberg, and finally at Würzburg, where he studied under
Julius Sachs. After holding various academic posts in Germany, he
became (in 1871) associated with the University of Amsterdam, first
as a lecturer and later as professor of botany and curator of the botanical gardens. He was particularly interested in variations among living
things and in evolution.

The mutation theory provided the much-needed explanation for
the variation observed in nature. Mutations are abrupt changes that
originate without any visible preparation and without any observable series of transitional forms. The theory was set forth in modern
form by de Vries in his book entitled *The Mutation Theory* (1901).

De Vries was searching for evidence to support the mutation
theory, particularly in the evening primrose, *Oenothera lamarckiana,*
with which he was familiar. This plant, originally introduced into
Europe from America, had spread widely over Europe and in de
Vries's time it grew in masses in the meadows near Amsterdam. Seed
from the recognized species usually perpetuated the species characteristics, but occasionally a new type appeared that had distinctly
new traits. One of these was a giant, another was a dwarf, and a
latifoliate form was also observed. De Vries determined that these
variants bred true, and he at first considered them to be new species.
It should be noted that this observational experimental work, introduced by de Vries as a method of studying evolution, was in definite

Figure 17.3
Hugo de Vries who developed the mutation theory. (From Gardner, *Principles of Genetics.* Copyright©, 1960, 1964, 1968 by John Wiley & Sons, Inc.)

contrast to the inference method practiced by some of de Vries's contemporaries.

De Vries's mutation theory was subjected to much criticism when it first appeared in 1901, particularly from Darwinists. Critics pointed out that the experimental lines he used might not have been pure and that the new traits could have resulted from recombinations. Some biologists considered de Vries's theory too different from traditional ideas to be acceptable. The theory of mutations was not really in conflict with Darwin's natural selection, however, and it later was shown to strengthen Darwin's case by providing for a source of variation different from that of the Lamarckian view, which Darwin had reluctantly accepted. Darwin had no place in his theory for profound or sudden modifications. The mutation theory became more acceptable when it was shown that small as well as large variations could occur abruptly and be permanent.

The original mutation theory was gradually refined and now is widely accepted as an explanation for the origin of variation. Now that the theory of the gene has been established, a mutation is defined as a change in a gene. The definition is sometimes extended to include chromosome aberrations as well as gene changes.

The Birth of a Science

The science of genetics developed rapidly after 1900 and soon took its proper place among the biological sciences. Once started, an explosive growth of knowledge occurred, to which many investigators contributed. Five will be cited as especially conspicuous during this period: an Englishman, William Bateson; a Frenchman, Lucien Cuénot; an American, W. E. Castle; a Dane, W. L. Johannsen; and a German, Carl Correns.

Bateson

William Bateson (1861-1926; Fig. 17.4), an experimental biologist at Cambridge University, became interested in Mendel's work immediately following its discovery in 1900. According to R. C. Punnett, Bateson's student and close associate, Bateson, on May 8, 1900, was first to announce Mendel's work in England. He was on a train en route to London to deliver a lecture entitled "The Problems of Heredity as a Subject for Horticultural Investigation" before the Royal Horticultural Society, when he read de Vries's account of Mendel's work. He immediately revised his speech to include Mendel's discoveries and dramatically gave England its first news of Mendel. A full English translation of Mendel's paper was published at the beginning of the 1901 *Proceedings of the Royal Horticultural Society*.

At the time Mendel's work was discovered, Bateson was engaged in experimental breeding, and he had already made significant contributions. In 1894, for example, he had published a book entitled *Materials for the Study of Variation* which suggested many problems similar to those with which Mendel was concerned. Bateson then designed and conducted experiments to answer the unsolved questions. These experiments were in progress when Mendel's work was discovered. Bateson subsequently designed new experiments to determine whether Mendel's laws held for other plants, in particular whether they applied to animals as well. Bateson had been breeding poultry, and he, along with C. C. Hurst, immediately organized experiments to see if the Mendelian principles applied in poultry. These animals breed rapidly, and large numbers of progeny can be obtained in a short time. It was soon obvious that Mendel's principles did apply to poultry and rabbits as well as to sweet peas. Bateson and his associates, including Mrs. Bateson and R. C. Punnett, then set out in earnest to repeat and supplement Mendel's experi-

Figure 17.4
William Bateson, English
experimental biologist
who confirmed Mendel's
results. (From Gardner,
Principles of Genetics.
Copyright© 1960,
1964, 1968 by John Wiley
& Sons, Inc.)

ments. Their work demonstrated that nine plant genera and four animal genera followed the Mendelian pattern.

In historical perspective, it seems possible that Bateson would have discovered Mendel's laws himself if Mendel's work had not come to light in 1900. In 1899, before the International Congress of Hybridization, Bateson had described his objectives and experiments that paralleled closely those of Mendel. Through Bateson's publications, *Mendel's Principles of Heredity—A Defense* (1902) and *Mendel's Principles of Heredity* (1909), the basic principles of genetics were clarified and confirmed. In 1906, at a hybridization conference sponsored by the Royal Horticultural Society, Bateson presented the word *Genetics* that he had coined as the name for the new science. This name was used in the 1909 edition of Bateson's text. He also coined the terms *homozygous* and *heterozygous,* and described the two kinds of pairing of genes. The word *allelomorph* from which the abbreviation "allele" has been taken, was also introduced by Bateson. A professorship of genetics was established at Cambridge in 1909, with Bateson as the first professor. The John Innes Horticultural Institute was established a year later, and Bateson relinquished his professorship to Punnett and became the first director of the new institute.

Cuénot

Lucien Cuénot (1866-1951), who had distinguished himself as an invertebrate zoologist, began genetic studies with mice immediately after the discovery of Mendel's work. In 1902 he announced a case of simple Mendelian inheritance in the mouse, and in 1903 he published a more detailed analysis of coat patterns and colors in mice. He not only discovered Mendelian ratios but followed the genetic investigations with chemical analyses designed to trace the steps between genes and traits. Basic and substantial contributions in rodent genetics were published by Cuénot in a series of papers completed in 1911. He then turned to studies of the genetics of cancer in mice and later published extensively in the broader aspects of biology.

Castle

W. E. Castle (1867-1962), in the United States, immediately applied the newly recognized Mendelian principles to mammals. In 1903, he published five papers dealing with such subjects as the heredity of the angora cat, albinism in man, the heredity of sex, and Mendel's law of heredity. This early work was followed by extensive investigations on coat characteristics in guinea pigs, rabbits, mice, rats, and other mammals. Castle later delved into other aspects of mammalian genetics including body size and form and became a leading authority in this field. In his later years his interests led to studies on domesticated animals, particularly horses.

Johannsen

Wilhelm L. Johannsen (1857-1927; Fig. 17.5), a Danish geneticist and plant physiologist, is closely identified with the early development of genetics as a science. His first genetic paper *"On Heredity and Variation"* appeared in 1896, and in 1898, he began investigations on barley and beans that have become classics. From this time on he devoted himself to the field of genetics. In 1905, he became director of the Institute of Plant Physiology at the University of Copenhagen.

In the same year (1900) that Mendel's paper was discoverd, Johannsen published a paper on "pure lines" that greatly stimulated interest in genetics. Through a series of ingenious investigations he showed a difference in the effects of selection when applied to populations of ordinary cross-fertilizing organisms as compared with self-fertilizing plants. Self-fertilization was found to be effective in altering the proportion of different types. When plants were self-fertilized over long periods, selection was no longer effective. The plants had become completely or nearly homozygous, and no genetic variation was left for selection to act on. All variation in a pure genetic line was thus attributed to the environment.

Johannsen later devoted his attention to the formulation of the principles of genetics and wrote one of the first textbooks in the field, *Elements of Genetics,* published in 1905. This book represented a landmark in the development of the new science. The word *gene,* which had been coined in 1903 by Johannsen, was used in this text. Johannsen also coined the terms *genotype* and *phenotype* and stressed the importance of making a clear distinction between genes and traits.

Correns

Carl Correns (1864-1933), one of the three men who discovered Mendel's work in 1900, experimented extensively with hybridization in maize, stocks, beans, peas, and lilies at the University of Tübingen during the 1890s. He had accumulated data from four generations in a garden pea experiment and had arrived at conclusions similar to those of Mendel. Searching the literature for work on the subject, he found Mendel's paper, which he cited in his publication (1900). In 1901, he published extensive data on maize hybrids and, in later years, he investigated many aspects of plant genetics including sex, the mechanics of reproduction, self-sterility, and leaf variegation dependent on cytoplasmic inheritance. This work (1909) involved the four o'clock plant of the genus Mirabilis and has now become classic in genetics. In 1902, he moved to the University of Leipzig and in 1913 he became a Director of the Kaiser Wilhelm Institute for Biology at Berlin-Dahlem where he continued his interest in cytoplasmic inheritance.

Chromosome Theory of Inheritance

Mendel formulated a precise mathematical pattern for the transmission of genes, but he had no concept of the biological mechanism involved. The location of the genes and their relation to biological structures and functions were to be determined by others. Transmission of genes from parents to offspring was obviously associated with the reproductive mechanism, but curious biologists wanted to define more specific mechanisms of inheritance. Which parts of the

Figure 17.5
Wilhelm L. Johannsen,
Danish botanist who
contributed to the science
of genetics during the
early part of the present
century. (From Gardner,
Principles of Genetics.
Copyright©, 1960,
1964, 1968 by John Wiley
& Sons, Inc.)

sex cells were involved, and how were the dynamic living processes associated with inheritance? While Mendel's experiments were in progress, the general aspects of reproduction in animals and plants were known, but only crude notions of cells and their behavior in reproduction had been formulated. Fortunately, the basic principles of cytology were established between 1865, when Mendel's work was completed, and 1900, when it was discovered.

The Mendelian pattern of inheritance requires that the genes be transmitted from cell to cell during division. But by what mechanism does each daughter cell receive all that is in the parent cell and become a complete cell rather than half a cell or some fraction of the parent cell? Wilhelm Roux (1850-1924; Fig. 11.5) speculated on this question and made models to see how such a division might be accomplished. The only procedure he could devise that would provide the required results necessitated lining up objects in a row and duplicating them exactly. He therefore suggested that nuclei must have strings of beadlike structures that line up and duplicate themselves during mitosis. If nuclei really have such structures, he reasoned, it might be possible to explain the mechanics of gene transmission from cell to cell. The constituents of the nucleus that seemed best designed to fill these requirements were the chromosomes. Boveri followed this lead and was largely responsible for developing the theory that genes are in chromosomes.

A young American, W. S. Sutton (1876-1916), also recognized the parallelism between the behavior of chromosomes and the Mendelian segregation of genes. As judged by the end products, genes behaved during reproduction and cell division as they would be expected to behave if they were located in chromosomes. It was eventually shown that genes definitely are located in chromosomes.

Controversy Between the Mendelians and the Biometrical School

Mendel's work with garden peas led to the concept of discontinuous variation. Peas resulting from the cross cited earlier in this chapter had vines that were either tall or dwarf. No intermediates were observed, and simple ratios were obtained between the well-marked traits. Analysis of Mendelian ratios required only simple arithmetic and most of the work in the early part of the present century, particularly that of Bateson and his associates, supported the Mendelian pattern. It apparently did not occur to these investigators that both continuous and discontinuous variation might occur in nature.

Mendel's results were in marked contrast to those of Kölreuter, Darwin, and other investigators who had studied quantitative traits in various plants and observed continuous variations. Traits studied by these latter investigators graded into each other and could not be separated into distinct classes. Simple ratios were not obtained, and complex mathematics were required to analyze the results of crosses. Continuous variation seemed to support Darwin's theory of pangenesis and to demonstrate the natural selection mechanism. Sir Francis Galton, the founder of biometry (the application of statistics to biological problems) along with later biometricians, Carl Pearson and W. F. R. Weldon, provided strong support for the statistical approach to genetics.

Arguments grew more emotional and more personal as time went on. During the first decade of the present century some of the most vitriolic exchanges in the history of science centered around this issue. (See preface to *Mendel's Principles of Heredity—A Defense* by W. Bateson, 1902.) Bateson, who did not have a broad statistical view of genetics, became the chief spokesman for the Mendelians. In

Figure 17.6
H. Nilsson-Ehle,
Swedish plant breeder
who developed the multiple
gene hypothesis to explain
the genetic mechanism for
quantitative inheritance.
(From Gardner, *Principles
of Genetics.* Copyright
1960, 1964, 1968 by John
Wiley & Sons, Inc.)

1914, he was the one who proclaimed a victory for the Mendelians. This war ended as most wars end—with losses on both sides and no real victory. Even as Bateson made his victory announcement, the basic data had been accumulated that eventually brought the biometricians and the Mendelians together.

Unfortunately, biometrical studies were eclipsed while (1900-1920) genetics was growing into a science. Actually both the Mendelians and the biometricians had a place in genetics, but it was not until continuous variation was explained in Mendelian terms as the "multiple gene hypothesis" that the place for each group was established. Yule (1873-1949) laid the foundation for the union in 1907 by suggesting that Mendelism could explain continuous variation if it was assumed that large numbers of genes could act in a cumulative manner. Experimental evidence for this type of inheritance was found by the Swedish plant breeder, Nilsson-Ehle (Fig. 17.6), in 1908 and by East in the United States in 1910, both of whom worked with cereal crops. R. A. Fisher (1890-1962; Fig. 17.7), in 1918, added the final touch required to unite the two schools of thought by showing that Mendelism can best explain the results of the biometricians.

Figure 17.7
Ronald A. Fisher, British
statistician and geneticist.
(From Gardner, *Principles
of Genetics.* Copyright©,
1960, 1964, 1968 by John
Wiley & Sons, Inc.)

Fisher thus helped to make biometry one of the most useful tools with which the geneticist may analyze and interpret quantitative data.

The Gene Concept

The concept of the gene (Table 17.1) was implicit in the conclusions of Mendel, who visualized a physical element (*anlage*) as the foundation for development of a trait. The Dutch biologist, de Vries, also set forth an early view of inheritance centered around a physical determiner. Johannsen coined the word "gene" to identify such an entity. Bateson promoted Mendel's view of paired genes to allelomorphs. Boveri and Sutton identified a gene as part of a chromosome and thus introduced the chromosome theory of inheritance. The theory of the gene as a locus or discrete unit on the chromosome was developed by Thomas Hunt Morgan (1866-1945; Fig. 17.8) from his studies on the fruit fly, *Drosophila melanogaster*.

William Bateson, the leading supporter of Mendelian inheritance in the early part of the century, was at first not impressed with the chromosome theory. In 1922, however, after visiting Morgan's laboratory at Columbia University, he recognized the importance of chromosomes as evidenced by the following quotation from his address before the American Association for the Advancement of Science at Toronto, in December, 1921: "...For the doubts—which I trust may be pardoned in one who had never seen the marvels of cytology, save as through a glass darkly—can not as regards the main thesis of the Drosophila workers, be any longer maintained. The arguments of Morgan and his colleagues, and especially the demonstra-

Table 17.1   Contributions to the gene concept

Investigator	Date	Contribution
Mendel	1866 (1900)	Paired physical elements of inheritance segregate.
de Vries	1901	Determiner capable of mutation.
Bateson	1902	Allele concept.
Boveri and Sutton	1902	Determiners are parts of chromosomes.
Johannsen	1903	Coined the word "gene."
Garrod	1909	A gene produces an enzyme.
Morgan	1910	A gene is a discrete chromosome locus.
Beadle and Ephrussi	1935	One gene produces one enzyme.
Goldschmidt	1937	Gene is a chemical rather than a physical unit.
Oliver	1940	A gene locus can be divided (pseudoalleles).
Avery et al.	1944	Genes are deoxyribose nucleic acid (DNA).
Hershey and Chase	1952	Specificity depends on DNA, not protein.
Watson and Crick	1953	DNA model with four organic bases.
Nirenberg and Ochoa	1961	DNA code for specifying protein synthesis.

Figure 17.8
Thomas Hunt Morgan,
American geneticist.
(From Gardner, *Principles
of Genetics*. Copyright©,
1960, 1964, 1968 by John
Wiley & Sons, Inc.)

tions of Bridges, must allay all skepticism as to the direct association of particular chromosomes with particular features of the zygote. The transferable characters borne by the gametes have been successfully referred to the visible details of nuclear configuration.

"The traces of order in variation and heredity which so lately seemed paradoxical curiosities have led step by step to this beautiful discovery. I come at this Christmas Season to lay my respectful homage before the stars that have arisen in the West." (*Science* 55:57, 1922)

Morgan

Morgan's early investigations were in the area of embryology, and his first publication in this field, *The Development of the Frog's Egg* (1887), was developed along morphological lines. He soon shifted to an experimental attack on problems of development, however, and became a leader in the rising school of experimental embryology. This in turn led naturally into the field of regeneration, with which he was occupied for several years, publishing a book entitled, *Regeneration* (1901). An ardent believer in experimentation, he turned next to a study of evolution and obtained many empirical facts described in his book, *Evolution and Adaptation* (1903).

Soon after the discovery of Mendelism, Morgan began genetic investigations, at first with mice and rats. He began his work on Drosophila in 1909. The adaptability of *Drosophila melanogaster* to breeding experiments had been shown by Castle, Moenkhaus, Lutz, and Payne. Morgan's extensive and thorough investigation of this

Figure 17.9
Calvin B. Bridges,
American geneticist and
cytologist. (From Gardner,
Principles of Genetics.
Copyright©, 1960,
1964, 1968 by John Wiley
& Sons, Inc.)

organism produced much information in support of the gene concept
and other aspects of genetics.

Morgan gathered around him a group of students who went on to
contribute extensively in the fields of cytology and genetics. Among
Morgan's students were such men as C. B. Bridges (1889-1938; Fig.
17.9), who was instrumental in demonstrating and analyzing structural
changes in chromosomes and in the cytological mapping of genes
on the chromosomes of Drosophila; H. J. Muller, a leader in chromosome studies of Drosophila, particularly balanced lethal arrangements,
and induced mutations; A. H. Sturtevant, who carried on the work
initiated by Morgan concerned with chromosome mapping, the
analysis of the gene, and position effects; and Curt Stern, who provided a cytological demonstration of crossing over and has contributed extensively to the mechanics of heredity and variation. These
scientists were greatly influenced by Morgan and in turn influenced
him.

In 1910, three basic contributions came from Morgan's laboratory:

(1) an account of the first spontaneous gene mutation observed in Drosophila (white eye); (2) an explanation of sex-linked inheritance; and (3) the first statement of the gene theory of inheritance, which included the principle of linkage and crossing over. Also in 1910, Morgan started to compile chromosome maps of Drosophila, utilizing advances made in chromosome studies up to that time. This project was carried on for several years by Morgan and his associates, particularly Sturtevant and Bridges, who located hundreds of genes on the four chromosomes of Drosophila, first from crossover data and later from cytological data interpreted from giant salivary gland chromosomes. Morgan, Sturtevant, Bridges, and Muller published an elaborate book, *The Mechanism of Mendelian Heredity* (1915), in which they summarized the progress that had been made in five years of experimentation with Drosophila. This was an epoch-making compilation in the field of genetics.

Morgan delivered a series of lectures at Princeton University in 1916; this was printed in book form under the title, *Evolution and Genetics*. In this book he summarized the evidence for evolution from the fields of comparative anatomy, embryology, paleontology, and genetics, and emphasized the genetic evidence that might explain the mechanism of heredity and variation. Mutant variations were considered to be the small heritable variations that Darwin had visualized as furnishing the material basis for organic evolution. Morgan distinguished these from noninherited environmental variations, which Darwin had not done. In reply to the objection that changes due to gene mutations are trivial and not significant in evolution, Morgan pointed out that minor superficial traits used by the geneticist and taxonomist may often be secondary results of more important physiological changes that have invisible but important effects on actual survival value of the organism. Also, genes for "useless" characteristics may be closely linked to those for "useful" ones. Systematists, of necessity, had distinguished many species on the basis of traits that played no part or only a minor part in the actual production of that species.

Morgan later updated the gene theory in his book, *The Theory of the Gene* (1928). Here he explained traits of an individual on the basis of paired physical elements (genes) in the germinal material. Morgan stated that genes were organized in a definite number of linkage groups, which separate when the germ cells mature, in accordance with Mendel's principle of segregation; and in consequence each gamete (egg or sperm) contains one set (n) of genetic

Figure 17.10
Richard B. Goldschmidt,
pioneer in the field of
sex determination and
physiological genetics.
(From Gardner, *Principles
of Genetics.* Copyright ©,
1960, 1964, 1968 by John
Wiley & Sons, Inc.)

material only. Physical units belonging to different linkage groups
were presumed to assort themselves independently, in accordance
with Mendel's principle of independent combinations. An orderly
interchange (crossing over) between the elements in corresponding
linkage groups also was hypothesized. The frequencies of crossing
over were found by Sturtevant and Morgan to indicate the linear
order of the genes in each linkage group and the relative position of
the physical elements with respect to each other.

Goldschmidt

Richard Goldschmidt (1878-1958; Fig. 17.10) had reviewed a broad
spectrum of literature on the subject and concluded in 1937 that the
gene must be a chemical entity operating through synthetic processes
in living systems. Although Goldschmidt was not a biochemist, he
was a biological theorizer of the first rank. Some of his theories have
not held up in the light of more recent accomplishments, but those
bearing on the chemical nature of the gene are being thoroughly
tested in the 1970s and are generally being substantiated.

Chemical Nature of the Gene

By the middle 1900s, biochemists and biophysicists were finding a place in the study of genetic material. A. E. Garrod had shown in 1909 that a gene produces an enzyme. G. W. Beadle and E. Ephrussi in 1935 and Beadle and E. L. Tatum in 1941 demonstrated the physiological role of genes and formulated the one gene-one enzyme hypothesis. This provided a basis for an understanding of the functional properties of genes and suggested some inadequacies in the classical gene concept of the Morgan school. The classical gene was a recombinationally indivisible unit of structure, a unit of mutation, and a unit of function. All three of these attributes of the gene were considered to be equivalent, and the gene was inseparable. Even though the beginnings in the biochemical findings suggested interacting functional properties of genes, the classical unit was still satisfactory because the gene was still an indivisible unit.

Subdivisions of the Gene

The discovery of C. P. Oliver in 1940 that the lozenge (*lz*) locus in Drosophila could produce "pseudoalleles" and the later M. M. Green and K. C. Green (1949) report that the *lz* locus could be divided into sites indicated that a gene could no longer be defined as a single unit of structure and function. Investigators then sought an ideal experimental system with which to investigate the functional aspects of the gene. The procaryotes (organisms lacking well-defined nuclei and meiosis, that is, viruses, bacteria, and blue-green algae) were chosen as the experimental material. Immediate triumphs in using procaryotes were: verification of the one gene-one enzyme hypothesis and identification of the genetic chemicals, deoxyribonucleic acid and ribonucleic acid.

J. Lederberg and E. L. Tatum (1946) demonstrated recombination in a bacterium (*Escherichia coli* K 12) and thus began bacterial genetics. As soon as a sexual phase was discovered, investigators could study segregation of genetic traits and map a bacterial (*E. coli*) chromosome. Although bacteria did not have the same systems on which genetics had been established in eucaryotes (organisms made up of cells with true nuclei), their gene function was related to metabolic control, which opened the field to biochemical genetics. The nature of genetic material and the role of DNA in specifying protein structure could thus be investigated in bacteria.

Molecular Genetics

Molecular genetics represents the blending of chemistry (biochemistry) and physics (biophysics) with biology. Dependent on modern tools and background in the physical sciences, this aspect of genetics could not develop before the middle of the twentieth century. The molecular geneticist studies the ultimate chemical units which make up genetic systems, and he is interested in their function as well as structure. Until recently, the protein component of the chromosome was suspected to be the essential material for carrying information from cell to cell and from generation to generation. This idea became untenable when the results of several experiments showed that genes are nucleic acid and that genes specify the synthesis of proteins.

When nucleic acid was extracted from one strain of a bacteria (pneumococcus which causes lobar pneumonia in mice), highly purified, and allowed to be taken up by organisms of another strain, the recipients developed certain characteristics of the donor bacteria. These transferred traits were perpetuated generation after generation, indicating that the gene complement (genotype) as well as the outward expression of genes or characteristics of the organism (phenotype) had been altered. Nucleic acid, obtained from killed pneumococcus organisms, highly refined and presumed to be pure, could be incorporated into and expressed through the genotype of recipient organisms. This phenomenon, called transformation, substantiated the idea that nucleic acid carried genetic information.

From an elaborate series of chemical analyses and enzymatic investigations involving the transforming substance, O. T. Avery and his associates (1944) identified a particular nucleic acid. (deoxyribose nucleic acid or DNA) as the effective chemical agent, the carrier of genetic information. The identification was further substantiated by the experiments of A. Hershey and M. Chase (1952) making use of radioactive tracers. These investigators labeled the DNA of a virus (bacteriophage) with radioactive phosphorus, and the protein envelope with radioactive sulphur. The results showed that the DNA and not the protein was genetically active.

Transduction

In other experiments by N. D. Zinder and J. Lederberg (1952), bacteriophage was found to transfer bacterial genetic information from one bacterial cell to another. The phenomenon was called transduction. The bacteriophage used in this experiment lived and reproduced in the typhoid organism *Salmonella typhimurium*. Virus particles were known to be composed mostly, if not entirely, of protein and nucleic acid. Following multiplication or reproduction of the virus particles within a bacterial cell, lysis or rupture of the cell occurred, and large numbers of new particles carrying some bacterial traits were freed from the host cell. In this stage the particles were infective and capable of entering other bacterial cells. With radio-active tracers, the different components of the test virus were followed through an invasion cycle. Significantly, it was observed in these experiments that when a virus entered a host cell, the protein part remained outside. Only the nucleic acid was present in a bacterium when the fundamental processes of duplication were being accomplished. Furthermore, each nucleic acid unit provided itself with a new protein envelope following the duplication process. Obviously the nucleic acid of the bacteria and not the protein was functional in inheritance.

Watson and Crick Model

All the information available about DNA, including the X-ray diffraction pictures by M. H. F. Wilkins, was brought together in 1953

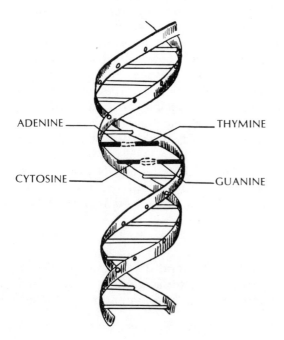

ADENINE —————————— THYMINE

CYTOSINE —————————— GUANINE

Figure 17.11
Watson and Crick Model
of DNA. The outside rims
are composed of phosphate
and sugar and the cross
links are pairs of organic
bases connected in the
center by hydrogen bonds.

by J. D. Watson and F. H. C. Crick in a model (Fig. 17.11). According to the Watson and Crick model of DNA, four organic bases—adenine (A), cytosine (C), guanine (G), and thymine (T)—are arranged in pairs forming cross links between two sugar phosphate rims of a double helix. Adenine pairs with thymine and cytosine with guanine. Through extensive chemical analysis of DNA, E. Chargaff and J. N. Davidson (1955) provided background for the Watson and Crick model by showing that the amount of adenine equals the amount of thymine and the amount of cytosine equals the amount of guanine.

RNA Viruses

Plant viruses (e.g., tobacco mosaic virus, TMV) were found to contain a different kind of information-carrying nucleic acid, ribose nucleic acid (RNA), and apparently they have no DNA. Fraenkel-Conrat (1956) and R. C. Williams reconstituted "hybrid" tobacco mosaic virus from nucleic acid and protein components from different sources. Some viruses, on the other hand, (e.g., those that infect the colon bacillus *E. coli*) seemed to have only DNA. Two different genetic information carriers have thus been developed among the viruses. Both DNA and RNA virus particles are on the borderline between living and non-living things. Some, at least, can be crystallized and maintained as chemicals on the laboratory shelf for months or years. When, however, the seemingly inert particles are introduced into an appropriate living host, such as a living tobacco plant in the case of TMV, they "come to life," produce characteristic symptoms on the host plant, and reproduce more virus particles.

Bacterial Genetics

F. Jacob and E. L. Wollman (1956) interrupted experimentally the sexual process in *E. coli* and showed that a piece of chromosome was donated from one bacterium to another. This established a physical basis for bacterial genetics. Jacob and Wollman (1958) showed further that the single linkage group (chromosome) in *E. coli* was circular and subject to genetic change by insertions of genetic material from different strains. S. Ochoa and A. Kornberg in the same year synthesized *in vitro* the polymers of ribonucleotides and deoxyribonucleotides demonstrating the chemical nature of the genetic material. M. Meselson and F. W. Stahl (1958) devised a density gradient centrifugation technique and with it demonstrated the manner in which the DNA, synthesized, in bacterial cells is distributed among the progeny. Each half of the molecule is conserved and used as a template for the formation of a new strand (i.e., semiconservative replication).

DNA, RNA, and Protein Synthesis

Several investigators in the late 1950s and early 1960s showed that cells of higher plants and animals have both DNA and RNA, and in such organisms each of these nucleic acids is specialized for a particular function. Although most of the experimental work has been done on procaryotes, the basic mechanisms are presumed to apply to other organisms as well. DNA's unique properties that permit its important function are: (1) it can replicate in a living system with catalytic action of one or more enzymes; (2) it has a coding device capable of carrying a tremendous amount of information; (3) it can genetically change or mutate; and (4) it can, in conjunction with RNA and protein synthesis, interpret information carried in a living system.

The DNA of higher organisms is carried from the parents to the zygote by the two gametes, egg and sperm. It makes new copies of its incorporated information as frequently as cell division occurs. Every new cell thus has a copy of all the information required for the architecture and functioning of the entire organism, even though only a small part may be used in an individual cell. The mechanism for the transfer of information is a coding device involving the arrangement of the organic bases in the DNA molecule.

Jacob (1961) and others showed that RNA is involved in the synthetic activity of proteins and is the prime agent for the interpreting of information carried by DNA. It contains four organic bases, three of which are the same as those in DNA: adenine, cytosine, and guanine. In RNA, however, uracil takes the place of thymine. Functionally, there are at least four kinds of RNA, (1) template RNA, which acts as the carrier of information in plant viruses, (2) RNA which is part of the cytoplasmic macromolecules called ribosomes, (3) messenger RNA, which acts as an intermediary between DNA and the ribosome (Jacob, 1961); and (4) transfer RNA (Crick, 1958), which transports amino acids to the ribosomes for protein synthesis (P. C. Zamecnik, 1958). Investigations by M. W. Nirenberg (1961), S. Ochoa, and others have provided an understanding of the mechanism through which information is carried by DNA, transferred to messenger RNA, and incorporated into protein synthesis.

Molecular biology is Cartesian biology (Chapter 14), based on the premise that biological mechanisms can be reduced to inanimate chemical and physical processes. Living organisms, however, are more than bundles of DNA. Since many interactions occur above the molecular level, areas other than molecular biology will continue to have a place in biology.

Figure 17.12
W. Weinberg, German
physician who developed
the Hardy-Weinberg
theorem. (From Gardner,
Principles of Genetics.
Copyright© 1960, 1964, 1968
by John Wiley & Sons, Inc.)

Population Genetics

Behavior of genes in a population and genetic and environmental
influences on the equilibrium of allele frequencies was recognized
as a problem by Mendel and others who investigated the principles
of genetics in the light of evolutionary theory. In 1908, an English
mathematician, G. H. Hardy, and a German physician, W. Weinberg
(1862-1939; Fig. 17.12), independently discovered a principle concerned with the frequency of genes (alleles) in a population that has
since formed the basis for population genetics. Before 1908 Mendelian
genetics had been amply demonstrated by W. Bateson and others.
Sample 3:1 and 1:1 ratios had been obtained from controlled pair
matings involving many different kinds of plants and animals. The
observed proportions of different phenotypes in natural populations,
however, suggested different gene frequencies than those characteristic of laboratory experiments. Hardy and Weinberg showed that
an equilibrium is established between the frequencies of alleles in a

population. The relative frequency of occurrence of each allele tended to remain constant, generation after generation. A mathematical relation now called the Hardy-Weinberg theorem was developed to describe the equilibrium between allele frequencies.

The significance of the Hardy-Weinberg equilibrium was not immediately appreciated. A rebirth of biometrical genetics was later brought about with the classical papers of R. A. Fisher, beginning in 1918, and those of Sewall Wright, beginning in 1920. Under the leadership of these mathematicians, emphasis was placed on the population rather than on the individual or family group, which had previously occupied the attention of most Mendelian geneticists. In about 1937, T. Dobzhansky and others began to interpret and to popularize the mathematical approach for studies of genetics and evolution. Questions concerning the behavior of genes in populations and the mechanics of evolution took on new significance.

Chronology of Events

B.C.

7000   Most animals and plants that have proved capable of domestication by man were already domesticated.

6000   Conscious effort by stock raisers and date growers to improve domesticated animals and plants was in evidence.

400-200   Hippocrates, Aristotle and other Greek philosopher-scientists speculated on principles of inheritance.

A.D.

1717   T. Fairchild, successful hybridization between carnation and sweet William plants.

1760   J. Kölreuter, quantitative inheritance.

1866   G. Mendel, *Experiments in Plant-Hybridisation.*

1883   W. Roux postulated that chromosomes in the nucleus are bearers of hereditary determiners.

1900   H. deVries, C. Correns, and E. von Tschermak discovered Mendel's paper.

1901   H. deVries observed "mutations."

1902   W. S. Sutton and T. Boveri, chromosome theory of inheritance.

1902-1909   W. Bateson, terms: genetics, allelomorph, homozygote, heterozygote, F_1, F_2 and epistatic genes; W. Bateson, L. Cuénot, W. E. Castle, W. L. Johannsen, and C. Correns all published papers supporting Mendelian inheritance.

1903-1909   Johannsen, terms: gene, genotype and phenotype.

1906   W. Bateson and R. C. Punnett, linkage (in sweet peas).

1908   The Hardy-Weinberg law.

1908-1910   H. Nilsson-Ehle, multiple gene hypothesis.

1909   G. H. Shull, hybrid corn.

              A. E. Garrod, paper on biochemical genetics, *Inborn Errors in Metabolism.*

1910	T. H. Morgan, white eye mutant, sex linkage and autosomal linkage, gene theory of inheritance.
1913	A. H. Sturtevant, linkage-distance concept and chromosome mapping.
1918	H. J. Muller, balanced lethal phenomenon in Drosophila which led to methods for detecting sex-linked lethal mutations.
1921	C. B. Bridges, triploid intersexes in Drosophila.
1927	H. J. Muller, induced mutations by X-rays.
1928	F. Griffith, virulence for pneumonia in mice transferred from dead virulent strains of pneumococci to live avirulent strains.
1930	R. A. Fisher, *The Genetical Theory of Natural Selection.*
1931	S. Wright, *Evolution in Mendelian Populations.* This book, along with Fisher's book, constitutes the mathematical foundation of population genetics.
	C. Stern, and H. B. Creighton and B. McClintock, cytological basis of crossing over.
1935	G. W. Beadle and B. Ephrussi, biogenetic production of eye pigments in Drosophila.
1937	R. B. Goldschmidt theorized that the gene was a chemical entity rather than a discrete physical structure.
	T. Dobzhansky, *Genetics and the Origin of Species.*
1940	C. P. Oliver, pseudodominance for the *lz* locus in Drosophila. In 1949, M. M. Green and K. C. Green described crossing over between sites within the *lz* locus.
1941	G. W. Beadle and E. L. Tatum, biochemical genetics of Neurospora including the one gene-one enzyme hypothesis.
1944	O. T. Avery, C. M. MacLeod and M. McCarty, transforming principle, DNA and not protein.
1946	M. Delbrück and W. T. Bailey, genetic recombination in bacteriophage.
	J. Lederberg and E. L. Tatum, genetic recombination in bacteria.
1950-1955	E. Chargaff, chemical structures of nucleic acids.
1952	N. D. Zinder and J. Lederberg, transduction in Salmonella.
	A. D. Hershey and M. Chase, only DNA and not protein enters the host cell in phage infection.
1953	J. D. Watson and F. H. C. Crick, model for DNA.
1955	S. Benzer, fine structure of the genetic material of phage T_4 of *E. coli.*
1956	F. Jacob and E. L. Wollman, a piece of chromosome was donated from one *E. coli* into another in mating.
	S. Ochoa and A. Kornberg synthesized *in vitro* polymers of ribonucleotides and deoxyribonucleotides.
	H. Fraenkel-Conrat and R. C. Williams reconstituted "hybrid" tobacco mosaic virus.
1958	F. Jacob and E. L. Wollman showed that the single circular linkage group in *E. coli* can be altered by insertions from different strains.
	M. Meselson and F. W. Stahl demonstrated semiconservative replication in bacterial cells.
	F. H. C. Crick predicted the discovery of transfer RNA.
	P. C. Zamecnik characterized amino acid-transfer RNA complexes.

1961   F. Jacob, proteins are synthesized on ribosomes and messenger
              RNA transports specifications from DNA to ribosomes.
              M. W. Nirenberg and S. Ochoa cracked the DNA, messenger
              RNA code.

References and Readings

Babcock, E. B. 1950. *The Development of Fundamental Concepts in the Science of Genetics. Portugaliae Acta Biologica Series,* 1949. Reprinted by American Genetics Association, Washington, D. C.

Baltzer, F. 1964. "Theodor Boveri." *Science* 144:809-815.

Bateson, W. 1909. *Mendel's Principles of Heredity.* Cambridge: At the University Press.

Boyer, S. H. 1963. *Papers on Human Genetics.* Englewood Cliffs, N. J.: Prentice-Hall.

Carlson, E. A. 1966. *The Gene: A Critical Review.* Philadelphia: W. B. Saunders.

Crow, J. F. 1959. "Darwin's influence on the study of genetics and the origin of life." *Biological Contributions, The University of Texas,* Austin. Publ. no. 5914.

Dobzhansky, T. 1951. *Genetics and the Origin of Species.* 3rd ed. New York: Columbia University Press.

Dunn, L. C. ed. 1951. *Genetics in the 20th Century.* (Chapter 3, The Knowledge of Heredity before 1900 by C. Zirkle; Chapter 4, The Beginnings of Mendelism in America by W. E. Castle.) New York: Macmillan Co.

_____. 1965. *A Short History of Genetics.* New York: McGraw-Hill Book Co.

Gardner, E. J. 1972. *Principles of Genetics.* 4th ed. New York: John Wiley and Sons.

Garrod, A. E. 1909. *Inborn Errors of Metabolism.* London: Oxford University Press. (Reprinted in H. Harris, *Garrod's Inborn Errors of Metabolism,* London: Oxford University Press, 1963, pp. 1-93.)

Goldschmidt, R. 1938. "The theory of the gene." *Sci. Monthly* 46:268-273.

Hecht, M. K. and W. C. Steere, eds. 1970. *Essays in Evolution and Genetics in Honor of Theodosius Dobzhansky.* New York: Appleton-Century-Crofts.

Iltis, H. 1932. *Life of Mendel.* trans. E. and C. Paul. New York: W. W. Norton and Co.

Iltis, H. 1947. "A visit to Mendel's home." *J. Heredity* 38:163-166.

Jaffe, B. 1958. *Men of Science in America.* New York: Simon and Schuster. (Chapter 16 on Morgan.)

Mendel, G. 1958. *Experiments in Plant Hybridization.* Cambridge: Harvard University Press. (Available in original German in vol. 42, *J. Heredity.*)

Morgan, T. H. 1926. *The Theory of the Gene.* New Haven: Yale University Press.

Muller, H. J. 1956. "Genetic principles in human populations." *Sci. Monthly* 83:277-286.

Peters, J. A. ed. 1959. *Classic Papers in Genetics.* Englewood Cliffs, N. J.: Prentice-Hall.

Punnett, R. C. 1950. "Early days of genetics." *Heredity* 4:1-10.

Roberts, H. F. 1929. *Plant Hybridization Before Mendel.* Princeton: Princeton University Press.

Sootin, H. 1959. *Gregor Mendel, Father of the Science of Genetics.* New York: Vanguard Press.

Stent, G. S. 1971. *Molecular Genetics.* San Francisco: W. H. Freeman and Co.

Stern, C., ed. 1950. "The birth of genetics." Suppl. *Genetics* 35:(5) pt. 2. (English trans. of letters from Mendel to Nägeli and papers of deVries, Correns, and Tschermak, the three men who discovered Mendel's paper in 1900.)

Stern, C. 1953. "The geneticist's analysis of the material and the means of evolution." *Sci. Monthly* 77:190-197.

————. 1962. "Wilhelm Weinberg." *Genetics* 47:1-5.

————. 1970. "The continuity of genetics." *Daedalus, Proc. Amer. Acad. Arts and Sciences* 99:882-908.

Stern, C. and E. R. Sherwood, eds. 1966. *The Origin of Genetics.* San Francisco: W. H. Freeman and Co.

Sturtevant, A. H. 1946. "Thomas Hunt Morgan." *Amer. Nat.* 80:22-23.

————. 1965. *A History of Genetics.* New York: Harper and Row.

Tschermak-Seysenegg, E. von. 1951. "The rediscovery of Gregor Mendel's Work." *J. Heredity* 42:163-171.

Vries, H. de. 1901-3. *The Mutation Theory.* Leipzig: Verlag von Veit and Co.

Watson, J. D. 1970. *Molecular Biology of the Gene.* 2nd ed. New York: W. A. Benjamin.

Weismann, A. 1893. *The Germ-Plasm; A Theory of Heredity.* trans. W. N. Parker and H. Rönnefeldt. New York: Charles Schribner's Sons.

Zirkle, C. 1964. "Some oddities in the delayed discovery of Mendelism." *J. Heredity* 55:65-72.

CHAPTER 18

BIOLOGY PAST AND PRESENT

O
UR nameless ancestors probably began to study living things because of a combination of curiosity and a desire to achieve control. But their efforts were at best sporadic and incoherent. Not until circumstances favored a sustained effort, as in the case of the early Greek philosopher-scientists, did the scientific method of objective observation and experimentation begin to emerge. These early beginnings and the humanistic philosophy that developed in the Renaissance generated biology as we know it.

Once established as a developing science, accomplishments in biology became dependent upon other affairs as well as appropriate times and places. Some developments could not occur until other events had taken place. A scholarly climate, a new availability of materials for dissection, the challenge of an unsatisfied need, and a brilliant young anatomist, named Vesalius, combined to produce *The Fabric of the Human Body*. Once human anatomy had been learned from actual observation and set forth in the pages of books, comparative anatomy took on added significance and comparative physiology developed under leadership of an Englishman (Harvey) who had studied at Padua and had followed in the Vesalian tradition. Harvey included actual measurements of blood flow and other bodily functions in his experimental methods. Small groups of curious observers who shared the spirit of investigation began to meet informally (and later more formally) to discuss their thoughts and observations. Thus the academy movement took root.

When microscopes became available, still another world was opened to view. Critical observers and interpreters then laid the foundation for studies of microorganisms and cells. Comparisons of reproductive and developmental processes followed, and biologists began to understand the mechanics of life. As the recognized numbers of living things grew and more and more plants and animals became known, systematizers inevitably began to classify them into meaningful groups. Systematics whetted the interest in finding more forms to classify. Explorers and comparers eagerly sought to satisfy that interest.

Prestige-hungry governments supported scientific expeditions to foreign lands. As various plants and animals were observed in their natural habitats, the stage was set for broad questions about how and why they reached their present state of development and why they were *where* they were. Fossils gained significance, and the grand view of natural relationships began to unfold. The unifying sciences of evolution and genetics could then take their places on the biological scene.

The Story, As Told by Historians

Only fragments of historical information are available to suggest the insights and accomplishments that were made in biology before the seventeenth century. Not all the events that occurred, but simply those that were in some way recorded, found their way into history. Chronological discussions, therefore, must rely on more or less sporadic records for the early years. Similarly, even when records are more complete, the individual historian inevitably exercises considerable selectivity. Furthermore, organization into periods, localities, and subjects simplifies presentations but does not reflect reality. Particularly in recent periods of time, significant activities in different fields have been occurring simultaneously, but it is impossible either to present them simultaneously or to be all-inclusive.

But we do know that many contributions of great importance to biology in the nineteenth and preceding centuries were made and recorded by amateur biologists. Gregor Mendel, for example, was a teacher who devoted summer vacations to satisfying his curiosity about inheritance in plants. Charles Darwin was essentially an amateur when he began his monumental work on organic evolution. Alfred Russel Wallace had been a land surveyor, an architect, a watchmaker, and a teacher when he set out for the Amazon with an entomologist friend, Henry Walter Bates. These and other amateurs established the foundations of today's biology.

In the latter part of the seventeenth century, through the eighteenth, and into the nineteenth century, written records became more feasible and thus more inclusive. Increasing amounts of data, transmitted with increasing care, permitted biology to reach an unprecedented peak of achievements during the last half of the nineteenth and the beginning of the twentieth century. Many plants and animals had been named and catalogued and awaited more intricate investigation. Soon after microorganisms were discovered they were associated with fermentation, decay, and disease. The organization of living things and their corresponding functional characteristics were explored and described. Mechanisms of cell division, gamete formation, and fertilization were defined and illustrated in detail. Developmental processes were documented for individual plants and animals. Broad relations of population dynamics were synthesized, and the concept of evolution was developed. Mendelian genetics was rediscovered and described in detail. Behavior patterns began to attract the interest of biologists. Sweeping generalizations could be made on the basis of cumulated literature.

Today's leading biologists, such as 1971 Nobel Prize recipient for physiology, Earl W. Sutherland, who is investigating the action of hormones, require advanced training, specialized instruments, and rely upon interactions among disciplines. By contrast with biology of earlier times, late twentieth-century biology is characterized by teamwork, high-powered tools, and generous applications of mathematics, chemistry, and physics. Regardless of the century or the state of technology, however, good ideas remain the cornerstone of all scientific investigations. And original ideas must come from the creative activities of curious individuals.

Population Studies

Thus Karl Sax, Garrett Hardin, and Paul Ehrlich, who are studying populations, are applying new viewpoints to known phenomena. When representatives of a single species are viewed in terms of life cycles, a population will have some characteristics similar to and some different from those of individuals within the group. A population, for example, grows, differentiates, and maintains itself, as does the individual. But a population also has characteristics that are unique to a group and different from those of individuals within the group. No single individual can adequately represent the entire population. Averages of measurements therefore must be used to describe a group, and characteristics of groups are best expressed as statistical functions. In trying to translate such generalities into useful specifics, biologists turn to laboratories and controlled conditions. Fruit flies in a population cage, for example, can be maintained in the laboratory with temperature, light, and humidity subject to manipulation as desired. Feeding and waste removal can be managed on a regular schedule. Random samples may be taken as needed to measure trends within the population and to compare different populations.

But in natural populations, the situation is more complex. All the contributing individuals can rarely be identified. Environmental conditions often cannot be defined in detail and certainly cannot be controlled to any great extent. Therefore, involved methods of sampling and statistical analysis are required. Even with the best computerized procedures, it is difficult to obtain meaningful interpretations of non-captive population characteristics. Differences between the total natural population and those individuals observed through the commonly used method of random sampling may be substantial. One study on the reproductive potential of deer, for example, was based on a sample (which could only be presumed to be random) of 500 deer that were observed in areas representative of the distribution of the species. The sample consisted of observations on progeny among the 500 deer. The actual population numbered in the thousands and extended over a wide geographical area. Improved sampling methods taking into account breeding patterns and life cycles are making possible valid estimates of reproductive performance or biotic potential under different environmental conditions.

Biotic Potential

As crowding becomes more acute on the earth's surface, reproductive or biotic potential becomes of prime concern. This potential equates with the inherent power of a population to increase in numbers when the age ratio is stable and environmental conditions are optimal. For natural populations, environmental conditions are usually less than optimal. Differences between the measured or observed rate of population growth and the potential is attributed to environmental resistance. Without environmental resistance, or man's artificially imposed limitations, each species would presumably expand until eventually all the space available and suitable was filled.

Examples of reproductive expansion without the usual restrictions have been observed when animals or plants have been purposely or inadvertently introduced into a new part of the world. Rabbits in Australia and in the San Juan Islands in Puget Sound; English sparrows in the United States; starlings in Western North America; Japanese beetles in the United States; Colorado potato beetles in Western United States; mongooses, goats, and pigs on the Island of Hawaii are examples of animals that increased at rapid and sometimes incredible rates in a new environment. On the other hand, some environments remain uninhabited by organisms that seem likely to survive well, if once introduced. Some snakes, for example, would undoubtedly find a happy home in the lush, warm Hawaiian Islands, but thus far none have been introduced in that part of the world.

Nature is exacting and efficient in controlling population density, but thus far man has found ways of circumventing, to some extent, nature's controls in his own population and in others that he purposely or inadvertently controls. The far-reaching significance of one aspect of man's intervention on nature's system of checks and balances was dramatized in Rachel Carson's book *Silent Spring* (1962). Effects of insecticides on the whole ecological pattern of plants and animals were vividly and forcefully portrayed. Insecticides were synthesized and widely disseminated before it was recognized that they could kill birds and other wildlife and might be harmful to people.

DDT, for example, has had a worldwide distribution since 1942. Its residues now appear in the lipids of most organisms, in air, and occasionally in meltwaters of Antarctic snows. Concentrations in living things have caused spectacular declines in populations of certain carnivorous and scavenging birds and fish and have threatened human food chains. Even though restrictions are now applied, DDT residues still persist. Fortunately, much of the toxic material is held in the biosphere rather than the living plants and animals. Miss Carson's book opened the way for objective appraisals of the ecological effects of pesticides, air pollution, water pollution, mining practices, harvesting the seas, energy utilization, radiation control, and disease control.

Pollution as a complex and rapidly growing problem for plants, animals, and man himself is being investigated by a diversity of scientists—including biologists. The population is increasing while the physical depositories for pollutants are limited. Interdependence of air, water, and waste pollution controls is finally being recognized. Taking the burden off the environment requires finding a way to recycle "used" goods, whether air, additives, or car bodies. Simple treating generally just changes the form to one that may be less harmful than another substance, but it does not eliminate their pollutant potentials. More attention must be given to the pollution problem if man and other living things are to survive on the earth. The answer for air and water pollutants is to prevent their entering the air or water in the first place. New techniques and cooperation with industry can do much to prevent the problem. Car bodies are being crushed and recycled.

Synthetic detergents have been a serious pollution problem. In the 1940s, detergents replaced soap in most domestic and industrial uses. Only after tons of detergents had been used was it realized that they accumulated in surface waters. Soap wastes were broken down by bacterial action, but detergents were indestructible by the agents operating in nature. The result was serious contamination of large bodies of water and pollution of domestic water supplies. New products are being developed that will serve man's needs without leaving unwanted residues.

A visit to any one of the large industrial cities with numerous automobiles, or even a flight over such an area, will illustrate that man is far behind in his effort to control air pollution. Much is being done to control the discharge of industrial plants and automobile exhaust. The problem of removing pollutants from the air and elsewhere seems insurmountable, but man may instead learn to keep them from ever entering the atmosphere. Alternative forms of transportation, such as electric trains, hold some promise. Man has learned to manipulate his environment, but his technology must do more to preserve nature's dynamic equilibrium.

The Human Population

Since World War II, the superabundance of people has become a serious threat to man's life on the earth. Man himself must be considered as a possible pollutant to his own environment. The worldwide human population increase has become so acute that other aspects of man's population problems have been crowded into the background. Knowledge gained from other species, both past and present, give the biologist some insight for his own situation.

The problem of the relation between population size and available space has been studied in other living things. Technological advances in food production may keep pace with a growing population for some time to come, but the limitations of land and fresh water suggest that the population cannot increase indefinitely without dire results. It seems obvious that if the present trend persists, education, cultural opportunities, and other "good things in life" will be inaccessible to substantial numbers of people. Only a limited number of people can be supported in the available space.

Large groups of people in the world are not now adequately supplied with the basic necessities of life. The world population is now more than 3 billion and, according to estimates of the United Nations, it is increasing at the rate of some 50,000,000 per year. The entire history of man on earth up to about 1830 was required to bring the population to one billion. In another century (1830-1930) the population doubled to about 2 billion. The third billion was added in about 30 years (1930-60), and the United Nations statisticians estimate that, at the present rate, it will take less than 15 years to add the fourth billion. If the present trend continues, some 6 billion people will be living on the earth at the end of the twentieth century. Opinions differ concerning the possibilities for meeting the needs of so many people. Most people, however, acknowledge that in a world with finite supplies of space, land, and water, a time will inevitably come when the limits of population that can be supported have been reached.

Even now a large proportion, perhaps two-thirds, of the people in the world are hungry. This is largely because of poor distribution and economic factors. Biological research has resulted in better yielding wheat, rice, corn, and other farm products. More effective use of certain habitats such as seas and other bodies of water may also increase productivity of organic material. Estuaries where fresh and salt water meet, for example, are capable of tremendous production of food. For comparison, an acre of semiarid land may produce one ton of organic material, whereas an acre of well-watered, intensely cultivated farm land in Florida or California may produce six tons. An acre of mangrove islands in Florida Bay, in contrast, produces 10 tons of organic material with little or no human intervention.

Many people are poorly clothed and inadequately sheltered. Large numbers of people in many countries are poorly educated. Moreover, in the increasingly complex culture of the most advanced countries, proportionately more adults are needed to properly educate the children and to provide adequately for the young and old in present-day society. Already there is overcrowding in many places. The problem of controlling the birth rate and improving the world living standard is so pressing that it must be considered in any evaluation of the future of man on earth. If we ignore the population problem, we are inviting for future generations the war, disease, and starvation that were envisioned by Malthus in 1798 and also the loss of the cultural gains society has won.

Behavior Patterns

The discovery of the genetic material, the universal code, and the interrelation of heredity and environment in the mid-twentieth century has brought new, broad, unifying principles and a new emphasis in biology. Mechanisms at the molecular and cellular level are challenging present-day and future biologists. Intricate timing mechanisms, for example, which bring all parts of a developing organism together in functional harmony may now be discovered. Complex systems of interactions may soon be defined. It is the whole organism that must be eventually understood and the whole is more than the sum of its parts because profound interactions are involved.

Behavioral characteristics of any animal develop under the joint, tightly interwoven influence of heredity and environment. DNA in the genome determines the individual's physiological, structural, and behavioral potentials, but not all of the potentials are inevitably realized in the developing individual. A fundamental question in the study of the relation between genes and behavior is whether heredity directly affects behavior in any way or merely sets the stage on which behavioral patterns may be molded by environmental factors.

Biologists and psychologists have taken somewhat different views on this issue in past decades, but now both groups recognize that heredity and environment are equally relevant. Environmental factors are interacting with inheritance mechanisms at every point in the developmental process. The problem is to disentangle the two and evaluate their relative importance in specific situations. Genetic programming simply provides that animals have specific learning capabilities at certain developmental periods. Zebra finches, for example, that have been isolated from species members before they are 35 days old can never learn to distinguish males and females of their own species. In other cases, the environment in which an animal matures can drastically affect its adult behavior. It is ranges of modifiability that are inherited, with the segregation of genes and the forces of natural selection accounting for observable individual differences in behavior patterns within a given population.

Internal mechanisms that activate a given behavior also mature and become more selective through learning. Recently discovered behavior patterns of food intake in cormorants, for example, reveal a gradual transformation from infantile behavior patterns to an increasing integration of the individual acts with the appropriate motivations. Until the third week of life the young birds beg and gape. Between the third and fifth week they begin to cudgel fish, and from the sixth week on they begin to catch their own. Begging drops out when they are six months old. Current research has shown that the behavior of any animal consists of patterns in time and may or may not be subject to modification through learning. Some basic behavior patterns are fully functional or at least "programmed" at the time of birth or hatching. Such behavior commonly involves so-called "fixed action patterns" or "instinctive movements." Once activated by the proper stimulus, these movements proceed through a preset pattern. For example, if the squirrel *Sciurus vulgares L.* is hand-raised in isolation on a liquid diet, its adult behavior repertoire will still include the species-characteristic, stereotyped method of storing nuts. Both inheritance and learning are known to be involved in the complex social patterns of some Hymenoptera, such as honeybees, that make use of chemical, optical, and sound signals. Studies of behavior patterns can now be carried out on a wide variety of animals because the basic research has been accomplished.

Aggression

Aggressive behavior is the result of interactions between the environment (and training) and the genetic makeup of the organism. Biologists are investigating these interactions in laboratories and in the field. In experiments with aggressive and relatively non-aggressive strains of mice, for example, babies from the non-aggressive strain increased in aggressive tendencies when raised by an aggressive-type mother. By contrast, aggressive-type babies retained their aggressiveness when raised by a non-aggressive mother. When baby mice were allowed to stay with their own mother, but a rat aunt took over many mothering activities other than supplying milk, results differed with the strain of mice used. Activity levels varied with the presence or absence of the rat aunt, regardless of strain of mice. But the level of aggression was lowered in one strain when a rat aunt was involved and not affected in another. Like many other behavior patterns, aggression may be modified through experience, though always only within genetic bounds. Aggression can be increased or decreased by purposeful education. Biologists are making a significant contribution by learning basic aspects of aggression.

Aggressiveness among nonhuman beings is a product of inheritance, maturation, various endogenous factors, and experience. Manifestations of aggression, however, depend upon the presentation of proper external stimuli, usually specific sign stimuli from other individuals of the same species. Modern man can learn aggression-limiting techniques by analyzing the behaviors of nonhuman beings. The first and foremost principle is the need to avoid population density-induced aggression. Even if technology could provide life's necessities for an increasing population, crowding per se leads to greater aggression. A dynamic balance between order and disorder must be established within living populations. Overcrowding is an environmental change that encourages expression of any innate qualities of aggression and prohibits peacefulness until proper adjustments are made and a new balance is established. Overcrowding, whether among human beings or animals, leads to a breakdown in social structure. Biologists now recognize that the innate aggressiveness necessary throughout evolution for the species to survive is the primary cause of this breakdown.

Differentiation in Embryology

The new dimension in biology that came with the use of the microscope opened many questions about embryonic development. Leeuwenhoek and his pupil, Hamm, discovered human spermatozoa in 1677. In 1694 another microscopist imagined he saw a miniature human figure in the head of the sperm. Von Baer, in 1828, observed the mammalian egg. With this basic information the story of differentiation began to unfold.

Fertilized eggs or zygotes of several animals are much alike in size and outward appearance, but they differ immensely in the coded information they carry, which determines the characteristics of the adult organism. A butterfly egg, for example, appears much like a grasshopper egg in size and shape, but the information carried in each gives rise to vastly different creatures. Zygotes of a mouse or man are microscopic entities, but each must carry the equivalent of many volumes of highly specialized printed material in order to designate the physical characteristics of the adult member of the particular species represented by the egg. All cells making up the body of an adult organism have come from a single fertilized egg by the process of cell division and all cells appear alike during their early embryonic stages; but, as cleavage proceeds, some groups of cells develop special characteristics (differentiate) and eventually give rise to the nervous system. Others, of the same basic origin, become skin cells, still others connective tissue cells, and so on. Such generalizations were recognized in the last century. But the processes involved in the differentiation of cells for particular functions and positions are still far from satisfactory definition.

Within the past few years, biologists have discovered mechanisms for turning certain genes on and off at appropriate times. Several possible mechanisms have been explored. The simplest process depends on a feedback which is responsive to the quantity of an end product in the cellular medium. This device is envisioned as operating like a thermostat on a central heating plant. When the temperature reaches a certain level, the furnace turns off. If, for example, bacteria *(Escherichia coli)* are supplied with an excess of a product they need but *can* make, production of that material stops and the organisms use that which is supplied. An example of the possible feedback mechanism in organisms made up of cells with true nuclei (eucaryotes) involves changes in the cellular content of the protein, histone, that regulates RNA synthesis. Attempts to offset the scourge of cancer have demonstrated that tumor cells which normally divide more rapidly than other cells in an organism have unusually low concentrations of growth-inhibiting histone. The present hypothesis is that histone may possess a variety of chemical structures, each of which is an inhibitor for a particular RNA. Different types of histone thus could selectively turn on and off the production lines stemming from units of DNA.

Another mechanism in organisms lacking well-defined nuclei (procaryotes) has a unit of several genes working together and includes a switch for turning production of an enzyme on or off. This "operon" (Fig. 18.1) model, proposed by Jacob and Monod from studies on bacteria (*E. coli*), requires a repressor substance that controls an enzymatic action. In the operon model a repressor controls an operator which is the switch for the structural genes that determine the primary structure (messenger RNA) for an enzyme. When the end product is needed, the switch is "on," and the relevant series of structural genes is engaged in production. When extraneous supplies of the end product are available, the repressor blocks the operator, and the structural genes are not producing. Enzyme repression occurs,

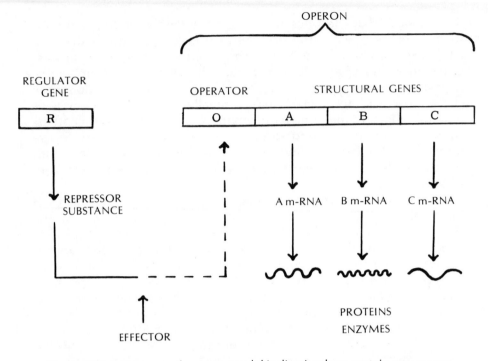

Figure 18.1. Diagram of operon model indicating how certain genes may
be turned on and off and thus suggesting an explanation for
one aspect of differentiation.

as expected, much more slowly than feedback inhibition. This model
has been substantiated for *E. coli* and some other procaryotes, but
little evidence is available for such a simple "operon" control scheme
in higher organisms.

A hormonal type of control of differentiation has been suggested on
the basis of the "puffs" observed in the giant chromosomes of
dipterous larvae. When these chromosomes were followed throughout the developmental cycle of Diptera larvae, it could be seen (as
evidenced by the puffs) that particular areas of the chromosomes
were active in producing RNA at different stages of the developmental sequence. The sequence of puffing was controlled, in part at
least, by a hormone in the larvae. This was interpreted as indicating
that different genes were influenced to become active at different
stages of larval development. The puffs eventually were found to consist of a protein (RNA polymerase) and RNA. This could mean that
protein synthesis and nucleic acid synthesis are coordinated proces-

ses. A particular sequence of puffs is now known to represent a corresponding pattern of gene activity, and a relation has been established between the activities of certain gene areas and particular cell functions. When Ulrich Clever injected the molting hormone, ecdysone, into larvae of the genus Chironomous, activity of specific gene loci was increased. Hormones thus are involved in some aspects of differentiation in some species, but they cannot account for the entire process.

Tissue culture techniques have also been useful in solving the riddle of differentiation. I. R. Königsberg, for example, has cultured animal cells from single progenitor cells (cloning) for the purpose of following single embryonic cells and their descendants. Cells cultured from presumptive muscle areas and treated in a manner known to favor cellular growth and differentiation multiplied rapidly during the first four days, each producing 50 to 60 descendants. During the fifth and sixth days, multinuclear cells appeared that showed the characteristics of muscle fibers. At the same time, fusion was occurring in the culture; single cells were apparently producing colonies of muscle cells. Extensive experiments demonstrated that a single cell (myoblast) could give rise to a colony. Each such colony contained many nuclei, showed the cross striations characteristic of skeletal muscle, and was capable of contracting spontaneously.

The mechanism of cellular differentiation apparently involves several processes, including a feedback mechanism, control of structural genes by operator and regulator genes, and hormonal control of a sequence of active sites along the chromosome. These processes are, for the most part, a step above the molecular level and require profound interactions within the organism.

Genetic Engineering

Advances in biology and medicine suggest that modifications may be made in the capacities and activities of man by direct interaction and manipulation of the body and mind. Enough is now known about gene action and genetic control to make some earlier speculations a near reality. The possibility, however remote, that man may re-create himself is dramatic and wins headlines in the popular press. Engineering the engineer arouses the imagination much more than engineering an engine. The real problem associated with altering and ultimately directing gene action is one that involves basic cell biology. How can enzyme action within a cell be added, removed, or altered?

Enzymes that operate within a cell are made according to DNA specifications within that cell. It is true that chance mutations may create new genes (alleles) that synthesize new enzymes in a cell. Thus far, however, no one has been able to control such mutational changes at will and thus alter enzymatic activity in a directional pattern.

Current progress in medical genetics is dramatic enough to be compared with the medical revolution of the last half of the nineteenth century that followed the establishment of the germ theory of disease. After years of basic accomplishments, the impact of genetics in medical practice is just beginning to be felt. This impact is bound to be substantial because the decline of infectious diseases has brought genetic diseases into new prominence. Infant mortality from congenital malformations is roughly the same as in 1900, about 5 per 1,000 live births. In 1900, however, the total infant mortality was 150 per 1,000 live births compared with 20 per 1,000 in 1970. While the incidence of congenital abnormalities has remained constant, these causes of death now account for about 25 percent of all infant mortality compared with only 4 percent in 1900. The incidence of congenital defects has not changed appreciably, but vaccines, antibiotics, and sanitation have decreased infectious diseases, and the relative importance of genetic diseases has increased.

Instead of treating a small number of infectious diseases that affect large numbers of people, the physician must learn how to handle an enormous number of genetic diseases, each of which affects a relatively few persons. More than 1,200 genetic diseases are known, some of which have an incidence of one per 100,000 in the general population. A physician may only see a few patients with genetic diseases in years of practice. Obviously, specialization and a system of counseling centers and referrals are needed. No medical center can afford to specialize in more than a few rare diseases. Cooperation among centers with expertise in particular diseases has proved to be an effective method of handling an otherwise formidable situation.

Cell Enzyme Activity

Genetic diseases may be controllable by adding, deleting, or replacing enzymes that operate with cells. Enzyme production is dependent on DNA specificity within the cell. In a procedure that has been described as the first attempt at "genetic engineering," DNA is being introduced into human cells by a virus. Stanfield Rogers of Oak Ridge National Laboratory has prepared the virus, and H. G. Terheggen of Cologne is treating two young girls in Germany who suffer from low blood levels of the enzyme arginase. They are receiving injection of live Shope papilloma virus for the purpose of inducing arginase activity in their cells.

Viruses are one agency known to carry DNA (or RNA) into host cells that they infect. Through transduction, the virus DNA takes over the genetic machinery of the host cell and enzymes are produced according to the specifications of the virus. The present objective is to induce the production into the cells of the German girls of a particular enzyme (arginase) that is needed for restoration of their health. This pioneering effort may foreshadow the day when artificial viruses, tailored to correct specific enzyme deficiencies, will be used to treat genetic diseases. Current experimental results indicate that animal viruses, bacterial viruses, and cell fusion techniques are all capable of introducing new functional genes into mammalian cells, although many of the genetic and regulatory processes remain unknown.

Although procedures for genetic engineering are only now being developed, the concept goes back to 1908, when Sir Archibald Garrod, physician to the British royal family, described several diseases as "inborn errors of metabolism." Among these diseases were alkaptonuria, cystinuria, porphyria, and albinism. Far ahead of his time, Garrod proposed that these diseases were dependent on recessive genes. Based on Garrod's discovery, it was possible to predict the probability of expression of a hereditary disease, but methods of early diagnosis, prevention, and treatment of these diseases are only now being developed with support from more recent discoveries of molecular genetics.

Prenatal Diagnosis of Genetic Disorders

In the late 1950s several obstetricians were attempting to improve the treatment for erythroblastosis fetalis. One approach was an analysis of amniotic fluid to determine whether the bilirubin content signifying destruction of red corpuscles was sufficient to justify delivery of the fetus prematurely by caesarian section. They observed that amniotic fluid contained cells and that the cells were those of the fetus and not those of the mother. Fritz F. Fuchs in Copenhagen observed Barr bodies, which appeared in the cells of the female fetus but not in those of the male. This showed that the sex of the fetus could be determined prenatally. When the fetus was a female with virtually no chance of inheriting a sex-linked gene such as the one for the disease, hemophilia, there was no cause for alarm. When, on the other hand, the fetus was a male and the chance was one-half that a defective sex-linked gene would be inherited, therapeutic abortions could be performed as warranted.

One of the diseases being seen in a new light because of advances in genetics has been known until recently as "mongolism." This poor descriptive term came into being because of the characteristic oblique opening of the eye with the inner corner covered by the epicanthal fold. Other facial features of the patients could have been chosen for a name. Symptoms associated with mental retardation and congenital heart disease are far more serious than the shape of the eye. The syndrome was first described by the clinician, Langdon Down, in 1866. But correlation of the condition with a chromosome irregularity had to await modern techniques. The chromosome identified as number 21 is replicated three times instead of being present as a single pair. The total number of chromosomes is thus 47 rather than 46, with the extra one being number 21. This chromosome irregularity has now been associated with physical abnormalities that have been recognized for a long time and was recently named Down's syndrome.

In the general population, this syndrome occurs about once in 750 people. It would be difficult to locate the fetus of the one in 750 that might be expected in the general population, but a maternal age differential makes diagnosis easier. The frequency is much greater among children of older mothers than among those of younger mothers. Expectancy of Down's syndrome in mothers below the age of 30 averages only about one in 2500 births. Since the incidence of Down's syndrome goes up rapidly with the age of the mother, screening of older mothers is more rewarding in terms of locating than testing all age groups in the general population. One case in every 100 mothers of age 40, and one in every 50 mothers of age 45 might be expected. Four cases of Down's syndrome were detected in one actual series of 104 amniocenteses performed by Henry L. Nadler's group at Northwestern University Medical School on pregnant women over 40. Therapeutic abortions were performed for all four. It has been suggested that all pregnant women of age 45 or over should be checked. On the average, about one in 50 of these would be expected to show a positive diagnosis for Down's syndrome.

Translocations between a chromosome from the so-called D group and one from the G group are more frequently encountered as the cause of Down's syndrome. Chromosome number 15 is the one from the D group that is involved in translocation. Disease-producing translocations follow a regular and predictable pattern of inheritance. Mothers carrying a translocation between a D and G chromosome have a 25 percent or one-fourth risk of having a baby with Down's syndrome. This risk is great enough to make it desirable for the fetuses of all expectant mothers known to have this kind of translocation to be tested. Sometimes the presence of a translocation can be determined by pedigree information and appropriate chromosome studies.

Phenylketonuria (PKU), for which most states require routine screening at birth, occurs once in about 10,000 live births. The test is accomplished by a simple color change in treated urine. Massachusetts tests 97 percent of the babies born in the state for PKU. In 1969, eight infants were found to have PKU. Infants giving a positive reaction are placed on a particular diet for the first few years of life. Care must be taken to avoid side effects and malnutrition of babies subjected to this diet. All eight of the infants detected in Massachusetts in 1969 were treated, and all have apparently escaped the mental retardation associated with PKU.

Massachusetts also requires a test on cord blood for level of galactose to detect galactosemia in newborn infants. In addition, the mother of a baby is requested to supply a sample of urine from the baby 3 or 4 weeks after she leaves the hospital. A kit containing a strip of filter paper is given to the mother along with an envelope pre-addressed to the state laboratory. Instructions are given for impregnating the filter paper with urine and mailing it. At the laboratory chromatographic tests are performed to detect enzyme abnormalities associated with histidinemia, hyperlysinemia, cystinuria, and several other genetic disorders.

Some genetic diseases have a built-in aid to screening because of the high incidence in certain population groups. Sickle-cell anemia, for example, is much more common among American blacks than among American whites. Some 97 percent of the victims of this disease in the United States are black. Eight to 10 percent of all American Negroes carry the recessive gene for sickling. Red blood corpuscles of those with sickle-cell anemia are collapsed and sickle shaped. They clog the capillaries and deprive the cells of the oxygen usually carried by the red corpuscles. About half of the sickle cell patients die before the age of 20, only a few live beyond 40, and most of these are crippled long before death. A solution of urea and invert sugar is now available for treatment. The urea breaks the bonds between hemoglobin molecules, reduces the sickling effect, and allows the corpuscles to return to normal shape.

In spite of the recent advances that have been made in techniques, especially in diagnostic techniques with reference to amniocentesis, much remains unknown about the hereditary abnormalities that are common in infants and growing children. No one can guarantee a perfect baby, even when all the tests that are now known have been run. Even more difficult is the question of what constitutes the ideal human being. If "engineers" are eventually competent to design and produce the kind of human beings they choose, what will be their makeup? Consideration must be given to the dynamic and interwoven complex in nature where the new or reconstituted individual must take his place. The "engineer" who undertakes such innovations must be super wise and appropriately humble.

Present-day biology is in the center stream of man's practical and intellectual heritage. It holds a key position in general education and has exciting research possibilities, particularly in those areas that help man to understand himself and to control his relations with his environment. Simultaneously, current ideas and activities are helping define dimensions of tomorrow's biology.

References and Readings

Ardrey, R. 1970. *The Social Contact*. New York: Atheneum.

Boyer, S. H., ed. 1963. *Papers on Human Genetics*. Englewood Cliffs, N. J.: Prentice-Hall. (Primary source selections.)

Durand, J. D. 1967. "The Modern Expansion of World Population." *Amer. Phil. Soc., Proc.* 3(3):136-159.

Harris, H. and K. Hirschhorn. 1970. *Advances in Human Genetics*. New York: Plenum Press.

Jacobson, C. B. and R. H. Barter. 1967. "Intrauterine Diagnosis and Management of Genetic Defects." *Am. J. Obst. Gynec.* 99:796-807.

Kass, L. R. 1971. "The New Biology: What Price Relieving Man's Estate." *Science* 174:779-788.

Lerner, I. M. 1968. *Heredity, Evolution and Society*. San Francisco: W. H. Freeman.

McKusick, V. A. 1970. "Human Genetics." *Ann. Rev. Genet.* 4:1-46. H. L. Roman (ed.). Annual Reviews, Inc., Palo Alto, Calif.

Moody, P. A. 1967. *Genetics of Man*. New York: W. W. Norton.

Scott, J. P. and J. L. Fuller. 1965. *Genetics and Social Behavior of the Dog*. Chicago: University Chicago Press.

Wallace, B. 1970. *Genetic Load, Its Biological and Conceptual Aspects*. Englewood Cliffs, N. J.: Prentice-Hall.

Woodwell, G. M., P. P. Craig, and H. A. Johnson. 1971. "DDT in the Biosphere: Where Does It Go?" *Science* 174:1101-1107.

General Readings

History of Biology

Adams, A. B. 1969. *Eternal Quest.* New York: G. P. Putnam's Sons.

Andrews, R. C. 1943. *Under a Lucky Star.* New York: Viking Press.

Asimov. I. 1960. *The Wellsprings of Life.* New York: Abelard-Schuman.

_____. 1964. *A Short History of Biology.* Garden City, New York: Natural History Press.

Barbour, T. 1943. *Naturalist at Large.* Boston: Little, Brown and Co.

Beebe, C. W. 1945. *The Book of Naturalists.* New York: Alfred A. Knopf.

Bodenheimer, F. S. 1958. *History of Biology.* London: W. Dawson and Sons.

Dawes, B. 1952. *A Hundred Years of Biology.* London: Gerald Duckworth and Co.

Ditmars, R. L. 1937. *The Making of a Scientist.* New York: Macmillan Co.

Doncaster, I. 1961. *In the Footsteps of the Naturalists.* London: Phoenix House.

Drachman, J. M. 1930. *Studies in the Literature of Natural Science.* New York: Macmillan Co.

Gabriel, M. L. and S. Fogel. 1955. *Great Experiments in Biology.* Englewood Cliffs, N. J.: Prentice-Hall.

Geisler, S. 1948. *Naturalists of the Frontier.* University Park, Texas: Southern Methodist University.

Glass, H. B., O. Temkin and W. Straus, Jr., eds. 1959. *Forerunners of Darwin 1745-1859.* Baltimore: Johns Hopkins Press.

Goldstein, P. 1965. *Triumphs in Biology.* Garden City, New York: Doubleday and Co.

Greenwood, T. 1886. *Eminent Naturalists.* London: Simpkin, Marshall, and Co.

Jordan, D. 1969. *Science Sketches.* Chicago: A. C. McClurg and Co.

Lenham, U. 1968. *Origins of Modern Biology.* New York: Columbia University Press.

Locy, W. A. 1910. *Biology and Its Makers.* New York: Henry Holt and Co.

_____. 1918. *The Main Currents of Zoology.* New York: Henry Holt and Co.

Mazzeo, J. A. 1967. *The Design of Life.* New York: Random House.

Meisle, M. 1924-29. *Bibliography of American Natural History.* Brooklyn, N. Y.: Premier Publishing Co.

Miall, L. C. 1912. *The Early Naturalists.* London: Macmillan and Co.

Moore, R. E. 1961. *The Coil of Life.* New York: Alfred A. Knopf.

Mornet, D. 1911. *Les Sciences de la Nature en France au XVIII Siècle.* Paris: Librairie Armand Colin.

Nordenskiöld, E. 1928. *The History of Biology, A Survey.* New York: Alfred A. Knopf.

Osborn, H. F. 1924. *Impressions of Great Naturalists.* New York: Charles Schribner's Sons.

Peattie, D. C. 1941. *The Road of a Naturalist.* Boston: Houghton Mifflin Co.

Raven, C. E. 1947. *English Naturalists from Neckam to Ray.* Cambridge: At the University Press.

Ritterbush, P. C. 1964. *Overtures of Biology: The Speculations of 18th Century Naturalists.* London: Yale University Press.

Rodgers, A. D. 1944. *American Botany.* Princeton: Princeton University Press.

Rostand, J. 1945. *Esquisse d'une Histoire de la Biologie.* Paris: Gallimard.

Singer, C. J. 1959. *A History of Biology.* New York: Abelard-Schuman.
Sirks, N. J. and C. Zirkle. 1964. *The Evolution of Biology.* Ronald Press.
Snyder, E. E. 1940. *Biology in the Making.* New York: McGraw-Hill Book Co.
Suner, A. P. 1955. *Classics of Biology.* New York: Philosophical Library.
Williams, H. 1961. *Great Biologists.* London: G. Bell and Sons.
Wood, L. N. 1944. *Raymond L. Ditmars.* New York: J. Messner.
Woodruff, L. L. 1923. *"The development of biology"* in The Development of the Sciences. New Haven: Yale University Press.

History of Medicine

Ackerknecht, E. H. 1955. *A Short History of Medicine.* New York: Ronald Press.
Allen, P. W. 1932. *The Story of Microbes.* Knoxville, Tenn.: Bookmill Co.
Atkinson, D. T. 1956. *Magic, Myth and Medicine.* Cleveland: World Publishing Co.
Baker, R. 1946. *Morton, Pioneer in the Use of Ether.* New York: Julian Messner.
Bettmann, O. L. 1956. *A Pictorial History of Medicine.* Springfield, Ill.: Charles C Thomas, Publisher.
Calder, R. 1958. *Medicine and Man.* New York: New American Library.
Camac, C. N. B. 1959. *Classics of Medicine and Surgery.* New York: Dover Publications.
Castiglioni, A. 1947. *A History of Medicine.* 2nd ed. New York: Alfred A. Knopf.
Clendening, L. 1942. *Source Book of Medical History.* New York: Paul B. Hoeber.
Cumston, C. 1926. *An Introduction to the History of Medicine.* New York: Alfred A. Knopf.
Eberson, F. 1948. *Microbes Militant: A Challenge to Man.* New York: Ronald Press.
Elgood, C. 1951. *A Medical History of Persia.* Cambridge: At the University Press.
Garland, J. 1949. *The Story of Medicine.* Boston: Houghton Mifflin Co.
Garrison, F. H. 1929. *An Introduction to the History of Medicine.* Philadelphia: W. B. Saunders Co.
Gordon, B. L. 1945. *The Romance of Medicine.* Philadelphia: F. A. Davis Co.
Graham, H. 1939. *The Story of Surgery.* New York: Doubleday, Doran and Co.
Guthrie, D. 1946. *A History of Medicine.* London: J. B. Lippincott Co.
Haggard, H. W. 1929. *Devils, Drugs, and Doctors.* New York: Harper and Brothers.
Haggard, H. W. 1934. *The Doctor in History.* New Haven: Yale University Press.
Hiller, L. A. 1944. *Surgery Through the Ages.* New York: Hastings House.
Keys, T. 1945. *The History of Surgical Anesthesia.* New York: Henry Schuman.
New York Academy of Medicine. 1939. *Landmarks in Medicine.* New York: Appleton Century.
Rapport, S. and H. Wright. 1952. *Great Adventures in Medicine.* New York: Dial Press.
Robinson, V. 1931. *The Story of Medicine.* New York: Tudor Publishing Co.

Schwartz, G. and P. Bishop, eds. 1958. *Moments of Discovery.* New York: Basic Books.

Sigerest, H. E. 1951. *A History of Medicine.* New York: Oxford University Press.

Singer, C. J. and E. A. Underwood. 1962. *A Short History of Medicine.* 2nd ed. New York: Oxford University Press.

Starobinsky, J. 1964. *A History of Medicine.* New York: Hawthorne Books.

Stevenson, L. G. 1953. *Nobel Prize Winners in Medicine and Physiology, 1901-1950.* New York: Henry Schuman.

History and Philosophy of Science

Arber, A. 1954. *The Mind and the Eye.* Cambridge: At the University Press.

Beveridge, W. I. B. 1950. *The Art of Scientific Investigation.* New York: W. W. Norton and Co.

Boynton, H., ed. 1948. *Beginnings of Modern Science.* New York: Classics Club.

Brown, G. B. 1950. *Science, Its Method and Its Philosophy.* New York: W. W. Norton and Co.

Butterfield, H. 1957. *The Origins of Modern Science.* London: G. Bell.

Conant, J. B. 1947. *On Understanding Science.* New Haven: Yale University Press.

_____. 1951. *Science and Common Sense.* New Haven: Yale University Press.

Dampier, W. C. 1944. *A History of Science.* New York: Macmillan Co.

Feigl, H. and M. Brodbeck, eds. 1953. *Readings in the Philosophy of Science.* New York: Appleton-Century-Crofts.

Forbes, R. J. 1950. *Man the Maker.* New York: Henry Schuman.

Frank, P. 1957. *Philosophy of Science.* Englewood Cliffs, N. J.: Prentice-Hall.

Glass, B. 1960. *Science and Liberal Education.* Baton Rouge, Louisiana: Louisiana State University Press.

Hall, A. R. 1954. *The Scientific Revolution.* New York: Longmans, Green and Co.

Hildebrand, J. H. 1957. *Science in the Making.* New York: Columbia University Press.

Ingle, D. J. 1958. *Principles of Research in Biology and Medicine.* Philadelphia: J. B. Lippincott Co.

Koyre, A. 1957. *From the Closed World to the Infinite Universe.* Baltimore: Johns Hopkins Press.

Libby, W. 1917. *Introduction to the History of Science.* Boston: Houghton Mifflin Co.

Mees, C. E. K. and J. R. Baker. 1946. *The Path of Science.* New York: John Wiley and Sons.

Moulton, F. R. and J. J. Schifferes, eds. 1945. *The Autobiography of Science.* Garden City, N. Y.: Doubleday, Doran and Co.

Needham, J. 1938. *Background to Modern Science.* New York: Macmillan Co.

Oppenheimer, J. R. 1954. *Science and the Common Understanding.* New York: Simon and Schuster.

Pledge, H. T. 1947. *Science Since 1500.* New York: Philosophical Library.

Sarton, G. 1927-48. *An Introduction to the History of Science.* 5 vol. Baltimore: Williams and Wilkins.

———. 1937. *History of Science and the New Humanism.* Cambridge, Mass.: Harvard University Press.

———. 1952. *Guide to the History of Science.* Waltham, Mass.: Chronica Botanica Co.

Singer, C. J. 1959. *A Short History of Scientific Ideas to 1900.* Oxford: Clarendon Press.

Singer, C. J., E. J. Holmyard and A. R. Hall, eds. 1954-58. *A History of Technology.* 5 vol. Oxford: Clarendon Press.

Smith, V. E. 1958. *The General Science of Nature.* Milwaukee: Bruce Publ. Co.

Taylor, F. S. 1949. *A Short History of Science and Scientific Thought.* New York: W. W. Norton and Co.

Thomas, D. L. and H. Thomas. 1959. *Living Biographies of Great Scientists.* New York: Garden City Books.

Whetham, W. C. D. 1924. *Cambridge Readings in the Literature of Science.* Cambridge: At the University Press.

Wiener, P. P. and A. Noland. 1957. *Roots of Scientific Thought.* New York: Basic Books.

Wightman, W. P. D. 1951. *The Growth of Scientific Ideas.* New Haven: Yale University Press.

Williams, H. S. 1909-10. *The Beginnings of Modern Science.* New York: Goodhue Co.

Wilson, G. 1937. *Great Men of Science.* Garden City, N. Y.: Garden City Publishing Co.

Wolf, A. 1935-39. *A History of Science, Technology and Philosophy in the 16th, 17th and 18th Centuries.* 2 vol. New York: Macmillan Co.

Woodruff, L. L. 1923. *The Development of the Sciences.* New Haven: Yale University Press.

INDEX